Hydrodynamics of Gas-Liquid Reactors

Hydrodynamics of Gas-Liquid Reactors

Normal Operation and Upset Conditions

B. J. AZZOPARDI

Faculty of Engineering, University of Nottingham, Nottingham, UK

R. F. MUDDE

Faculty of Applied Sciences, Delft University of Technology, Delft, The Netherlands

S. LO

CD-adapco, UK

H. MORVAN

Faculty of Engineering, University of Nottingham, Nottingham, UK

Y. YAN

Faculty of Engineering, University of Nottingham, Nottingham, UK

D. ZHAO

Faculty of Engineering, University of Nottingham, Nottingham, UK

A John Wiley & Sons, Ltd., Publication

Library of Congress Cataloging-in-Publication Data

Hydrodynamics of gas-liquid reactors : normal operation and upset
conditions / B.J. Azzopardi ... [et al.].
      p. cm.
   Includes bibliographical references and index.
  ISBN 978-0-470-74771-1
1.   Chemical reactors–Design and construction. 2.   Chemical reactors–Fluid
dynamics–Mathematical models. 3.   Gas-liquid interfaces. I. Azzopardi, B. J.
(Barry J.)
   TP157.H93 2011
   660'.2832–dc22

                                                          2011004546

A catalogue record for this book is available from the British Library.

Print ISBN: 9780470747711
ePDF ISBN: 9781119970323
oBook ISBN: 9781119970712
ePub ISBN: 9781119971405
Mobi ISBN: 9781119971412

Set in 10/12pt Times Roman by Thomson Digital, Noida, India
Printed and bound in Singapore by Markono Print Media Pte Ltd

Contents

List of Figures

List of Tables

Preface

One way of identifying areas of importance in Chemical Engineering is to look at the subject groups or working parties of national or transnational professional bodies. Amongst the Working Parties of the European Federation of Chemical Engineers, there are two that are relevant to the present text, one on Chemical Reaction Engineering and the other on Multiphase Flow. Although each has its own programme of events, they come together once in a while to air matters of common interest. Multiphase flow, the simultaneous flow of more than one phase, has applications in: electrical power generation – nuclear and fossil fired; oil and natural gas production and refining; distillation and absorption and in heat transfer. This has produced a body of knowledge that can be drawn on for the benefit of the design of chemical reactors.

The relative efforts in the two facets of reactor design have been very succinctly illustrated in Figure A, by Professor Octave Levenspiel [Levenspiel, O. (1999) Chemical Reaction Engineering, *Ind. Eng. Chem. Res.* **38**, 4140–4143]. This text aims at strengthening the weaker link of the chain shown.

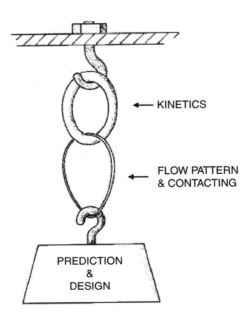

Figure A   Reprinted with permission from Industrial & Engineering Chemistry Research, **38**, 11, Chemical Reaction Engineering, Levenspiel © 1999, Americal Chemical Society

The preparation of this present work brought together a team with complimentary skills. All are firmly based in multiphase flows and they research into chemical reactor applications in addition to other industrial applications. The work presented here is an expression of the realization that there is the need for more and more complex modeling methods. There are also requirements to consider not just mixing but also separation of gas and liquid and to look into the upset conditions, which might lead to accidents, as well as a steady state. All of these have found a place in this book from the point of view of multiphase flows, in a combination of well proven engineering concepts and modern developments, both numerically and experimentally.

Nomenclature

A	constant in Equation 2.45 [–]
A	constant in film thickness correlating equation [–]
A	constant in mass transfer equation (Chapter 4) [–]
A	cross-sectional area of vessel (Chapter 8) [m^2]
A	constant in Antoine equation (Chapter 8) [–]
A_D	drag force coefficient [–]
A_H	Hamaker constant [–]
A_p	projected area of one wire [m^2]
a	transverse dimension of elliptical bubble [m]
a	constant in Equation 2.45 [–]
a	parameter in upper limit log normal equation [–]
a	constant in Equation 5.14
a'	specific interfacial area (also known as interfacial area density) [1/m]
B	constant in film thickness correlating equation [–]
B	constant in mass transfer equation (Chapter 4) [–]
B	ratio of latent heat to gas constant in Antoine equation (Chapter 8)
b	axial dimension of elliptical bubble [m]
b	constant in Equation 2.45 [–]
b	constant in Equation 5.14
C	constant of integration in falling film analysis [kg/m s]
C	constant in mass transfer equation (Chapter 4) [–]
C'	constant of integration in falling film analysis [m/s]
C_D	drag coefficient [–]
C_I	impeller–impeller distance [m]
C_L	lift coefficient [–]
C_M	virtual mass coefficient [–]
C_o	volume fraction correction factor [–]
C_t	response coefficient [–]
C_1	coefficient in Equation 3.6
C_2	coefficient in Equation 3.6
C_3	coefficient in Equation 3.6
c	constant in Equation 2.24
c	constant in Equation 2.45
c	constant in Equation 5.14
c	streaming speed (Chapter 6) [m/s]
c	velocity of sound (Chapter 9) [m/s]
c_g	concentration in gas phase
c_l	concentration in liquid phase

c_p	specific heat capacity [J/kg °C]
c_r	constant in radial profile equations [–]
c_1	constant in Equation 2.14
c_2	constant in Equation 2.14
c_3	constant in Equation 2.15
D	impeller diameter [m]
D_c	column diameter [m]
D_l	liquid diffusion coefficient
D_t	tube diameter [m]
d	particle (or bubble or drop) diameter [m]
d_{bL}	large diameter bubble size [m]
d_e	equivalent sphere diameter [m]
d_{max}	diameter of largest drop [m]
d_o	diameter of the orifice [m]
d_w	diameter of wire [m]
d^*	dimensional bubble size [$= d(\rho_l g/\sigma)^{0.5}$] [–]
d_{32}	Sauter mean diameter [m]
e	constant in Equation 2.45 [–]
E	aspect ratio of elliptical bubbles [–]
$\dot{E}$	dissipation rate of turbulent kinetic energy [m²/s³]
F	function in Equation 2.45
F	function defined by Equation 4.18 [–]
F_B	Bassett (or history) force [N]
F_b	buoyancy force [N]
F_D	drag force [N]
F_i	inertial force (Chapter 2) or interaction force (Chapter 5) [N]
F_L	lift force [N]
F_M	virtual mass force [N]
F_m	momentum force [N]
F_s	surface tension force [N]
F_T	turbulent drag force [N]
F_v	fractional free volume for knitted mesh pads [–]
F_W	wall lubrication force [N]
f	friction factor [–]
G	physical property group (Chapter 8) [m/s]
g	acceleration due to gravity [m/s²]
H	height of gas lift column [m]
H	height of liquid in vessel [m]
H_o	original height of liquid in vessel [m]
h	specific enthalpy [J/kg]
h_{cr}	critical film thickness [m]
K	pre-constant in Equation 8.4 [–]
K	rate [s⁻¹]
K_b	constant in Equation 2.24 [–]
K_e	constant in Equation 2.46 [–]
K_1	constant in Equation 3.9

K_2	constant in Equation 3.9
K_3	constant in Equation 3.9
k_{br}	inertia breakup time constant $[s^{-1}]$
k	turbulent kinetic energy $[m^2/s^2]$
k_l	mass transfer coefficient
L	mesh pad thickness $[-]$
L	axial length of film $[m]$
L	vent duct length (Chapter 9) $[m]$
L_B	Batchelor length scale $[m]$
L_f	length scale in falling film analysis $[-]$
L_k	Kolmogorov length scale $[m]$
M	moment of particle distribution $[m]$
M	mass of vessel contents (Chapter 8) $[kg]$
$\dot{m}$	mass flux (Chapter 8) $[kg/m^2\ s]$
m	power in Equation 8.38
N	non-equilibrium parameter (Chapter 9), two alternative definitions are given in Equations 9.37 and 9.49
N	rotation rate (Chapter 3) $[rad/s]$
N	daughter number of particles (Chapter 5) $[-]$
n	particle number density (Chapter 5) $[-]$
n	constant in Equation 2.24 $[-]$
n	exponent in radial profile equations (Equations 2.48 and 2.51) $[-]$
n	number of bends in wave plate $[-]$
n	number of layers in mesh pad $[-]$
n	power in Equation 8.38 $[-]$
Q	volumetric flow rate of gas through an orifice $[m^3/s]$
Q	volumetric flow rate on a falling film reactor $[m^3/s]$
$\dot{Q}$	total heat input rate $[W]$
$\dot{q}$	specific heat input rate $[W/kg]$
$\bar{q}$	mean specific heat input $[W/kg]$
P_{cl}	coalescence probability $[m]$
P_g	gassed power $[W]$
Po	power number $[-]$
P_0	power number in ungassed case $[W]$
$P(d)$	log-normal particle distribution $[-]$
p	pressure $[N/m^2]$
p_g	pressure on gas side of bubble interface (Chapter 2) $[N/m^2]$
p_g	partial pressure of evolved gas (Chapter 8) $[N/m^2]$
p_l	pressure on liquid side of bubble interface (Chapter 2) $[N/m^2]$
p_v	partial pressure of evolved gas (Chapter 8) $[N/m^2]$
R	radius $[m]$
R	radius of bend in wave plate $[m]$
r	radial coordinate $[m]$
S	cross-sectional area of vent (Chapter 8) $[m^2]$
S_0	cross-sectional area of the vent at zero overpressure $\left(S_0 = \dot{Q}v_{lg}M/\Delta h_v \dot{m}V\right)$ $[m^2]$
S	local intensity of dispersion $[-]$

s	entropy [J/kg K]
s	spacing of plate in wave plate [m]
T	temperature [°C]
T	tank diameter [m]
t	time [s]
t_b	bubble departure time [s]
t_e	contact time [s]
U_R	slip ratio [–]
u	macroscopic velocity [m/s]
u_b	bubble velocity [m/s]
u_{gs}	gas superficial velocity [m/s]
u_o	gas velocity in orifice [m/s]
u_r	radial velocity of drops in wave plate bend [m/s]
u_{tr}	gas superficial velocity at the homogeneous/heterogeneous transition [m/s]
u^*	dimensionless velocity [$= V_T(\rho_l/g\sigma)^{0.25}$] [–]
u^*	friction velocity (Chapter 4) [m/s]
V	volume of a bubble of diameter d [m^3]
V	velocity of gas in wave plates [m/s]
V	velocity of tip of impeller [m/s]
V	volume of vessel (Chapter 8) [m^3]
V	volume [m^3]
V_f	face velocity for knitted mesh pad
V_g	volume of gas generated
V_L	large bubble velocity [m/s]
V_l	volume of liquid associated with a bubble [m^3]
V_T	bubble terminal velocity [m/s]
$\dot{V}$	rate of vapour production [m^3/s]
v	velocity [m/s]
v_{rel}	relative velocity [m/s]
v_g	gas velocity
v_l	liquid velocity
$v(d)$	volume fraction of drops in sizes d to $d + \partial d$ [–]
W	width of falling film [m]
W	characteristic velocity (Chapter 9) [m/s]
X	Lockhart–Martinelli parameter [–]
x_g	vapour mass fraction [–]
y	distance from interface [m]
Z	dimensionless distance from bottom of vessel [–]
z	vertical axis [m]
α	heat transfer coefficient [W/m^2 °C]
α	angle of bend in wave plate
β	ratio of front to back dimensions for a flattened elliptical bubble [–]
β	fraction of wire perpendicular to the flow [–]
Δh_v	latent heat of evaporation [J/kg]
δ	parameter in upper limit log-normal equation
δ	film thickness [m]

δ	lateral distance cleared of drops [m]
ε	mesh pad voidage [–]
ε_g	void fraction [–]
$\bar{\varepsilon}_g$	calculated void fraction (Chapter 8) [–]
ε_{go}	homogeneous void fraction based on vessel contents (Chapter 8) [–]
ε_{bL}	void fraction contribution of large bubbles [–]
ε_{gtr}	void fraction at homogeneous/heterogeneous transition [–]
ε_{min}	permittivity of an empty bubble column
ε_{max}	permittivity of a bubble column filled with liquid
ϕ	thermal or adiabaticity factor (ratio of heat capacity of vessel and content to that of the vessel [–]
$\phi_l{}^2$	two-phase multiplier [–]
γ	isentropic exponent [–]
η	critical pressure ratio [–]
η_l	liquid viscosity [Pa s]
η_g	gas viscosity [Pa s]
η_F	grade efficiency [–]
η_o	overall efficiency [–]
η_w	collection efficiency for one wire [–]
η_1	efficiency for one bend of vane pack [–]
μ_w	attenuation coefficients of the column wall
μ_{liq}	attenuation coefficients of the liquid
μ_g	attenuation coefficients of the gas
λ	thermal conductivity [W/m °C]
θ	wetting angle
ρ_g	gas density [kg/m^3]
ρ_l	liquid density [kg/m^3]
ρ_m	mixture density [kg/m^3]
ρ_p	mesh pad density [kg/m^3]
ρ_s	solids density [kg/m^3]
σ	surface tension [N/m]
σ	ratio of downstream to upstream cross-sectional area [–]
σ_α	turbulent Prandtl number [–]
τ	shear stress [kg/m s]
τ_{br}	breakup time [s]
τ_k	stress tensor [1/s]
τ_I	interfacial shear stress [kg/m s]
τ_i	interaction time [s]
τ	relaxation time [s]
υ_{lg}	difference between gas and liquid specific volumes [m^3/kg]
ω	rate of rotation [radians/s]
ω	parameter defined by Equation 9.25
γ	distribution moment [–]
γ	isentropic coefficient [–]
$\dot{\gamma}$	shear rate [1/s]
v	kinetic viscosity [m/s^2]

ψ dimensionless group used in level swell calculations [–]

ψ modified Stokes number ($\rho v d_a^2/18\eta d_w$) [–]

Ar Archimedes number [–]

$E\ddot{o}$ Eötvös number [–]

Fr Froude number for stirrer ($N^2 D/g$) [–]

Fr_g Froude number (= u_{gs}^2/gD_c) [–]

K_f liquid number – inverse of Morton number [–]

Ka Kapitza number – $K_f^{0.33}$ [–]

Mo Morton number ($g\eta_l^4/\sigma^3\rho_l$) [–]

Nu Nusselt number [–]

O^* overpressure number ($c_p\Delta TMv_{lg}/(V\Delta h_v)$) [–]

Pr Prandtl number [–]

Po power number [–]

Re Reynolds number [–]

Re_g Reynolds number based on column diameter and gas superficial velocity [–]

Re_m mixture Reynolds number [= $(\rho_g u_{gs}+\rho_l u_{ls})D_c/3\eta_l$.] [–]

Re_w wire Reynolds number (=$\rho_l V_f d_w/\eta_g$) [–]

S' dimensionless vent area (S/S_o) [–]

Sc Schmidt number [–]

Sh Sherwood number [–]

Ta Tadaki number (= $Re\ Mo^{0.23}$) [–]

We Weber number [–]

We_o Weber number based on orifice diameter and velocity [–]

Ω capillary number (Chapter 5) [–]

Subscripts

br breakup

c continuous phase

cl coalescence

$coll$ collision

cr critical

d dispersed phase, drop

eff effective

HEM homogeneous equilibrium model

k phase indicator

max maximum

o upstream value

r relative

set at set pressure

tot total

1

Introduction

There are many reactions where the reactants consist of a liquid and a gas. Examples are oxidations, hydrogenations and halogenations. Although in some instances it might be possible to evaporate the liquid and operate entirely in the gas phase, this is not always desirable. The high temperatures involved are demanding on the materials of construction, the strength of the materials, compression costs and subsequent separations. For other reactions it might be advantageous to evaporate one of the products to drive the reaction to completion more rapidly. Examples here are condensation polymerisations where removal of water is a very positive advantage.

There is an entire subject area of chemical reaction engineering that concerns itself with the times, volumes, and so on, required to achieve the particular extent of reaction. However, what is often passed over in such texts is the actual contacting of the phases. Now, this contacting is crucial to the success or otherwise of the reactor operation. The present book, which focuses on this area, aims to fill that gap.

The book is divided into three parts, the interrelation between which is shown in Figure 1.1. Part One considers the methods that might be used for design calculation for the main types of reactors. It considers the methods at three levels. Firstly, there are engineering calculations. These are often methods carried out by hand or by the use of simple spreadsheets. They are laid out for bubbly flows without mechanical agitation (usually known as bubble columns) in Chapter 2, bubbly flows with mechanical agitation (sparged stirred tanks) in Chapter 3 and film flows in Chapter 4. However, it will be seen that these engineering methods are only adequate to a first level, particularly for bubbly systems. For this reason the next level of complexity, Multiphase Computational Fluid Dynamics, is presented in Chapter 5. Even this cannot handle some of the micro-scale processes that are occurring, such as coalescence. This is why Chapter 6 presents methods more suitable for modelling at this small scale.

Hydrodynamics of Gas-Liquid Reactors: Normal Operation and Upset Conditions, First Edition.
B. J. Azzopardi, R. F. Mudde, S. Lo, H. Morvan, Y. Yan and D. Zhao.
© 2011 John Wiley & Sons, Ltd. Published 2011 by John Wiley & Sons, Ltd.

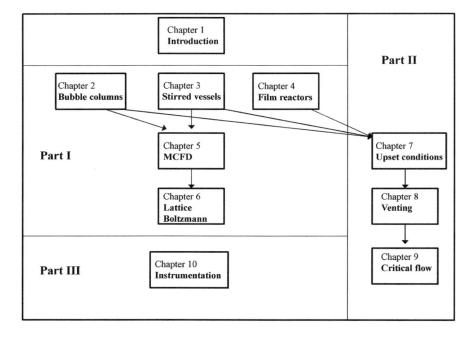

Figure 1.1   *Chart showing structure of the book and the interrelation of the chapters*

All the above material considers normal operation where the system stays at or about the design conditions. However, for many multiphase reactors (in addition to single-phase reactors) the occurrence of runaway reactions has to be anticipated, and measures put in place to protect against its consequences. Part Two, Chapters 7–9, considers these upset conditions. Chapter 7 presents the types of reactions where there have been incidents in the past and the reasons for those malfunctions occurring. The behaviour of the vessel contents and the process of removing material from the vessel are laid out in Chapter 8. Chapter 9 presents methods for the calculation of the limitation to flow that can occur at the relief point and the piping beyond.

The final Part of the book, Chapter 10, considers the methods of measurement for the many parameters that are relevant to these reactors.

This book is intended for learning and for familiarisation of the concepts presented. Therefore, where appropriate, chapters end with a number of exercises that test the acquired knowledge.

Part One

2

Bubble Columns

Hydrodynamics of Gas-Liquid Reactors: Normal Operation and Upset Conditions, First Edition.
B. J. Azzopardi, R. F. Mudde, S. Lo, H. Morvan, Y. Yan and D. Zhao.
© 2011 John Wiley & Sons, Ltd. Published 2011 by John Wiley & Sons, Ltd.

2.1 Introduction

Bubble columns are pools of liquid through which gas is bubbled. Liquid is fed in and then taken off at different points. The simplest versions are almost batch systems with low feed rates and take off of liquid. However, there can be more complex versions, particularly where the liquid is circulated through other, gas-free, regions. The design of bubble column reactors requires two types of calculations. Those from chemical reactor engineering will define the residence time required. This, together with the feed rate of the liquid, will give the volume of vessel necessary. The second type of calculations is to do with the hydrodynamics, and allows the details of the gas distributor to be designed so as to achieve the required phase distribution and so effect the mass transfer of species between the gas and liquid phases. In some instances the residence time and hydrodynamic requirements are incompatible. For those, bubble columns with recirculation are possible suitable variants that provide a larger volume for residence time whilst only a smaller part for the two-phase mass transfer. In all types of columns, the gas has to be removed without carrying over any liquid. Methodologies for this are considered.

In all types of bubble columns, consideration has to be given to getting the gas in and taking it out. This chapter will concentrate on these. Section 2.2 describes the different types of bubble columns. The injection of gas and motion in the columns is covered in Section 2.3 and gives information on the design of distributors, whilst Section 2.4 considers the arrangements to ensure that only gas emerges through the relevant outlet.

2.2 Types of Bubble Columns

A conventional bubble column consists of a cylindrical vessel (axial vertical) with an arrangement to introduce gas into the bottom of the vessel, as illustrated in Figure 2.1a.

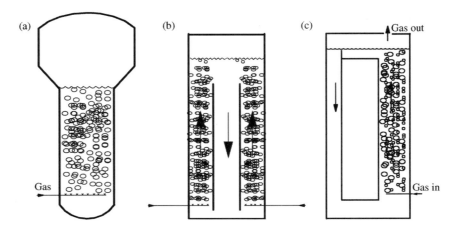

Figure 2.1 *Typical arrangements for bubble columns. (a) Standard bubble column; (b) gas-lift type with internal down-comer; and (c) gas-lift type with external down-comer*

Above the level of the gas-liquid mixture there might be a section with a larger diameter than the main part of the column. This is provided to encourage disengagement of drops from the gas by gravity before it emerges from the vessel. Secondary drop separation equipment (discussed in Section 2.4) might be positioned in this part of the vessel. Alternative arrangements using a gas-lift type of operation are illustrated in Figure 2.1b and c. These have sections into which gas is injected, and where it travels upwards carrying liquid up with it. There are other sections with no gas injection where the liquid can flow downwards. This recirculation occurs because of the difference in the density of the gas-liquid mixture and that of the liquid alone. The gas injection could be into an outer annulus with return down the centre as shown in Figure 2.1b. Alternatively, the injection could be into the centre with return down the annulus. These types are known as internal recirculation types. The arrangement shown in Figure 2.1c is usually termed an external down-comer type. There are some operations where the gas injection part (often known as the riser) and the down-comer are of the same diameter. The gas could be injected part of the way down the down-comer.

2.3 Introduction of Gas

In determining the way in which the gas is introduced into the bubble column, there are a number of questions to be posed. At what rate must mass be transferred between the gas and the liquid phases? Is the degree of mixing adequate? How much pressure drop can be afforded in the gas circuit? How can an even distribution of liquid about the cross-section be ensured?

In order to design the gas distributor system, it is necessary to appreciate how the phases are distributed about the bubble column. Because of the highly deformable nature of the gas-liquid interface, there are a large number of possible configurations. To be able to handle the problem it is conventional to group these possibilities into a small number of what are termed flow patterns, that is blanket descriptions focusing on their main features. At very low gas flow rates, the flow is described as homogeneous. This consists of uniformly spaced bubbles of uniform size, which rise at the same velocity. At higher gas flow rates, the flow is often described as heterogeneous and bubbles can be of a wide range of sizes, and this where the different sized bubbles rise at different velocities. This description is reasonable for larger diameter columns. However, for small diameter columns ($\leq$100 mm), some of the larger bubbles can occupy the greater majority of the column and the flow is given the name of slug flow. This has alternating regions of large bubbles and liquid slugs containing smaller bubbles. The conditions for the occurrence of each flow pattern are usually recorded on what is termed a flow pattern map. This could be a plot of gas superficial velocity, that is the velocity the gas would have if it alone occupied the column, versus column diameter. A typical map [1] is illustrated in Figure 2.2. A variant to the homogeneous flow pattern was reported by Rietema and Ottengraf [2]. They used a viscous liquid and multiple needle injectors. The streams of bubbles were much closer halfway up the column than at bottom or top, an effect caused by circulation of the liquid. This arrangement was called a bubble street.

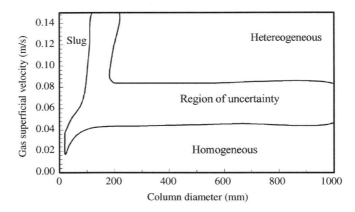

Figure 2.2 *Flow pattern map for bubble columns [1]*

2.3.1 Methodology of Gas Injection

The method of injection of gas into the liquid within the bubble column is usually designed to produce uniformly spaced small bubbles. Papers reporting research in this area describe a wide range of arrangements. At one extreme, a single hole at the bottom of the column is used. This assumes that the gas stream will be broken up into smaller bubbles. Though this might occur for low viscosity liquids such as water, it has been observed that for a very high viscosity liquid ($>25\,000$ times that of water) the gas formed an unstable coherent column up the middle of the bubble column at higher gas injection rates.

To understand better the important parameters in column behaviour, it is helpful to examine their effect on the relationship between the void fraction, the fraction of the two-phase mixture that is gas, and the flow rate of the gas as the gas superficial velocity. Important parameters are column height and diameter, the distributor, that is the number and size of holes, the gas density, the liquid viscosity and the presence and concentration of chemicals that affect the behaviour of bubbles by, for example, hindering coalescence.

From experiments on columns of 150, 225 and 400 mm diameter, Groen [3] showed that there is little effect of this parameter on the overall void fraction/gas superficial velocity relationship. The effect of column height was studied by, for example Růžička *et al.* [4], who showed that void fraction decreased as the height of the two-phase mixture increased.

In contrast, in the heterogeneous pattern, little difference is seen between the data of Letzel *et al.* [5] (200, 0.5 mm holes, column 1.2 m height) and that of Cheng *et al.* [6] (6 mm holes, 10 m column with measurements made at the 5 m level).

The effect of the distributor is illustrated in Figure 2.3. Here three sets of data from experiments on 225 mm columns using air and water are plotted. In the experiments of Groen [3], the air was introduced through a porous plate with 40% porosity and hole sizes of 40 µm. In the other two cases the gas was introduced through 5 mm diameter nozzles [7]. Either one nozzle, placed at the centre of the bottom of the column, or 25 nozzles, equally spaced about the bottom of the column were used. As seen, although there is little difference between the one and 25 nozzle data, those for the smaller holes give a

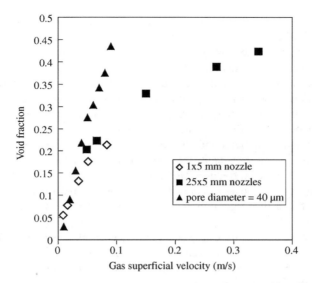

Figure 2.3   *Effect of gas distributor on void fraction/gas superficial velocity relationship*

significantly steeper void fraction/gas superficial velocity gradient. A number of other sources confirm this.

The effect of gas density (or system pressure) is seen to be significant in some circumstances. Figure 2.4 shows how increasing the pressure increases the void fraction [5].

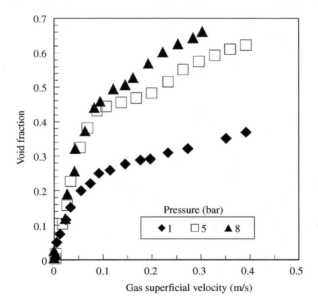

Figure 2.4   *Effect of pressure on void fraction/gas superficial velocity relationship. Air-distilled water*

However, although there is a large increase from 1 to 5 bar (1 bar $= 10^5$ Pa), the increase diminishes with further increases in pressure; there is very little difference between the void fraction data for 5 and 13 bar. What is clear is that there is very little difference in void fraction with pressure at low gas superficial velocities. These data sets are characterised by the large change in slope, which is attributed to the transition from homogeneous to heterogeneous flow. The increase in pressure seems to extend the occurrence of the homogeneous regime. In contrast, there is a smaller effect of pressure when the liquid viscosity has a value above that of water (eightfold) [8].

Another important parameter that has a significant effect on void fraction is liquid viscosity. Increase of this parameter over the value typical for water, that is 0.001 Pa s, causes a decrease in the void fraction. However, it is seen [9], Figure 2.5, that a 550 times increase in viscosity only causes a decrease in void fraction by a factor of 2.5. It is probable that the cause of this is the larger bubbles that are formed, decreasing the possibility of the formation of the small bubbles flowing along straight lines, which characterise homogeneous flow. It is reported that this effect occurs for viscosities of only 0.008 Pa s. Support for this point of view comes from the trend in the void fraction/gas superficial velocity data shown in Figure 2.5. There is not the significant change of slope that characterised the homogeneous/heterogeneous transition.

The presence of some chemicals, even in small amounts, can also affect the void fraction significantly [10]. Alcohols or salts have been seen to increase void fraction. The higher the number of carbon atoms in the alcohol, the greater the increase in void fraction. This is particularly so at higher gas superficial velocities. The effect at lower velocities, in the homogeneous regime, is much weaker. Similar effects are reported for the effect of the concentration of the alcohol employed. It is argued that the increase in void fraction is linked

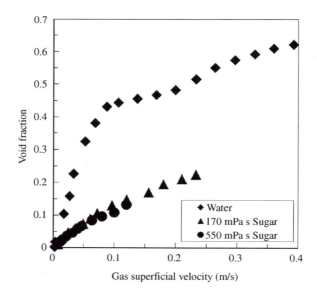

Figure 2.5 *Effect of viscosity on the void fraction/gas superficial velocity relationship. Pressure = 5 bar*

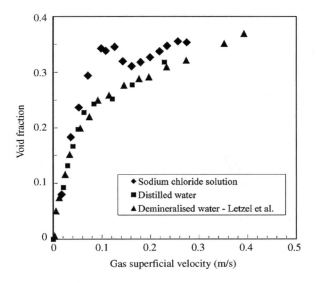

Figure 2.6   *Effect of additives on the void fraction/gas superficial velocity relationship*

to suppression of coalescence of bubbles by the presence of alcohol, that is the extent of the homogeneous region is increased. It is noted that this effect is present for both larger and smaller gas injection orifices, although the increase in void fraction is more pronounced for the latter. A similar increase in void fraction is seen with the addition of salts, Figure 2.6. This illustrates results for small gas orifices. An increase in void fraction is also seen for larger orifices but it is less pronounced and there is less distinction between the actual salts used. These data, as well as those for alcohol, are characterised by the very obvious peak. The region of negative gradient in the void fraction/gas superficial velocity data is identified by some researchers as a transition region.

The above examples have all related to a batch bubble column, that is one with no external circulation. The gross motion of liquid can have a noticeable effect. Figure 2.7 illustrates this with data from a 192 mm column operating with steam-water at 46 bar [11]. This and other available data show that once the recirculation velocity of the liquid exceeds a critical value, the effect diminishes.

2.3.2   Bubble Formation and Size Change

When a gas is passed through an orifice into a pool of liquid it naturally breaks up into individual bubbles. Systematic experiments [12] have shown that the higher the gas flow rate the larger the bubble size with a power law dependence of about 0.4. Decreasing the surface tension, as occurs when alcohols are added to water, causes bubble sizes to decrease. Bubble sizes increase as the concentration of *electrolyte* dissolved in water is increased. They also increase as the sizes of the orifices from which they are formed are larger. The bubbles are also larger in more viscous liquids. The height of liquid above the orifice has only a small effect on the bubble size.

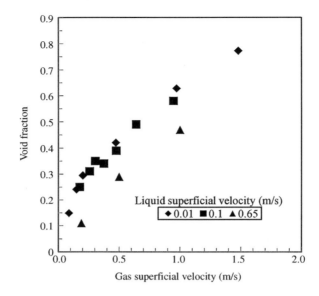

Figure 2.7   *Effect of liquid circulation velocity on void fraction*

The forces which have an effect on the formation of bubbles from an orifice are buoyancy, F_b, and gas momentum, F_m, which favour detachment and surface tension, F_s, drag, F_d, and inertial, F_i, forces which hinder detachment [13]. The *buoyancy force* is given by:

$$F_b = \frac{\pi d^3 (\rho_l - \rho_g)g}{6} + \frac{\pi d_o^2}{4}(p_g - p_l) \tag{2.1}$$

where

d is the spherical equivalent bubble diameter
d_o is the diameter of the orifice
g is the acceleration due to gravity
ρ_l, ρ_g are the densities of the liquid and gas, respectively.

The second term allows for a difference in pressure inside and outside the bubble with p_g and p_l being the gas and liquid pressures, respectively. The *gas momentum force* is the product of the momentum flux through the orifice and the orifice area

$$F_m = \frac{\pi d_o^2 \rho_g u_o^2}{4} \tag{2.2}$$

with u_o being the gas velocity through the orifice, that is $= Q/(\pi d_o^2/4)$. The *surface tension force* is:

$$F_s = \pi d_o \sigma \tag{2.3}$$

where σ is the surface tension. The drag force is defined as:

$$F_d = \frac{\pi d^2}{4} c_D \frac{\rho_g u_b^2}{2} \tag{2.4}$$

where

u_b is the bubble velocity
c_D, the drag coefficient, is taken to be 24/Re$_o$ + 1.

The *inertia force* results from the bubble volume increases during the detachment stage. The bubble accelerates up to the point of detachment due to the continuous increase in bubble buoyancy. It is specified as:

$$F_i = (\rho_g V + \rho_l V_l) \frac{u_b}{t_d} \qquad (2.5)$$

where

V is the bubble volume
V_l is the liquid volume that moves with the bubble
$\rho_l V_l$ is called the added mass in some analyses
t_d is the time to detachment.

The simplest analysis considers only the buoyancy and surface tension forces and assumes equal pressure in the gas and liquid phases and that $\rho_l \gg \rho_g$, and gives the bubble diameter explicitly as:

$$d = \left(\frac{6d_o \sigma}{\rho_l g} \right)^{1/3} \qquad (2.6)$$

However, this is far too simplistic and for realistic gas flow rates, because, as noted above, surface tension plays a minor role in determining the size of bubbles.

The more thorough analysis takes all forces into account, that is:

$$F_b + F_m = F_s + F_d + F_i \qquad (2.7)$$

which, using the definitions of Equations 2.1–2.5, results in:

$$d^3 = \left(\frac{6d_o \sigma}{(\rho_l - \rho_g)g} \right) \left(1 - \frac{We_o}{4} \right) + \frac{81 \eta_l Q}{\pi (\rho_l - \rho_g) g d} + \left(\frac{135}{4\pi^2} + \frac{27\rho_g}{\pi^2 \rho_l} \right) \left(\frac{\rho_l Q^2}{(\rho_l - \rho_g) g d^2} \right) \qquad (2.8)$$

with

$$We_o = \frac{\rho_g u_o^2 d_o}{\sigma} \qquad (2.9)$$

Equation 2.8 requires iteration to solve for d. However, Gaddis and Vogelpohl [13] showed that the simplified versions for the surface tension, viscous drag and inertia controlled cases results in:

$$d = \left[\left(\frac{6d_o \sigma}{\rho_l g} \right)^{1.33} + \left(\frac{81 \eta_l Q}{\pi \rho_l g} \right) + \left(\frac{135 Q^2}{4\pi^2 g} \right)^{0.8} \right]^{0.25} \qquad (2.10)$$

which is a good fit to much data. Note that if only the first term in Equation 2.10 is considered, it reduces to Equation 2.6. Considering only the second term results in an

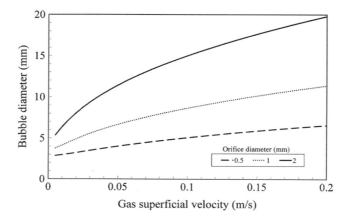

Figure 2.8 *Bubbles sizes predicted by the equation of Gaddis and Vogelpohl [13] for air-water for a 0.16% open area – effect of gas flow rate and hole diameter*

equation similar to that deduced by Davidson and Schuler [14], but predicting 2.7% smaller bubble diameters. The third term of Equation 2.10 is similar to those for inviscid liquids derived by Davidson and Schuler [15], but with a marginally different constant. Figure 2.8 illustrates the dependence of bubble size on superficial gas velocity and orifice diameter for an air-water system whilst Figure 2.9 shows the corresponding information for a liquid with viscosity 550 times that of water.

Another problem that arises is that bubbles are influenced by the preceding one. In the limit, the string of bubbles links together to form a jet. Lin and Fan [16] showed how there is a critical gas velocity in the orifice for jetting to occur. They studied the effect of liquid

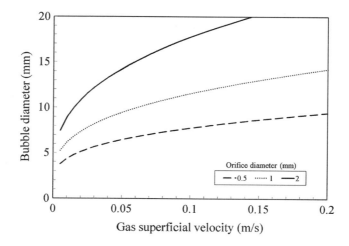

Figure 2.9 *Bubbles sizes predicted by the equation of Gaddis and Vogelpohl [13] for air-viscous liquid (dynamic viscosity = 0.55 Pa s) for a 0.16% open area – effect of gas flow rate and hole diameter*

viscosity and system pressure on the critical value. They noted that the gas density effect was to the power of -0.38. Idogawa *et al.* [17] identified the power as -0.8. To a first approximation it might be suggested that there is a critical gas momentum.

Bubbles can be formed by the breakup of larger bubbles. Three mechanisms have been identified: (i) due to collisions between bubbles as a result of the eddying motion of the flow; (ii) due to instabilities of large bubbles and (iii) by the tearing off of the edges of larger bubbles, particularly spherical cap bubbles. In more viscous liquids very small bubbles can be formed by the bursting of bubbles at the bubbly liquid-gas interface. These tiny bubbles have negligibly low rise velocities and are good flow followers for the liquid. Consequently, they can be distributed throughout the bubble column by backmixing of the liquid. They also accumulate and can quickly turn clear liquids milky.

The size of bubbles can increase through coalescence. This is often conceived of as two bubbles coming together, the thinning of the film of liquid between them and its subsequent rupture. However, three types have been identified [18, 19]: coming together of bubbles due to turbulence; different sized bubbles approaching each other due to velocity differences; and one bubble being drawn into the wake of a preceding one.

Figure 2.10 shows coalescence between spherical cap bubbles [20]. In the first frame the lower bubble is seen to be distorted by being in the wake of the upper one. Here the liquid is an ionic liquid (1-ethyl-3-methylimidazolium ethylsulfate, ([EMIM][EtSO$_4$])). Coalescence between the spherical cap and intermediate bubbles appears far less common. Instead, the intermediate bubbles in the wake of the leading spherical cap bubble are thrust to one side and then resume their station in the tail of the combined spherical cap bubble.

In experiments on very viscous liquids in a large diameter column (225 mm), large bubbles (equivalent diameter of about 120 mm) were released alternately on either side of the column. It was seen that if the releases were close enough the trailing bubble was drawn across the column diameter into the wake of the leading one, and coalescence followed.

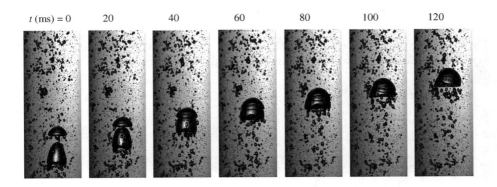

Figure 2.10 *Sequence of stills from high-speed video illustrating the coalescence of two spherical cap bubbles. Liquid = [EMIM][EtSO$_4$]. Gas superficial velocity = 3.6 mm/s. Note the intermediate sized bubbles, in the wake of the upper spherical cap bubble, that are displaced by the arrival of the second spherical cap bubble and which eventually reposition themselves in the wake of the combined spherical cap bubble (Reprinted from Kaji, R., et al., Studies of the interaction of ionic liquid and gas in a small diameter bubble column, Ind. Eng. Chem. Res.,* **48***, 7938–7944. Copyright (2009) with permission from American Chemical Society.)*

The low pressure in the wake region of the upstream bubble was sufficient to draw the trailing one across the pipe. Similar behaviour has been seen in a high-speed video taken of a 125 mm bubble column with a gas distributor consisting of a plate with 25, 1 mm diameter holes [21]. The liquid is the ionic liquid referred to above. Once again, bubbles were seen to move considerable lateral distances and move into the wake of a preceding bubble and thence coalesce.

2.3.3   Bubble Movement

2.3.3.1   Bubble Shape

Bubbles, if small enough, will have a spherical shape, Figure 2.11a. Larger bubbles become elliptical though of fairly constant shape, as shown in Figure 2.11b. The next change results in a difference between the front and the back of the bubble, the lower side is flatter [22], Figure 2.11c. Even large bubbles are approximately elliptical but wobble about a mean shape. Beyond this, bubbles are termed spherical cap bubbles. These have the shape of part of a sphere as illustrated in Figure 2.11d. There can be small bubbles torn off the trailing edge if the liquid is of low viscosity and waves can occur on the front surface. These waves have been shown to have considerable influence on mass transfer.

The volume of elliptical bubbles, Type **b**, is given by:

$$\frac{\pi d_e^3}{6} = \frac{4\pi}{3}(ba^2) \tag{2.11}$$

where

d_e is the equivalent sphere diameter.

Whilst for Type **c** bubbles it can be written as:

$$\frac{\pi d_e^3}{6} = \frac{2\pi}{3}(ba^2 + \beta ba^2) \tag{2.12}$$

The aspect ratio of the elliptical bubbles, E, is b/a, whilst for those with a flattened front the aspect ratio is defined as $E = (b + \beta b)/2a$. For both, the relationship to the equivalent sphere diameter is:

$$\frac{d_e}{2a} = E^{0.333} \tag{2.13}$$

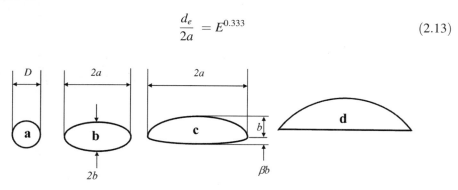

Figure 2.11   *Shapes of bubbles*

A number of relationships have been proposed for the aspect ratio. Some are simple functions of the equivalent diameter, for example:

$$E = \frac{1}{1 + c_1 Eo^{c_2}} \tag{2.14}$$

where the Eötvös number, $Eo = g d_e^2 (\rho_l - \rho_g)/\sigma$, is the ratio of gravitational to surface tension forces. Values of $c_1 = 0.163$ and $c_2 = 0.757$ have been proposed by Wellek *et al.* [23] based on work on non-oscillating drops in contaminated liquids, but it has also been seen to be applicable to oscillating bubbles in low viscosity liquids. Others are also functions of the terminal velocity of the bubbles and the two relationships need to be solved simultaneously. The simpler version takes the form:

$$E = \frac{1}{1 + c_3 We} \tag{2.15}$$

where the Weber number, $We = \rho_l V_T^2 d_e / \sigma$, is the ratio of inertial to surface tension forces. A more comprehensive method has been proposed by Fan and Tsuchiya [24]:

$$\begin{array}{ll} E = 1 & Ta < 0.3 \\ E = \{0.77 + 0.24 \tanh[1.9(0.4 - \log_{10} Ta)]\}^2 & 0.3 < Ta < 20 \\ E = 0.3 & 20 < Ta \end{array} \tag{2.16}$$

In this the composite dimensionless group, termed the Tadaki number, $Ta = \mathrm{Re}\, Mo^{0.23}$, is employed, which introduces the effect of the viscosity of the liquid. In this, the Reynolds number is, $\mathrm{Re} = \rho_l V_T d / \eta_l$ and the Morton number is, $Mo = g \eta_l^4 / \sigma^3 \rho_l$. These equations have been tested against experimental data by, for example Celata *et al.* [25]. Although the mean errors are reasonably small there is still considerable scatter. It is noted that Equations 2.15 and 2.16 involve the terminal velocity of the bubbles and so these equations will have to be solved simultaneously with those for velocity presented below.

2.3.3.2 Bubble Motion

The behaviour of bubbles of small sizes, of the order of 1 mm diameter, can be very clearly seen in a fairly simple domestic or leisure experiment. If a carbonated drink is very carefully poured into a smooth-walled container very few bubbles will be seen to be present. However, if a salted peanut (or other nut) is dropped into the liquid, then regular streams of bubbles are formed at the cavities where the salt crystals join the peanut. These rise in straight lines. Similar behaviour is observed in small bubbles formed at orifices.

However, when larger bubbles become distorted, they are seen to have curvilinear paths, as illustrated [26] in Figure 2.12. It has been observed that the motion can be of a zig-zag type or more of a spiral type, according to the original distortion undergone by the bubble at formation. Small distortions favour zig-zag motions whilst larger distortions cause spiral motions.

2.3.3.3 Bubble Velocity

When considering the terminal velocities achieved by bubbles rising in liquids, three different regions have been identified, corresponding to the most important forces that control the velocity. With increasing bubble size these are termed the viscous, surface tension and inertial regions. As will be shown below, the extent of these regions depends on

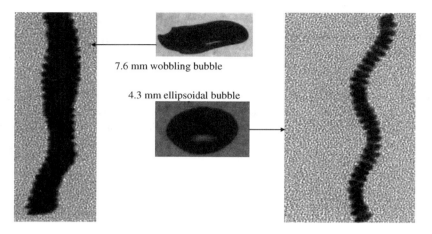

Figure 2.12 *Example of the shapes of real bubbles and their motion as illustrated by multiple images (Reprinted from Žun, I., The principles of complexity in bubbly flow, Multiphase Sci. Tech., 17, 169–190. Copyright (2005) with permission from Begell House, Inc.)*

the physical properties of the bulk liquid but also on the presence and concentration of additional chemicals, often present at low concentrations, which might have an effect on the micro-motion at the bubble interface.

A simple equation for the bubble terminal velocity, V_T, was put forward by Harmathy [27]

$$V_T = 1.53 \left[\frac{g(\rho_l - \rho_g)\sigma}{\rho_l^2} \right]^{0.25} \tag{2.17}$$

As can be seen, this does not involve the bubble diameter. In the viscous dominated region, the terminal velocity can be obtained from the balance of the drag and buoyancy forces [22].

$$c_D \frac{\pi d^2}{4} \frac{\rho_l V_T^2}{2} = \frac{\pi d^3}{6} g(\rho_l - \rho_g) \tag{2.18}$$

which can be rewritten in dimensionless terms as:

$$c_D = \frac{4Ar}{3\,\mathrm{Re}^2} \tag{2.19}$$

where the Archimedes number,

$$Ar = \frac{d^3(\rho_l - \rho_g)\rho_l g}{\eta_l^2} \tag{2.20}$$

A drag law equation is then required. A modification of the relationship proposed by Schiller and Nauman has been found to give accurate results for liquids of both low and higher viscosities [22].

$$c_D = \frac{16}{\mathrm{Re}} (1 + 0.15\,\mathrm{Re}^{0.687}) \tag{2.21}$$

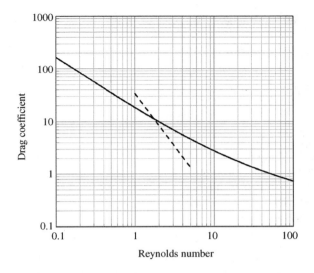

Figure 2.13 *Drag law Equation 2.21 (——) and force balance Equation 2.19 (------)*

The two relationships between drag coefficient and Reynolds number can be solved simultaneously for a given bubble diameter and physical properties. The curves are illustrated in Figure 2.13 for a liquid of viscosity 0.05 Pa s, density 900 kg/m^3 and a gas density of 5 kg/m^3. The bubble size used in the Archimedes number is 2 mm.

For the inertial dominated region, Mendelson [28] has proposed an equation using the similarity of the bubbling process to surface waves over deep water. This results in a combination of the Weber and Froude numbers. The terminal velocity can be calculated from:

$$V_T = \sqrt{\left(\frac{2\sigma}{\rho_l d} + \left[\frac{\rho_l - \rho_g}{\rho_l}\right]\frac{gd}{2}\right)} \tag{2.22}$$

Allowing for distortion of bubbles from the spherical to the elliptical, Tomiyama *et al.* [22] created a more complex formula

$$V_T = \frac{\sin^{-1}\sqrt{1-E^2}-E\sqrt{1-E^2}}{1-E^2}\sqrt{\left(\frac{8\sigma}{\rho_l d}E^{1.33} + \left[\frac{\rho_l - \rho_g}{\rho_l}\right]\frac{gd}{2}\frac{E^{0.67}}{1-E^2}\right)} \tag{2.23}$$

Lin *et al.* [8] published a non-dimensional equation that covers all three regions. It has been tested against experiments over a wide range of temperatures (affecting liquid viscosities) and pressures, and excellent agreement is shown. Temperature has a significant effect on the viscosity of the liquid employed. The equation relates dimensionless velocity [$u^* = V_T(\rho_l/g\sigma)^{0.25}$] to a dimensional bubble size [$d^* = d\,(\rho_l g/\sigma)^{0.5}$] and accounts for different types of liquids: whether aqueous or organic, single or multi-component and pure or contaminated.

$$u^* = \left\{ \left[\frac{d^{*2}}{Mo^{0.25}K_b} \left(\frac{\rho_l - \rho_g}{\rho_l} \right)^{1.25} \right]^{-n} + \left[\frac{2c}{d^*} + \left(\frac{\rho_l - \rho_g}{\rho_l} \right) \frac{d^*}{2} \right]^{-0.5n} \right\}^{-1/n} \tag{2.24}$$

where

$n = 0.8$ for contaminated liquids
$n = 1.6$ for purified liquids
$c = 1.2$ for single component liquids
$c = 1.4$ for multi-component liquids
$K_b = \max (K_{b0}Mo^{-0.038}, 12)$
$K_{b0} = 14.7$ for aqueous liquids
$K_{b0} = 10.2$ for organic liquids.

A similar superposition technique has been employed by Jamialahmadi *et al.* [29] who used an equation similar to that of Mendelson for the larger bubble contribution

$$V_{TL} = \sqrt{\left(\frac{2\sigma}{(\rho_l - \rho_g)d} + \frac{gd}{2} \right)} \tag{2.25}$$

and a direct equation for the viscous drag contribution

$$V_{TS} = \frac{(\rho_l - \rho_g)gd^2}{18\eta_l} \left[\frac{3\eta_l + 3\eta_g}{2\eta_l + 3\eta_g} \right] \tag{2.26}$$

which are combined via

$$V_T = \frac{V_{TL}V_{TS}}{\sqrt{V_{TL}^2 + V_{TS}^2}} \tag{2.27}$$

It was noted above that the motion of the bubble depended on its size, shape and initial disturbance. This also has an impact on the terminal velocity of the bubbles. Figure 2.14 shows the velocity/size data for air bubbles in distilled water. The data are distinguished according to the type of motion. As seen in Figure 2.14, there can be a factor of two difference in the velocity according to whether the bubble moves in a zig-zag or spiral motion. It is shown that Equation 2.22 provides an upper limit and Equation 2.23 a lower limit to the data. The velocity predicted by Equation 2.17, which is 0.25 m/s, passes through the midst of the measured data.

The behaviour of these equations is seen in Figure 2.15. Here a liquid of viscosity 0.05 Pa s, density 900 kg/m^3 and surface tension of 0.02 N/m, and a gas of density 5 kg/m^3 have been used. As seen, the viscous and inertial equations (Equations 2.19–2.22) show good asymptotic agreement with the Fan–Tsuchiya equation. For this liquid there is not the region that is independent of bubble size as proposed by Harmathy [27]. If the effect of liquid viscosity is examined systematically, with all other properties being kept the same as above,

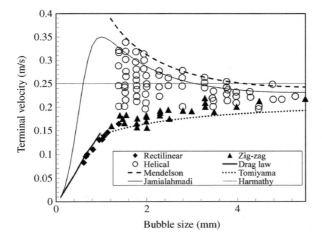

Figure 2.14 *Bubble velocity dependence on bubble size and type of motion. Experimental data from Tomiyama et al. [22]*

then it is seen in Figure 2.16 that Equation 2.17 is fairly reasonable in the bubble size range 1–5 mm for liquid viscosities around that of water. The figure shows that as the liquid viscosity increases the surface tension dominated region disappears and the viscous region blends into the inertial region. As the viscosity is increased, the terminal rise velocity of very small bubbles drops towards zero, they become very good flow followers of the liquids. They are also difficult to remove, as they hardly rise in still liquid.

2.3.3.4   *Effect of Multiple Bubbles*
Under homogeneous conditions, the presence of many similar sized bubbles causes an effect akin to hindered settling in sedimentation. The bubble velocity is reduced from that of an isolated bubble by a factor, often $(1-\varepsilon_g)$ raised to some power. In contrast, when there are

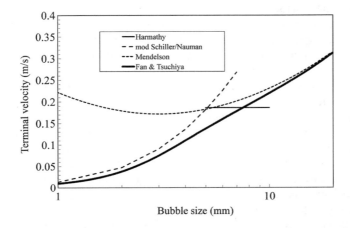

Figure 2.15   *Velocities calculated by different methods – liquid viscosity = 0.05 Pa s*

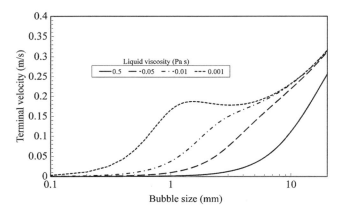

Figure 2.16   *Effect of liquid viscosity on the bubble velocity calculated using the method of Lin et al. [8]*

large bubbles present, it has been observed that larger bubbles rise more rapidly in the presence of smaller bubbles. This is caused by the small bubbles altering the shape of the nose of the large spherical cap, or Taylor bubble, and making it move faster [30].

2.3.4   Void Fraction Prediction

In selecting equations to predict the void fraction, overall correlations can be used. However, it is often better to use more physically based methods. Above, two major flow patterns were identified, homogeneous and heterogeneous flow. The former is identified as containing uniform small bubbles whilst for the latter there is a much larger range of bubble sizes. The description can be taken further and as the gas flow rate increases it can be considered that gas goes to form more small bubbles until a transition point is reached. Thereafter, all gas goes to form larger bubbles. Therefore, before going into the detail of void fraction equations, at this point the coordinates of the transition, the gas superficial velocity and void fraction need to be considered. Methods to identify these and routes to specifying them are given.

The two main approaches for the identification of the transition are seeking characteristics in the void fraction/gas superficial velocity data or interrogation of a time varying signal, such as the wall pressure. There are a number of variants in extracting information from void fraction data. One is illustrated in Figure 2.17. Here the ratio of gas velocity (gas superficial velocity divided by void fraction) to bubble rise velocity is plotted against void fraction. Also plotted is a theoretical curve. Initially, the values derived from the experimental data lie reasonably well on the theoretical curve. Suddenly there is a strong deviation. This is taken to be the transition. For this exercise, the bubble rise velocity is calculated using the equation of Harmathy. The theoretical curve is that suggested by Richardson and Zaki [31] $\{= u_{gs}/[V_T(1-\varepsilon_g)^{1.39}]\}$. Note that there are a number of variants to this approach and different values can be obtained for the transition according to the rise velocity and the model equation used. The specific approach used here might not be very applicable to viscous liquid cases. As seen in Figure 2.17, the equation used over-predicts the bubble rise velocity.

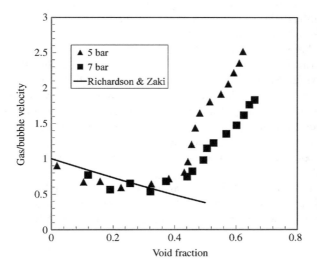

Figure 2.17 *Drift flux method to identify void fraction and gas superficial velocity at the homogeneous/heterogeneous transition*

An alternative approach is to consider the scale of bubble aggregation, that is the difference between gas superficial velocity and the multi-bubble corrected bubble velocity normalised with the gas superficial velocity $[=1-V_T\varepsilon_g(1-\varepsilon_g)^{1.39}/u_{gs}]$. As shown in Figure 2.18, when plotted against gas superficial velocity this shows a minimum, which corresponds to the homogeneous/heterogeneous boundary.

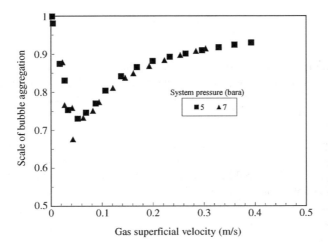

Figure 2.18 *Scale of bubble aggregation for air-water in a 150 mm diameter column at different pressures*

The dependence of the gas superficial velocity at transition has been examined systematically [32]. Although there is an effect of column diameter and liquid height on this value, this effect seems to disappear for diameters greater than ~0.3 m and liquid heights of more than 4 m (for columns of diameter ~0.15 m). Examination of the effect of the size and position of the orifices in the gas distributor plate shows that there is little effect for orifice diameters > 1.5 mm. Smaller orifice diameters result in a higher transition velocity. For orifices in the range 5–10 mm diameter, the transition velocity is not affected if the spacing between orifices is > 30 mm. Smaller spacings diminish the transition velocity. Again, available research is almost exclusively on air-water systems, but where the effect of liquid viscosity is examined it is found that, for viscosities greater than that of water, the critical void fraction at transition decreased with increasing viscosity to the power of ~0.25–0.4 [33].

Available equations for predicting the transition void fraction and gas superficial velocity have been tested against experiment by Ribeiro [34]. The most accurate was an equation he presented himself. However, this has 20 constants. Equations have been proposed by Wilkinson *et al.* [35],

$$\varepsilon_{gtr} = 0.5e^{-193\frac{\sigma^{0.11}\eta_l^{0.5}}{\rho_g^{0.61}}} \tag{2.28}$$

$$u_{tr} = 2.25\left(\frac{\sigma\varepsilon_{gtr}}{\eta_l}\right)Mo^{0.273}\left(\frac{\rho_l}{\rho_g}\right)^{0.03} \tag{2.29}$$

and Reilly *et al.* [36]

$$\varepsilon_{gtr} = \frac{0.59B^{1.5}\rho_g^{0.48}\sigma^{0.06}}{\rho_l^{0.5}} \tag{2.30}$$

$$u_{tr} = \frac{0.352\varepsilon_g(1-\varepsilon_g)\sigma^{0.12}}{\rho_g^{0.04}} \tag{2.31}$$

where B is a parameter that depends on the fluid. On the whole, these are reasonably accurate but the latter can have large overall errors due to a few outlier points, mainly from high viscosity liquids. If we examine the values obtained from Figures 2.17 and 2.18, it is seen that these methods agree very well, and reasonably well, respectively, with the predictions of Reilly *et al.* [36].

Void fraction can be defined as the ratio of the gas superficial velocity to the actual velocity, that is the terminal velocity of the bubbles. However, the equations provided in the last section are for isolated bubbles. As noted, the presence of other bubbles alters the velocities: small bubbles are hindered and so their velocities are lowered; for large bubbles other bubbles can cause augmentation of the isolated bubble velocity. In general terms, the slip velocity can be written as:

$$u_s = \frac{u_{gs}}{\varepsilon_g} - \frac{u_{ls}}{1-\varepsilon_g} = V_T F(\varepsilon_g) \tag{2.32}$$

where

$F(\varepsilon_g)$ is a function describing bubble interaction.

Wallis [37] proposed different functions for small, $(1-\varepsilon_g)$, and large, $1/(1-\varepsilon_g)$, bubbles. For zero liquid flow rate this results in

$$\varepsilon_g = 0.5 - \sqrt{0.25 - \frac{u_{gs}}{V_T}} \quad \text{for small bubbles} \tag{2.33}$$

$$\varepsilon_g = \frac{u_{gs}}{u_{gs} + V_T} \quad \text{for large bubbles} \tag{2.34}$$

For a finite liquid flow rate this latter equation becomes

$$\varepsilon_g = \frac{u_{gs}}{(u_{gs} + u_{ls}) + V_T} \tag{2.35}$$

Alternative functions have been provided for $F(\varepsilon_g)$. Some of these have the void fraction raised to a power $\neq 1$ and for those instances iteration is required to solve for the void fraction. For example, Richardson and Zaki [31] suggested $F(\varepsilon_g) = (1-\varepsilon_g)^{n-1}$, where n depends on the Reynolds number for an isolated bubble, Re_T. The proposed relationship between n and Re_T is given in Table 2.1.

Figure 2.19 shows the dependence of the void fraction on gas superficial velocity and bubble size calculated using this approach for an air-water bubble column. These curves were derived using Equation 2.14 for the bubble shape and the Tomiyama correlation, Equation 2.23, for the terminal velocity. The Reynolds number of a single bubble at terminal velocity was computed using the equivalent diameter. Note that from the Equation 2.32, assuming zero liquid flow rate and using the Richardson–Zaki form of $F(\varepsilon_g)$, the maximum gas fraction and corresponding gas superficial velocity are readily obtained and found to be $\varepsilon_{g\,max} = 1/n$.

Another refinement is to allow for velocity profiles in the overall sense that there is an upflow of gas and liquid at the centre and downflow at the wall. This leads to a correction factor on the denominator of Equation 2.35, that is

$$\varepsilon_g = \frac{u_{gs}}{C_o(u_{gs} + u_{ls}) + V_T} \tag{2.36}$$

Table 2.1 *Dependence of exponent of Richardson and Zaki [31] on Reynolds number (Data from Richardson, J. F. and Zaki, W. N., Sedimentation and Fluidization: Part I, Trans. I. Chem. Engrs., **32**, 35–53. Copyright (1954) Institute of Chemical Engineers.)*

Range of Reynolds numbers	n
$Re_T < 0.2$	4.65
$0.2 < Re_T < 1$	$4.35\,Re_T^{-0.03}$
$1 < Re_T < 500$	$4.45\,Re_T^{-0.1}$
$500 < Re_T$	2.39

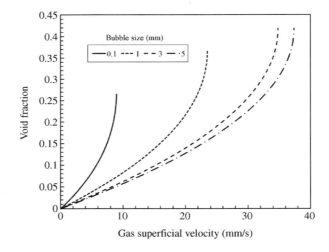

Figure 2.19   *Effect of gas superficial velocity and bubble size on void fraction in the homogeneous region*

This is called the drift flux equation and was originated by Zuber and Findlay [38]. Here, C_o is the radial distribution parameter, $\langle \varepsilon_g(r)u_g(r)\rangle / \langle \varepsilon_g(r)\rangle \langle u_g(r)\rangle$, with $\langle \cdot \rangle$ cross-sectional averaging.

The behaviour of these equations is illustrated in Figures 2.20 and 2.21. In the former the data were taken on a 125 mm column using deionised water. The bubbles were created using

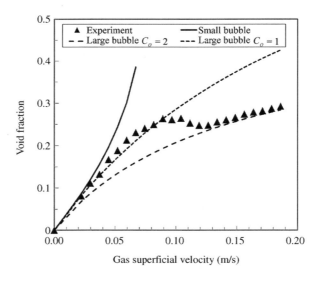

Figure 2.20   *Gas superficial velocity dependence of void fraction distilled water-air on a 127 mm column with a distributor of 0.2% open area and 0.5 mm diameter holes*

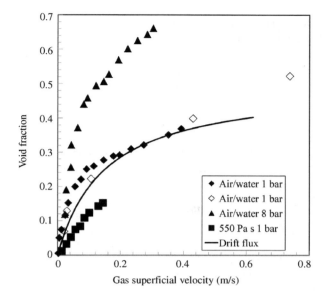

Figure 2.21 *Comparison of predictions of drift flux method with experimental data*

a plate containing 121, 0.5 mm holes and the data show the typical initial peak characteristic of such small holes. Equation 2.33 predicts well at low gas velocities, whilst Equation 2.35 shows the correct trends and values at the higher gas velocities. A very early version of such equations used $C_o=2$, $u_{ls}=0$ and $V_T=0.3$. Figure 2.20 illustrates how this predicts atmospheric pressure air-water data at higher gas velocity fairly well, it is less suitable for lower velocities and also for both higher pressure air-water and for more viscous liquids.

Others have proposed variants on Equation 2.36. Hills [39] suggested

$$\varepsilon_g = \frac{u_{gs}}{1.35(u_{gs}+u_{ls})^{0.93}+V_T} \tag{2.37}$$

and Clark and Flemmer [40] proposed

$$\varepsilon_g = \frac{u_{gs}}{1.95u_{gs}+0.93u_{ls}+V_T} \tag{2.38}$$

In some studies, the constant C_o has been taken as a function of the density ratio. Kataoka and Ishii [41] took $C_o=1.2-\sqrt{(\rho_g/\rho_l)}$ whilst Cumber [42] suggested $C_o=(\rho_l/\rho_g)^{0.05}$. The proposal of Beattie and Suguwara [43] differs from other reports in that it is based on a simple model. It follows the idea that C_o is identified through the ratio of maximum to average velocities and considers a two-phase 'universal' velocity profile to give $C_o = 1 + 2.6\sqrt{f}$. They gave an implicit expression for the friction factor but indicated that it could be approximated by:

$$f = \frac{0.05525}{Re_m^{0.237}} + 0.0008 \qquad (2.39)$$

where

$$Re_m = (\rho_g u_{gs} + \rho_l u_{ls})D_c/3\eta_l$$

This approach links to observations in turbulent pipe flow. The mean axial velocity profiles in both single- and two-phase flow have been found to fit an equation of the form

$$\frac{u}{u_{max}} = \left(\frac{y}{R}\right)^n \qquad (2.40)$$

where

u_{max} is the maximum velocity in the channel
y is the distance from the wall
R is the pipe radius.

Data from the drop-laden core of annular flow show that the values of the power law exponent, n, were larger than the equivalent single-phase data values. However, when plotted against friction factor (determined from the pressure gradient or from extrapolation of Reynolds stress data to provide wall shear stresses) the values of n lie on the same line as single-phase data measured by Nikuradse and Nunner [44a,b] in smooth and rough walled pipes. Over the range of data in this figure, the ratio of maximum to mean velocities is 1.15, 1.22 and 1.74 for $n = 0.098, 0.143$ and 0.43 whilst C_o as calculated from the expression of Beattie and Sugawara comes out at 1.13, 1.22 and 1.65. The predictions from the above variants have been compared with experimental information from the database in which the points were identified as bubbly flow [45]. This contained 614 points with data from air-water, steam-water from pressures up to 166 bar and nitrogen-mercury. The range of density ratios was from 6.6 to 11 560 and the pipe diameters were from 0.009 to 0.168 m. The resulting distribution of errors was reasonably Gaussian. Mean errors of 6% (Zuber and Finlay), 11% (Hills), 4% (Clark and Flemmer) and 6% (Beattie and Suguwara) were obtained. The spread of errors was similar for all four methods.

Apart from that of Wallis, these methods do not distinguish between homogeneous and heterogeneous flows. Moreover, even Wallis does not allow for the simultaneous presence of large and small bubbles. An approach that takes into account the effect of flow patterns might be expected to be more accurate. Krishna *et al.* [46] used the concept that up to the homogeneous/heterogeneous transition, all gas goes into small bubbles. Beyond the transition, additional gas goes entirely into large bubbles. The void fraction is given by:

$$\varepsilon_g = \varepsilon_{bL} + \varepsilon_{gtr}(1 - \varepsilon_{bL}) \qquad (2.41)$$

where

$$\varepsilon_{bL} = \frac{u_{gs} - u_{gtr}}{V_L} \text{ if } u_{gs} > u_{gtr} \text{ otherwise } \varepsilon_{bL} = 0 \qquad (2.42)$$

is the void fraction contribution of large bubbles (all in S.I. units) and V_L is the large bubble velocity which is obtained from

$$V_L = 0.71\sqrt{gd_{bL}}\left(2.73 + 4.505\left[u_{gs}-u_{tr}\right]\right)\sqrt{\frac{1.29}{\rho_g}} \qquad 0.125 > d_{bL}/D_c$$

$$V_L = 0.802e^{\frac{-d_{bL}}{D_c}}\sqrt{gd_{bL}}\left(2.73 + 4.505\left[u_{gs}-u_{tr}\right]\right)\sqrt{\frac{1.29}{\rho_g}} \quad 0.125 < d_{bL}/D_c < 0.6 \quad (2.43)$$

$$V_L = 0.35\sqrt{gD_c}\left(2.73 + 4.505\left[u_{gs}-u_{tr}\right]\right)\sqrt{\frac{1.29}{\rho_g}} \qquad d_{bL}/D_c > 0.6$$

In these equations, the first term is the basic large bubble velocity whilst the second and third are correction factors to account for the effect of bubble proximity and of gas pressure on bubble velocity. Note, this last is correct for water but the effect is smaller for higher liquid viscosities. Indeed, at a viscosity of 0.55 Pa s there is hardly any effect of pressure. The large diameter bubble size is calculated from:

$$d_{bL} = 0.069(u_{gs}-u_{gtr})^{0.376} \qquad (2.44)$$

The void fraction and gas superficial velocity at the homogeneous/heterogeneous transition are obtained from the equations of Reilly *et al.* (see above). The method gives good predictions for air-water data at pressures of 1–13 bar. Examples are shown in Figure 2.22. For the homogeneous region, Equation 2.33 is used. The figure also illustrates the use of alternative equations for the critical void fraction and gas superficial velocity proposed by Wilkinson *et al.* These are seen to under predict the air-water data. The equations have also been used to predict air-sugar solution data (liquid viscosity $= 0.55$ Pa s) as illustrated in Figure 2.23. The version using the transition parameters from Wilkinson *et al.* gives a much better prediction, not surprisingly as they contain liquid viscosity explicitly. It must be noted that the equations given above utilise nearly 20 constants. There have been many more equations published. Ribeiro and Lang [47] have gathered them together and list 37.

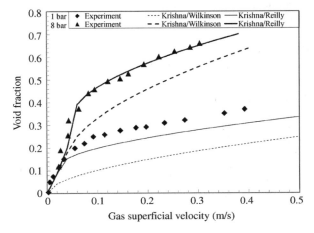

Figure 2.22 *Comparison of predictions of model of Krishna et al. [46] with air-water data*

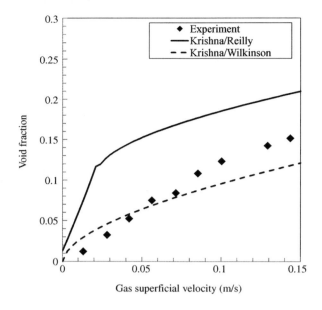

Figure 2.23 *Comparison of predictions of models of Krishna et al. [46] with viscous liquid data*

The predictions of two examples are shown in comparison with experimental data in Figures 2.24 and 2.25. Both use dimensionless groups and have the general form

$$F = A \left(\frac{We}{Re}\right)^a Mo^b \left(\frac{\rho_g}{\rho_l}\right)^c \left(\frac{\eta_g}{\eta_l}\right)^e \qquad (2.45)$$

where

We is the Weber number
Re is the Reynolds number
Mo is the Morton number.

For the Hikita equation,

$$\varepsilon_g = F$$
$$A = 0.672$$
$$a = 0.578$$
$$b = -0.131$$
$$c = 0.062$$
$$e = 0.107$$

whilst for the Hammer equation

$$\varepsilon_g = F/(1 + F)$$
$$A = 0.4$$
$$a = 0.87$$
$$b = -0.27$$
$$c = 0.17$$
$$e = 0$$

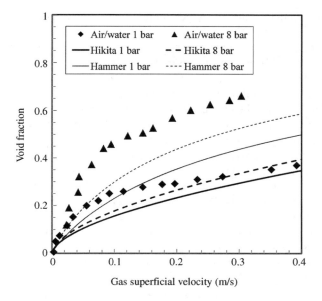

Figure 2.24 *Comparison of predictions of correlations (see Riberio and Lage [47]) with air-water data*

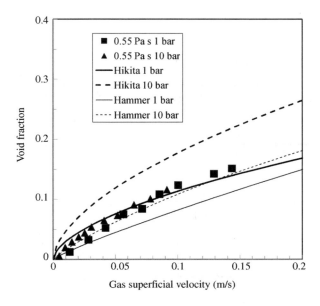

Figure 2.25 *Comparison of predictions of correlations (see Riberio and Lage [47]) with viscous liquid data*

As seen, they both under-predict the effect of pressure for the water data and over-predict it for the viscous liquid data.

Gas lift loop reactors are useful because they operate, as a result of the relatively high liquid flow, usually in a more or less uniform flow regime. A similar procedure as described above can be used to estimate the gas fraction in the loop as a function of the gas superficial velocity. An additional complication is that the liquid circulation rate also needs to be estimated, as this directly influences the bubble residence time and thus the gas fraction. The liquid circulation is a consequence of the difference in density of the bubbly mixture in the riser and down-comer, which is full of liquid. The riser density is, due to the gas injection, lower than in the riser and gravity will push the riser contents up, while pulling the downer fluid downwards. As the gas leaves the riser at the top of the loop, the driving force is maintained and the acceleration induced by gravity will be counteracted by friction of the flowing mixture with the walls. A one-dimensional mechanical energy balance going around the loop will connect the driving density difference between the riser and down-comer to the energy dissipation due to friction. Of course, the driving density difference is a function of the riser gas fraction (and if applicable the average gas fraction in the down-comer section). The gas fraction, in its turn, is influenced by the residence time of the gas bubbles. This depends on the bubble velocity, which is the sum of the bubble slip velocity and the liquid circulation velocity. Thus, both the mechanical energy balance and the gas fraction in the riser depend on the unknown gas fraction and the unknown liquid circulation velocity. Figure 2.26 shows a schematic of these equations and their inter-relationship.

For a gas lift of height H with a bubble-free down-comer, the mechanical energy balance can be written as (ignoring the gas density with respect to the liquid density):

$$\varepsilon_g \rho_l H = K_e \frac{\rho_l (u_{gs} + u_{ls})^2}{2} \tag{2.46}$$

where

u_{gs} denotes the gas superficial velocity fed to the air lift
u_{ls} denotes the superficial liquid velocity that represents the liquid circulation in the riser.

K_e contains all contributions to the friction, such as wall friction, bends, expansions, contractions, and so on, in addition to possible geometry changes that alter the local liquid flow rate (e.g. if riser and down-comer have a different cross-sectional area).

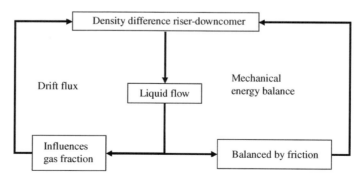

Figure 2.26   *Information flow for gas lift systems*

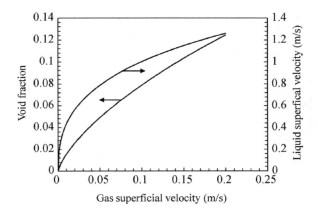

Figure 2.27 *The effect of gas superficial velocity on void fraction in the riser and on liquid circulation, expressed as liquid superficial velocity. Column height and diameter were selected as 5 and 0.2 m, respectively. Bubble size assumed to be 3 mm. Fluids were air and water at atmospheric pressure*

The slip velocity of the bubbles is again the difference between the actual bubble velocity and the local liquid velocity. These can be related to the gas superficial velocity and the liquid superficial velocity using Equation 2.32 with the Richardson and Zaki function:

$$\frac{u_{gs}}{\varepsilon_g} - \frac{u_{ls}}{1-\varepsilon_g} = V_T(1-\varepsilon_g)^{n-1} \qquad (2.47)$$

Note that the K_e is also a function of the velocities.

With these two equations, both the gas fraction and the liquid circulation can be calculated.

Figure 2.27 gives an example for an air lift with external down-comer. The height of the air lift is 5 m, the column diameter 0.2 m and the bubbles are assumed to be of 3 mm spherical equivalent diameter.

2.3.5 Detailed Behaviour of the Flow

When considering the flow in more detail, the aspects of axial, radial and temporal variations will be examined in addition to the distribution of the bubble sizes within a particular flow. There are two ways in which *axial variation* can be present. The first is due to the acceleration of bubbles to their terminal velocity. The second way in which axial variation might be important is through a change in pressure. As the flow rises through the column, pressure decreases due to a diminution in the static head. There is a consequent decrease in gas density with concomitant increase in gas superficial velocity. Obviously, this will affect the void fraction. Examples can help quantify the effect. For a column filled with a gas-water mixture to a height of 1.3 m and operating at a void fraction of 0.25 and atmospheric pressure at the gas outlet, there will be a 10% difference in pressure, gas density and gas superficial velocity from the top to the bottom of the column. If the operating pressure was 10 bar, the effect would be 1%. Another reason why there might be axial variations is because of end

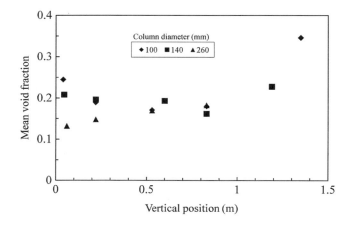

Figure 2.28 *Axial variation of cross-sectionally averaged void fraction – data from Kumar et al. [48]*

effects, particularly at the top of the column. Measurements at different planes by Kumar *et al.* [48] showed the extent of end effects as illustrated in Figure 2.28.

Data on the *radial* void fraction profile can be obtained from point probes (electrical or optical), γ or X-ray computed tomography, electrical resistance or capacitance tomography or wire mesh sensor tomography. Point probes give local time varying data and radial information, obtained by repeated tests with the probe moved from position to position. The computed tomography approaches will give time averaged data, whilst the other tomography methods can give data resolved in space and time. There is an extensive literature on this subject. Relevant material can be found in a review by Parasu Veera and Joshi [49]. Two types of radial profiles have been reported. These are described as the wall peak and the centre peak. The former show maxima close to the wall, correctly this is a ring with a dip in the middle. This seems to be found only in columns/pipes of ≤ 60 mm diameter. For large diameter units the norm is the void fraction that is maximum at the centre of the column, as reported, for example, in the much cited paper by Hills [50]. Initially a simple power law equation was proposed to fit the data. However, a subsequent examination of bubble column data indicated that it might be that the void fraction close to the wall was not zero. In terms of the cross-sectionally averaged void fraction, this can be written as:

$$\varepsilon_g = \bar{\varepsilon}_g \left(\frac{n+2}{n+2-2c_r}\right)\left(1-c_r\left[\frac{r}{R}\right]^n\right) \tag{2.48}$$

Wu *et al.* [51] examined their own and other researchers' data and proposed that n and c be specified by:

$$n = 2188\mathrm{Re}_g^{-0.598}\,Fr_g^{0.146}Mo^{-0.004} \tag{2.49}$$

and

$$c_r = 0.0432\,\mathrm{Re}_g^{0.2492} \tag{2.50}$$

where

$$Re_g \; \frac{(\rho_l - \rho_g)u_{gs}D_c}{\eta_l} \quad \text{and} \quad Fr_g = \frac{u_{gs}^2}{gD_c}$$

For the liquid velocity there is a similar equation [52]

$$u_l(r) = u_{lo}\left(1 - 2.65n^{0.44}c_r\left[\frac{r}{R}\right]^{2.65n0.44c_r}\right) \tag{2.51}$$

The profiles can be described by power laws in dimensionless radii. The exponent, n, has been found to vary between 1.4 and 11 for the void fraction and from 2 to 2.3, or even greater, for the velocity.

Most studies concentrate on either void fraction or velocity. In Figures 2.29 and 2.30 void fraction and velocity data [53] taken on the same facility are displayed. The experiments were carried out with liquid superficial velocities of 0 and 0.175 m/s and so correspond to bubble column and gas lift mode of operation. The void fraction data show a peak at the column centre. There are indications here that there is a non-zero void fraction at the wall. The velocity data similarly show a peak at the centre. For the bubble column version, there are regions of negative (downward) velocity at the wall. Obviously, this does not occur for the gas lift version. The almost identical profiles between the void fraction data from the two cases and between the velocity data when the bubble column data are shifted by the mean liquid velocity used in the gas lift case show that the flows are very similar. The equations of Wu are also shown. However, in this case the value of n given by Equation 2.49 has had to be divided by four to give the fit shown here. The predictions of Equation 2.48 have been

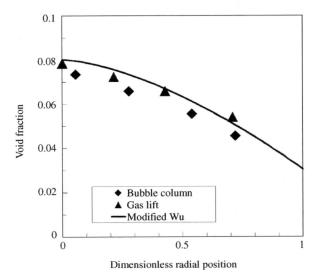

Figure 2.29 *Radial void fraction profiles measured by Mudde and Saito [53]. Also shown are predictions of modified equation of Wu et al. [51]*

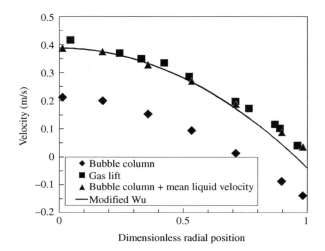

Figure 2.30 *Radial profiles of liquid axial velocity measured by Mudde and Saito [53] for both a standard bubble column and a column with recirculation. Also shown are standard column data transposed by liquid superficial velocity and predictions of the equation suggested by Wu and Al-Dahhan [52]*

compared with the data obtained using wire mesh sensor tomography on a 192 mm diameter 10 m tall bubble column operating with steam-water at 46 bar. Again good agreement could be achieved, but the parameter *n* had to be multiplied by a factor of 3.6, Figure 2.31.

The *variation with time* can be determined from point probes and tomography measurements. The complex nature of the flow can be seen in Plate A, which is formed by the phase distribution across a column diameter at successive steps in time. The Plate also shows frames from 'movies' of the variation of gas and liquid about a cross-section. Both this approach and point probes can provide time varying signals of the phase occupying the

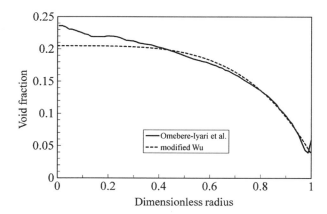

Figure 2.31 *Radial void fraction profile from 46 bar steam-water experiments: gas superficial velocity = 0.09 m/s; liquid superficial velocity = 0.01 m/s. Column diameter = 194 mm*

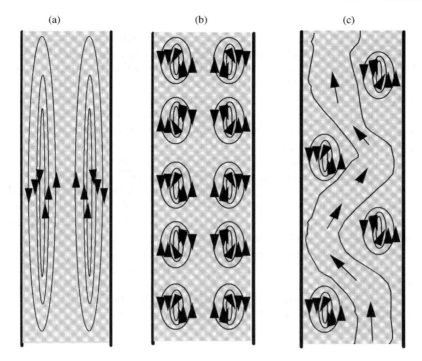

Figure 2.32   *Proposed circulation models for bubble columns: (a) vortex street [2]; (b) counter rotating vortices [50]; and (c) more complex pattern with moving vortices [51]*

sample space, which might appear quite random. Not surprisingly, the actual recirculation of liquid in the column becomes difficult to describe. Figure 2.32 shows some of the increasingly more complicated descriptions that have been suggested. Figure 2.32a is the simple bulk circulation, which was clearly seen by Reitema and Ottengraf [2]. However, a short, fat column with small nozzles and a viscous liquid is probably required for this to occur. A more complex picture of pairs of counter-rotating vortices, Figure 2.32b, was suggested by Joshi and Sharma [54]. The wealth of measurements, observations and computational fluid dynamics calculations made over the last few years have led to the description shown in Figure 2.32c. This indicates that there are packages or vortices that can move about [55]. These explain the downward motion of bubbles that are seen to occur when the flow is observed through transparent column walls. Please see colour plate A for further information.

2.3.6   Gas-Liquid Mass Transfer

When considering the transfer of chemical species from the gas to liquid phases, an important parameter is the surface area of the bubbles per unit volume of reactor space occupied by the two-phase mixture, the specific interfacial area. Before examining this, it is necessary to consider the bubble sizes present in the column. Although an equation was provided above for the drop size at creation, Equation 2.10, the *size of the bubbles within the body of the bubble columns* will also be influenced by processes such as coalescence and breakup. The actual sizes of the bubbles in the bulk of the column are important to determine the specific interfacial area, the surface area for mass transfer. Bubble sizes can be determined from photography (at low bubble concentration), point probes (usually two probes in line to measure transit times and contact times), multi-point probes and wire mesh sensor tomography. Because of the non-spherical shapes of many bubbles, several dimensions might be required to define the size, that is many measurements must be made. This is feasible for photography and the wire mesh approaches. However, the double-probe technique can only give one measurement. Even if the bubble is spherical, as there cannot be a guarantee that the probes go through a diameter, the measured length is usually a chord and methods are required to convert chords into diameter. The multi-point probes can give a more three-dimensional description of non-spherical bubbles. It is noted that the probe (single or multiple) has to be moved from one position to others to get cross-sectionally representative data. The wire mesh sensor approach addresses the entire cross-section at one time and statistically meaningful data can be obtained. Figure 2.33 shows data for increasing gas superficial velocity where it is obvious that the second, larger bubble size, peak is growing systematically. This is probably due to coalescence.

The gas-liquid interfacial area can be determined from the techniques given above for bubble size measurements as well as by use of chemical reaction methods. The *specific interfacial area*, a', which is very important in mass transfer is the bubble surface area per unit volume of column. The total bubble surface area, a, is given by:

$$a = \pi \sum_{all\ i} d_i^2 \qquad (2.52)$$

The total volume of bubbles can be linked to the mean void fraction

$$\varepsilon_g \frac{\pi D_c^2}{4} H = \frac{\pi}{6} \sum_{all\ i} d_i^3 \qquad (2.53)$$

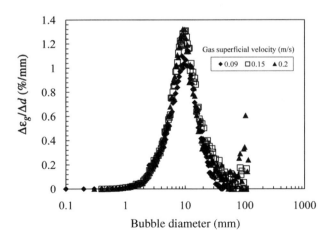

Figure 2.33 *Bubble size distribution measured with wire mesh sensor tomography: 46 bar steam-water experiments; liquid superficial velocity = 0.01 m/s. Column diameter = 194 mm*

The specific interfacial area, a', is then

$$a' = \frac{a}{\frac{\pi D_c^2}{4} H} = 6\varepsilon_g \frac{\sum\limits_{all\ i} d_i^2}{\sum\limits_{all\ i} d_i^3} = \frac{6\varepsilon_g}{d_{32}} \qquad (2.54)$$

where

d_{32} is the Sauter mean diameter.

The relationship between specific interfacial area bubble size and void fraction is shown in Figure 2.34. Obviously, it requires bubbles smaller than 5 mm to give a significant area. Now as the bubble size distribution is essentially bimodal, it is instructive to examine the proportions of large and small bubbles. An idea of the effect of the fraction of large bubbles can be seen in Figure 2.35. Of course this is an arbitrary division and so it is more effective to look at the results from a void fraction predication method that specifies the sizes of small and large bubbles and the proportion of each. Here the method of Krishna *et al.* [46], Equations 2.28–2.31 and 2.41–2.44, is used and results are given in Figure 2.36. The calculations were carried out for an air-water mixture at a number of system pressures.

The transfer of a chemical species from the gas phase to the liquid phase is usually calculated by means of a mass transfer coefficient, that is the ratio of flux of mass transferred to the driving force. The mass transfer coefficient will have contributions from both the gas and liquid sides of the interface. However, for most bubbles (> 1 mm diameter) the gas within the bubble can be considered to be well mixed so that the gas-side coefficient is assumed to be infinite. What follows below concentrates on the liquid side coefficient, k_l. There are several physical processes that occur in this mass transfer. Danckwerts [56] proposed the surface renewal approach. Later work looks at the interaction of the liquid motion surrounding the bubble to the effect of waves on the interface [57]. Most work on mass transfer proposes empirical correlations to determine values of this parameter, and are mainly based on $k_l a$, but some of the present methods are for k_l. Ribeiro and Lage [47] have

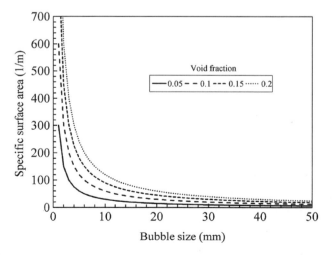

Figure 2.34 *Dependence of specific interfacial area on bubble size and void fraction*

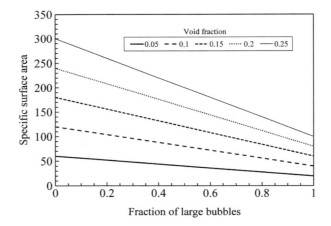

Figure 2.35   *Effect of proportions of large and small bubbles on specific interfacial area*

listed 35 equations. Many are in the form of dimensionless groups, but not all. Some usually have the gas superficial velocity explicitly raised to a power of between 0.68 and 1. Many employ the column diameter as a length scale whilst others bring in the bubble size (obviously less easy to use as it requires an ancillary equation).

The equation of Akita and Yoshida [58] is an equation of the first type that is usually recommended.

$$\frac{k_l a D_c^2}{D_l} = 0.6 \left(\frac{\eta_l}{D_l \rho_l}\right)^{0.5} \left(\frac{g D_c^2 \rho_l}{\sigma}\right)^{0.62} \left(\frac{g D_c^3 \rho_l^2}{\eta_l^2}\right)^{0.31} \varepsilon_g^{1.1} \tag{2.55}$$

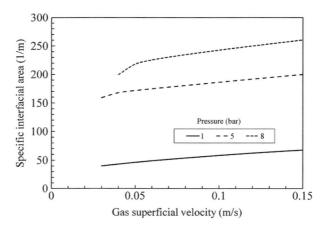

Figure 2.36   *Effect of superficial gas velocity and system pressure on specific interfacial area calculated using the void fraction method of Krishna et al. [46]*

where

D_l is the liquid phase diffusion coefficient.

The equation of Jordan and Schumpe [59], coming more from an organic liquid database, has been suggested for these fluids. This is of the second type, based on bubble size.

$$\frac{k_l a d^2}{D_l} = 0.6 \left(\frac{\eta_l}{D_l \rho_l}\right)^{0.5} \left(\frac{g d^2 \rho_l}{\sigma}\right)^{0.34} \left(\frac{g d^3 \rho_l^2}{\eta_l^2}\right)^{0.27} \left(\frac{u_{gs}}{\sqrt{gd}}\right)^{0.72}$$
$$\times \left[1 + 13.2 \left(\frac{u_{gs}}{\sqrt{gd}}\right)^{0.37} \left(\frac{\rho_g}{\rho_l}\right)^{0.49}\right] \tag{2.56}$$

2.3.7 Design of Gas Introduction Arrangement

There are two aspects that need to be considered in the design of a gas distribution system for bubble columns: smooth operation and pressure drop. The first of these has implications for the size of bubbles sought. In some applications, small bubbles in steady streams are required; for example if there are materials in the liquid that could be damaged by strong turbulence or there is violent breakage of large bubbles at the top surface of the gas-liquid mixture. Here it might be desirable to be in the homogeneous flow regime. Obviously, from the material presented in the sections above, porous plates or plates with multiple small orifices (diameter <0.5 mm) would be the best choice. However, there is a requirement that the orifices are not too close together, as this might encourage coalescence of bubbles. Two calculations are then required. One is to determine the (spherical equivalent) bubble size. In this situation Equation 2.10 would be used with the volumetric flow rate being calculated dividing the total gas flow rate by the number of orifices. The second calculation is to determine the area of influence of the orifices, and to compare this with the projected area of the bubble, which could well have been distorted from the spherical to a more elliptic shape. Equation 2.16 would give the aspect ratio of the bubbles, with Equation 2.11 relating it to the spherical equivalent diameter. To provide the terminal bubble velocity required in Equation 2.16, Equation 2.24 can be used.

If pressure drop is the major consideration, distributors with larger holes would probably be employed. This arises from cases where it is necessary to minimise the pumping power for the gas, as the capital and operating costs of any compressor could be crucial to the profitability of the reactor. The driving requirement is to achieve the necessary mass transfer. Distributor plates with larger holes or spiral or 'spider' spargers might be used. The latter has the form of a central vertical feed pipe and several horizontal, radial pipes with holes in them. The sparger to hole diameter ratio is chosen to ensure that gas emerges uniformly through the holes. Use of larger diameter holes could lead to dripping of liquid into the gas chamber below the plate or into the sparger. This phenomenon is similar to that of weeping of distillation trays. Methods developed for that field could be useful to check if liquid will drip down.

2.3.8   Worked Example

A bubble column of 2 m diameter is to run with 750 kg/h of gas. A distributor plate with 15 000, 1.0 mm diameter holes has been designed. Determine the bubble size, shape and velocity and the void fraction. The physical properties of the gas and liquid are:

gas density $= 2 \, \text{kg/m}^3$
liquid density $= 1000 \, \text{kg/m}^3$
liquid viscosity $= 1 \, \text{mPa s}$
surface tension $= 73 \, \text{mN/m}$

Calculate the gas superficial velocity, u_{gs}.

$$u_{gs} = \frac{4\dot{M}}{3600 d_c^2 \pi \rho_g} = \frac{4x750}{3600x2x2x\pi x2} = 0.033 \, \text{m/s}$$

The volumetric flow rate through each orifice, Q:

$$Q = \frac{\dot{M}}{3600 N \rho_g} = \frac{750}{3600 \times 15000 \times 2} = 6.9 \times 10^{-6} \, \text{m}^3/\text{s}$$

Calculate the bubble size using Equation 2.10.

$$d_e = \left[\left(\frac{6 d_o \sigma}{\rho_l g} \right)^{1.33} + \left(\frac{81 \eta_l Q}{\pi \rho_l g} \right) + \left(\frac{135 Q^2}{4 \pi^2 g} \right)^{0.8} \right]^{0.25}$$

$$d_e = \left[\left(\frac{6 \times 0.001 \times 0.073}{1000 \times 9.81} \right)^{1.33} + \left(\frac{81 \times 0.001 \times 6.9 \times 10^{-6}}{\pi \times 1000 \times 9.81} \right) + \left(\frac{135 \times (6.9 \times 10^{-6})^2}{4 \pi^2 \times 9.81} \right)^{0.8} \right]^{0.25}$$

$$d_e = 0.0071 \, \text{m}$$

Note that the three terms in the square bracket are in the ratio $0.07 : 0.0076 : 1$.

Calculate the Eötvös number.

$$Eo = \frac{g d_e^2 (\rho_l - \rho_g)}{\sigma} = \frac{9.81 \times 0.0071^2 (1000 - 2)}{0.073} = 6.82$$

Calculate the shape, E, using Equation 2.14.

$$E = \frac{1}{1 + 0.163 Eo^{0.757}}$$

$$E = \frac{1}{1 + 0.163 x 6.82^{0.757}} = 0.59$$

This is then employed to determine the terminal velocity of these bubbles using Equation 2.23:

$$V_T = \frac{\sin^{-1}\sqrt{1-E^2} - E\sqrt{1-E^2}}{1-E^2}\sqrt{\left(\frac{8\sigma}{\rho_l d_e}E^{1.33} + \left[\frac{\rho_l-\rho_g}{\rho_l}\right]\frac{gd_e}{2}\frac{E^{0.67}}{1-E^2}\right)}$$

$$V_T = \frac{\sin^{-1}\sqrt{1-0.59^2} - 0.59 \times \sqrt{1-0.59^2}}{1-0.59^2}$$

$$\times\sqrt{\frac{8 \times 0.073}{1000 \times 0.0071}0.59^{1.33} + \left[\frac{1000-2}{1000}\right]\frac{9.81 \times 0.0071}{2}\frac{0.59^{0.67}}{1-0.59^2}}$$

$$V_T = 0.199\,\text{m/s}$$

To determine the exponent, n, the Reynolds number needs to be calculated.

$$\text{Re}_T = \frac{\rho_l V_T d_e}{\eta_l} = \frac{1000 \times 0.199 \times 0.0071}{0.001} = 1418$$

From Table 2.1 the equation is:

$$n = 2.39$$

The void fraction is obtained by solving Equation 2.47. Simple iteration is required.

$$\varepsilon_g(1-\varepsilon_g)^{n-1} - \frac{u_{gs}}{V_T} = \varepsilon_g(1-\varepsilon_g)^{1.39} - \frac{0.033}{0.199} = 0$$

This results in a value of $\varepsilon_g = 0.25$.

2.4 Disengagement of Liquid from Gas

2.4.1 Mechanisms of Drop Formation

The bubbles rising up through the liquid in bubble columns will burst when they reach the top surface of the bubbly mixture. This process generates drops, which could be carried up by the gas. The quantity and size of the drops produced will now be considered. The design of the bubble column vessel, to minimise the carryover of liquid with the gas, will be examined with particular emphasis being placed on the secondary separators, which might be installed to increase the efficiency of liquid retention.

Two major mechanisms of drop creation from bubbles bursting at a surface have been reported [60, 61]. These are illustrated in Figure 2.37. One is associated with the bursting of isolated bubbles. The other mechanism occurs when there are multiple bubbles at the gas-bubbly flow interface and is more relevant to bubble columns with a plethora of bubbles. This is characterised by the formation of a thin skin of liquid that bursts, creating a shower of small drops.

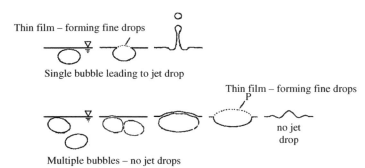

Figure 2.37   *Mechanism of drop formation from single and multiple bubbles (Reprinted from Günther, A., Wälchli, S., and von Rohr, P. R., Droplet Production from Disintergrating Bubbles at Water Surfaces, Single Vs. Multiple Bubbles, International Journal of Multiphase Flow, **29**, 5. Copyright (2003) with permission from Elsevier.)*

2.4.2   Drop Capture

The processes that are employed for phase separation include settling under the action of gravity or centrifugal forces and impaction onto solid surfaces. Other processes might be possible but are not normally used.

In mist eliminators, the sizes of drop to be removed from the gas phase can vary over a wide range, depending on their mode of origin and history. As a rough approximation, they can be classified according to their size; those above about 10 μm are regarded as sprays, below this range droplets are referred to as mists or aerosols. Sprays and mists are not uniform but always consist of drops of a distribution of sizes. The drop size distributions encountered in practice are frequently very complex. Droplet size distributions could be bimodal, that is even relatively fine distributions may be accompanied by a coarse fraction.

The type of mist eliminator employed will depend on the drop size distribution of the approaching liquid and the efficiency required. In many instances a combination of types will be used. For mists with drops > 100 μm, gravity separators or cyclones are suggested. A unit capable of handling finer drops would be located following this, should this be necessary. For finer drops, whether this is the main part of the mist or the tail of a coarser mist, wave plate or mesh pad mist eliminators or axial flow cyclones are used.

Wave plate mist eliminators or separators (also known as vane packs) consist of a series of corrugated plates set side by side as shown in Figure 2.38. Their effectiveness depends on the inability of drops to follow the gas as it zigs and zags through the plates. The drops, which tend to follow a straighter line, impinge on the walls and drain away. To assist drainage of the liquid, additional features are provided in the form of recessed or protruding channels, Figure 2.38. Alternatively, every second channel is blocked off to the gas, but is interlinked to accept the liquid. The detailed design depends on the manufacturer, each of which has considerable expertise having installed many such units.

The wave plate units are operated with vertical upflow or horizontal flow. For upflow, the collected liquid has to drain counter-currently to the gas flow. Therefore, a limitation is counter-current flow limitation or flooding when the gas commences to carry liquid up with it. In the case of horizontal gas flow, liquid drainage is obviously perpendicular to the gas

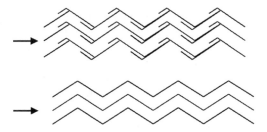

Figure 2.38   *Typical wave plate arrangements*

flow. In both instances local accumulation of liquid can lead to re-entrainment and loss of efficiency.

Important parameters of the wave plate geometry are the bend angle, the plate spacing and the approach velocity of the gas. Apart from grade efficiency, pressure drop though the mist eliminator has to be considered.

Knitting continuous filaments to form a tube (like a sock) is the first stage in producing mesh pad mist eliminators. After knitting, the tubes are flattened to form a double layer. Several of these then constitute the mesh pad, which with suitable support grids and so on, can be fitted into the required vessel. In many cases the double knitted layer is crimped before being made up into the pad, which allows for greater packing and is used to control the porosity of the pad. Crimping is corrugating the mesh pad to lessen even further the possibility of drops passing straight through. Filaments usually range from 0.15 to 0.4 mm diameter and are made from metals, polymers such as polypropylene, nylon or PTFE (polytetrafluoroethylene), or glass fibre. A single layer of knitted mesh is illustrated in Figure 2.39a. However, a drop approaching the pad will see several such layers, each slightly displaced and face a rather tortuous path through the pad, as illustrated in Figure 2.39b. Typical mesh pads have voidages in the high 90%. The packing can also be expressed in terms of packing density, ρ_p, which is linked to the fractional free volume, F_v, by $\rho_p = (1 - F_v)\rho_s/(2 - F_v)$.

(a)   (b)

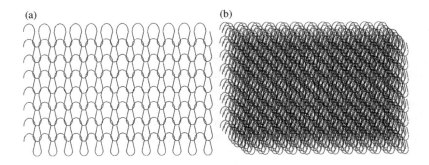

Figure 2.39   *Wire mesh pads for demisting: (a) one layer; and (b) multiple layers. Adapted from Feord et al.*

Wave plate units can be more compact than the equivalent mesh pad. Therefore, they are employed where space or weight are important considerations, for example, in high-pressure applications, where compactness could reduce vessel costs considerably. Another reason to employ wave plates could be if the process fluids are corrosive and exotic metals are required. Decreasing vessel size then becomes an important consideration to reduce costs. Axial flow cyclones are essentially tubes with inserts to cause the flow to swirl. Droplets are thrown onto the walls. Arrangements are required to remove the film and to ensure that there is no re-entrainment of liquid. Some details of this type of unit are found in the report by Verlaan [62].

In specifying the size of unit required, three elements need to be taken into consideration. The first is to determine the face velocity of the drop-laden gas approaching the unit. This has implications for both the efficiency of collection and pressure drop across the unit. The usual approach is to specify the gas momentum flux required and determine the face area necessary to achieve this. Equipment suppliers have the values of this parameter, which are based on tests, their considerable experience of successful installations and backup through modelling and correlations. A second element, which is important, is the possibility of re-entrainment of liquid. Although it is possible to capture smaller drops by increasing the gas velocity and hence incurring a greater pressure drop, there is an upper limit beyond which the efficiency falls as the liquid deposited on the collector surfaces is transformed once again into drops that can pass out of the unit. The third element to be considered is uniform presentation of the drop-laden gas to the mist eliminator. Given that these units, particularly in retrofit cases, are shoehorned into a minimal volume, it is important to check that flow resistances in the vessel do not force all the gas through a small part of the mist eliminator.

In the sizing of the vertical type of phase separators, where the drop size distributions might not normally be known, a maximum velocity approach is used. If this velocity were to be less than the terminal settling velocity of the drops, they would be separated under gravity. If they were below a pre-set value for the mesh pads or wave plates, these units would operate successfully. Normally the velocity is expressed as a constant times the dimensionless group $(\rho_l/\rho_g - 1)^{0.5}$. An alternative way of looking at the information would be to say that the velocity head, gas density times velocity squared, must be kept below a certain limit. Manufacturers have tables of these limit values often based on considerable experience. For vertical vessels, the limit value of the velocity head can be used to calculate gas velocity, which can be combined with volumetric gas flow rate to yield the vessel diameter.

The main complication of drop removal is that the effectiveness of many separators is decisively affected by the individual drop size. For a given configuration of mist eliminator, the fractional separation efficiency represents the *probability* of a drop with a given size being *retained* in the separator. The curve obtained by plotting the fractional separation efficiency as a function of particle size is called the *fractional separation curve*. This can be combined with the particle size distribution to give the effectiveness of a specific separator with a particular laden gas flow. If a given contaminant is to be fully separated, the entire fractional separation curve must lie on the left of the oversize cumulative distribution curve. On the other hand, if the fractional separation curve lies on the right of the oversize cumulative distribution curve, no separation occurs at all, that is the separator is totally ineffective. Frequently the two curves overlap, and partial separation takes place

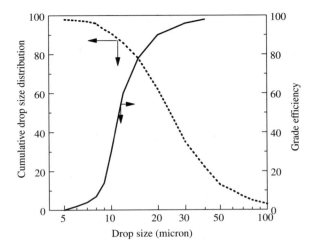

Figure 2.40 *Particle size distribution curve and fractional separation efficiency curve*

(Figure 2.40). The overall separation efficiency η_o is calculated from the fractional separation efficiency η_F and the oversize value $v(d)$:

$$\eta_o = \int_{d_{min}}^{d_{max}} \eta_F \, v(d) \, \mathrm{d}d \qquad (2.57)$$

The integration is performed from the smallest to the largest particle diameter. However, for practical purposes, it is sufficient if the particle size distribution and the fractional separation efficiency are available in graphical form. The approximate overall separation efficiency is then calculated from the following equation:

$$\eta_o = \sum \eta_F v_i \qquad (2.58)$$

For the data shown in Figure 2.40 we can calculate the overall efficiency to be 84.7%.

2.4.3 Wave Plate Mist Eliminators

Typical grade efficiency curves for wave plates are shown in Figure 2.41, where the effect of gas velocity can be seen. In addition to data from plain wave plates (closed symbols), the figure also shows data from a unit with drainage channels or hooks (open symbols). These results were taken on a carefully designed wind tunnel, which provided (nearly) saturated air to minimise evaporation of the drops. The drop flow was sampled before and after the mist eliminator and sorted into sizes. From this, the grade efficiency could be determined. Pressure drop across the unit is another important design parameter. A correlation has been proposed by Wilkinson [63] that predicts a large bank of data from plain wave plates to within +26 to −37% with all data being encompassed by $\sim \pm 53\%$. More interestingly, Burkholz [64] has suggested that pressure drop correlates well with the drop diameter for which the collection efficiency is 50%. Figure 2.42 shows that data from both plain wave

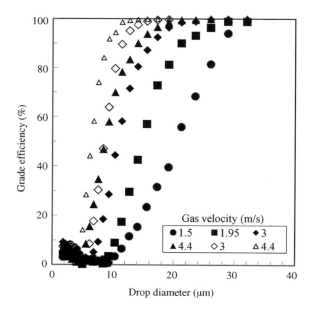

Figure 2.41   *Effect of gas velocity and wave plate geometry on grade efficiency for wave plate mist eliminators. Plain wave plates (closed symbols) and data from a unit with drainage channels or hooks (open symbols)*

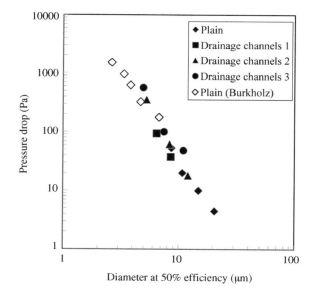

Figure 2.42   *Correlation of pressure drop with drop size whose collection efficiency is 50%*

plates and those with drainage channels both lie on one line when plotted in this way. The data are well correlated by the line $\Delta p = 38.2 d_{50}^{-3.02}$, where Δp is in kPa and d_{50} is in μm.

A number of workers have proposed models for wave plate mist eliminators. Several treat the system as a series of bends [53]. The flow through one zig is assumed to be as through a bend of angle α and radius R. During the time, t, taken for the gas to traverse the bend, the drop travels a distance, δ, radially outwards

$$\delta = u_r\, t = u_r\, \alpha\, \frac{R}{V} \qquad (2.59)$$

where

u_r is the radial velocity
V is the gas velocity.

From this we can say that a width δ will be cleared completely of drops. The efficiency can then be written as:

$$\eta_1 = \frac{\delta}{s} = \frac{\rho_d\, V\, d_d^2}{18\, \eta_g}\, \alpha \qquad (2.60)$$

where

ρ_d and d_d are, respectively, the density and size of the drops.

Assuming that the mist is remixed across the gap, the efficiency for a wave plate with n bends is:

$$\eta = 1 - (1 - \eta_1)^n \qquad (2.61)$$

The results of this approach are compared with experimental data in Figure 2.43. The theory over-predicts the efficiency, particularly at smaller drop sizes.

An alternative analysis, which looks at the flow in much more detail, is that utilising computational fluid dynamics. It has been used for wave plate mist eliminators [62, 65]. The method as used by the latter workers comprises the following steps. Firstly, the turbulent gas flow field is computed using the commercial computational fluid dynamics (CFD) code CFX4.1. The standard $k - \varepsilon$ turbulence model is used, although results are available to show how more complicated turbulence models affect predictions [66, 67]. Next, the trajectory of a droplet is calculated by assuming that it is a hard sphere, which experiences only a drag force due to the gas stream.

In order to allow for the influence of turbulent fluctuations on the droplet motion, the gas flow 'seen' by a droplet is constructed from the sum of a mean velocity, obtained from the CFD calculation, and a fluctuating velocity. It is the manner in which the fluctuating velocity is calculated that determines the particular type of turbulent dispersion model. Here we follow Wang and James and use a modification of the standard eddy interaction model of Gosman and Ionnides [68]. The key difference is that the gas velocity *and* turbulent eddy lifetime and length-scale are updated when a particle crosses a computational mesh boundary. Further details of the calculation method, the accuracy of the numerical solutions and other assumptions made are given by Wang and James [65b]. Of particular note are the following: (i) no droplet-droplet interaction takes place, (ii) the droplets exert no influence

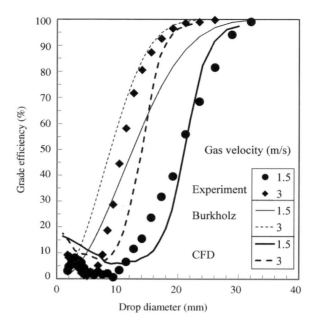

Figure 2.43 *Comparison of grade efficiencies predicted by the simple equation of Burkholz [64] and by computational fluid dynamics with experimental data from a plain wave plate*

on the mean gas flow and (iii) turbulence modulation due to droplets is ignored. Under these assumptions, the motion of a droplet spray with a given size distribution at the inlet to the mist eliminator can be simulated by calculating the track of a representative droplet from each size interval.

By recording where a droplet deposits, the collection efficiency of any element of the mist eliminator, in addition to the overall collection efficiency, can be calculated. Similarly, the evolution of any characteristic of the distribution, for example the Sauter mean diameter, can be found. Predictions from this method are shown in Figure 2.43. For the plain wave plate unit, this model is seen to give good predictions, even predicting the turn up in efficiency measured at very small droplet sizes. In the case of the hooked wave plates, the computational approach over-predicts, particularly at smaller droplet sizes. Note that Equation 2.59 gives better predictions even though the analysis does not allow for drainage channels (hooks).

The occurrence of re-entrainment of drops from the wall films in wave plate mist eliminators causes a decrease in efficiency [62, 69, 70]. For vertical wave plates (gas up flow, downwards drainage of the liquid) the mechanism for the decrease in efficiency has been attributed to flooding of the draining film by the upward shear of the gas [62]. Flooding is the condition at which an upwards gas flow starts to prevent the downflow of liquid. For air-water at ambient conditions, re-entrainment was expected to occur when a critical value of the dimensionless re-entrainment number $(U_g^4 \rho_g^2 / \rho_{lg} \sigma)$ was exceeded. This group is the fourth power of the dimensionless velocity, usually called the Kutateladze number, a parameter much used in the analysis of flooding processes. For horizontal systems it has been proposed [70] that it is the centrifugal forces on the film as it goes around a corner that causes re-entrainment. A simple model that successfully describes the limiting gas flow determined experimentally was suggested.

2.4.4   Mesh Mist Eliminators

A typical grade efficiency for a mesh pad [71] is shown in Figure 2.44. Increasing the thickness of the pad produces increased separation efficiency, but at the price of a high pressure drop through the pad. Considering the efficiency of one wire and then combining the effects of several layers is one method of modelling of the efficiency of mesh pads. For example, the efficiency of collection of one wire [72] can be expressed by:

$$\eta_w = \frac{\psi^3}{(\psi^3 + f_1\psi^2 + f_2\psi + f_3)} \tag{2.62}$$

where

$$f_1 = -0.0133 \ln (\mathrm{Re}_w) + 0.931$$
$$f_2 = 0.0353 \ln (\mathrm{Re}_w) + 0.36$$
$$f_3 = -0.0537 \ln (\mathrm{Re}_w) + 0.398$$

$$\mathrm{Re}_w = \frac{\rho_g V_f d_w}{\mu} \tag{2.63}$$

$$\psi = \frac{St}{2} = \frac{\rho_d d_d^2 V_f}{18\,\mu d_w}$$

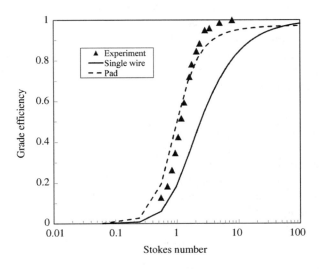

Figure 2.44   *Comparison of predicted and measured efficiencies for wire mesh pad mist eliminators*

where

V_f is the approach (or face) velocity of the gas.

The overall pad efficiency is then given by:

$$\eta_p = 1 - (1 - \eta_w A_p)^n \tag{2.64}$$

where

A_p as the projected area of wire in one layer of mesh
n is the number of layers.

For $n > 15$ we can use the approximation

$$\eta_p = 1 - e^{-\eta_w A_p n} \tag{2.65}$$

and $A_p n$ is

$$A_p n = \frac{4}{\pi} \beta \frac{L(1-\varepsilon)}{d_w} \tag{2.66}$$

where

β is the fraction of wire area perpendicular to the flow
L is the pad thickness
ε is the pad voidage.

Comparisons of the predictions of the above methods against experiment are given in Figure 2.44. Although the agreement is reasonable, the method shows sensitivity to β and there is not a single value of β that works best. Obviously, β is a difficult parameter to determine and more thorough methods of analysis are required. One computationally intensive method is that in which the mesh is described in detail, and where it can be broken down into a series of straight lines and semicircles [73]. The motion of the drops is then calculated until they impinge on a wire. The behaviour of deposited liquid is also tracked through the calculation of drops coalescing and dripping off when they get too large.

Mesh pad units are often used with an upflow of gas and a counter-current draining of liquid. Such a system is susceptible to the flooding process found in packed columns. Too high a gas flow will prevent liquid draining away. The additional liquid held up in the mesh has a strong effect on the pressure drop across such units [74, 75] as shown in Figure 2.45. These data show the characteristic change in the pressure drop/gas flow rate plot when flooding occurs. The break point in the data gives the limiting gas flow rate. Dickson and Morrison [74] note that a small increase in this value can be obtained by tilting the mesh pad. They also report that the drops emerging from the top of the mesh pad at conditions beyond flooding are much larger than those caught by the mesh pad, Figure 2.46. This means that the mesh pad is acting as a coalescer.

In many instances the distribution of droplet sizes can be described by the upper limit log-normal equation, a modified normal distribution [76]. Its basic equation is:

$$\frac{dv}{dy} = \frac{\delta}{\sqrt{\pi}} e^{-\delta^2 y^2} \tag{2.67}$$

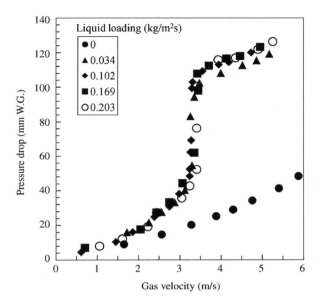

Figure 2.45 *Pressure drop across mesh pad mist eliminator showing increase in pressure drop associated with flooding, data of Dickson and Morrison [74]*

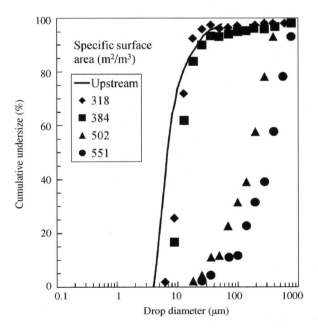

Figure 2.46 *Cumulative drop size distributions before and after mesh pad mist eliminator showing coalescing effect, data of Dickson and Morrison [74]*

where

$$y = \ln\left(\frac{aD}{D_{max} - D}\right) \tag{2.68}$$

If we are interested in the volume distribution or cumulative volume distribution, these can be written, in terms of diameter, as

$$\frac{dv}{dD} = \frac{\delta D_{max}}{\sqrt{\pi}D(D_{max} - D)} e^{-\delta^2\left[\ln\left(\frac{aD}{D_{max}-D}\right)\right]^2} \tag{2.69}$$

and

$$Cumv = \int_0^D \frac{dv}{dD}\,dD = \int_0^D \frac{\delta D_{max}}{\sqrt{\pi}D(D_{max} - D)} e^{-\delta^2\left[\ln\left(\frac{aD}{D_{max}-D}\right)\right]^2}\,dD \tag{2.70}$$

Here the parameters a and δ characterise the shape of the distribution. In Mugele and Evans [76], an example from a spray generated at a nozzle gives values of $a = 2.62$, $\delta = 1.131$. An examination of the literature for droplet size measurement in a gas-liquid pipe flow shows that more representative values are $a = 2$ and $\delta = 0.8$. The largest drop size, d_{max}, can be linked to the upper limit log-normal parameters and the Sauter mean diameter by:

$$d_{max} = d_{32}\left(1 + \frac{1}{e^{4\delta^2}}\right) \tag{2.71}$$

The methods outlined above can be applied to check the performance of separators. Drop fraction and efficiency can be combined as in Equations 2.55 or 2.56 to give the overall efficiency of the bend. The calculation will yield how much of each drop size passes into the inlet nozzle where the calculation can be repeated. This gives the drops passing into the separator body where any greater than the cut-off diameter for gravitational settling are eliminated. The calculation is then repeated with the mist eliminator. Equations 2.57–2.59 or 2.60–2.62 provide the efficiency for wave plates and mesh pads, respectively.

One important factor in the design of separators involving a mist eliminator is the assumption that the gas-drop mixture is presented evenly to all parts of the inlet face of the mist eliminator. Although in many cases this is not a problem, engineers dealing with such units must be aware of the potential problem. Obviously, if maldistribution occurs, part of the mist eliminator could be over loaded and part would receive very little feed. One approach that can help in this matter is the use of computational fluid dynamics, as discussed in Chapter 5. Commercial codes are available to carry out these calculations. Normally, the mist eliminator is not modelled in detail in these more global calculations. Instead, it is considered as a porous body with a prescribed flow rate/resistance relationship. Fewel and Kean [77] gave an example of such an approach in their paper.

Questions

These can be evaluated by hand calculations or a fairly simple spreadsheet.

2.1 Identify the range of bubble sizes for which the surface tension force is dominant. For this, the equation of Harmathy is a reasonable approximation. Use the viscous and inertia methods to determine the size limits. Liquid density $= 850\,kg/m^3$, gas density $= 5\,kg/m^3$, liquid viscosity $= 0.5\,mPa\ s$, surface tension $= 0.02\,N/m$.

2.2 How does the size of bubbles affect their terminal velocity in water? How does increasing the viscosity by $\times 10$, $\times 100$, $\times 1000$ change this relationship? Assume all other physical properties remain constant.

2.3 The mixture of liquid reagents introduced into a 1 m diameter bubble column reactor has a viscosity of $50\,mPa\ s$, a surface tension of $0.02\,N/m$ and a density of $850\,kg/m^3$. The reactor operates at a system pressure for which the gas has a density of $5\,kg/m^3$. The design requires that the bubble column operates in the homogeneous flow pattern. How many holes are required in the distributor so that coalescence does not occur immediately on bubble creation? Gas is introduced at a rate of $0.3\,kg/s$.

2.4 How does void fraction vary with gas superficial velocity? How does specific interfacial area, as calculated from Krishna's void fraction model, vary with gas superficial velocity for a liquid of viscosity $\times 80$ that of water? The surface tension is $0.02\,N/m$ and the gas density is $9.5\,kg/m^3$. Compare with predictions of empirical equations.

2.5 The mass transfer coefficient, $k_l a$, for a 1 m diameter bubble column operating at 10 bar is $0.1\,m^3/s$. What gas velocity is required to double the mass transfer coefficient? See Table 2.Q.1 for properties of the fluids employed at process conditions. Use the Krishna approach via void fraction as well as a more empirical correlation.

2.6 What is the overall efficiency of a wave plate mist eliminator placed in the head space at the top of a bubble column? This has a diameter of 1.5 m. Liquid density $= 850\,kg/m^3$, gas density $= 5\,kg/m^3$, liquid viscosity $= 0.5\,mPa\ s$, surface tension $= 0.02\,N/m$. The bubble column operates at a gas superficial velocity of 0.4 m/s. A vane pack with seven, $60°$ bends 0.3 m tall with 30 plates at a spacing of 8 mm has been proposed. If carry-up of liquid from the bubbling surface is determined to be 6 t/h (1 tonne $= 10^3\,kg$) and the drops have a Sauter mean diameter of $40\,\mu m$, what is the overall efficiency?

2.7 The unit referred to in Question 2.6 is to be operated with the same liquid and gas but at a pressure for which the gas density $= 12\,kg/m^3$. The bubble column operates at a gas superficial velocity of 0.35 m/s. A knitted mesh pad with a face area of 0.4×0.5 m has been proposed. A thickness of 0.1 m is recommended. The wire thickness is 0.25 mm, the pad voidage is 0.98. If carry-up of liquid from the bubbling surface is determined to be 3 t/h and the drops have a Sauter mean diameter of $70\,\mu m$, what is the overall efficiency (use $\beta = 0.666$)?

Table 2.Q.1 Properties of the fluids employed at process conditions

Parameter	Vapour	Liquid
Density (kg/m³)	6.7	611
Viscosity ($\mu N\ s/m^2$)	18	148
Surface tension (N/m)	0.022	

References

[1] Shah, Y.T., Kelkar, B.G., Godbole, S.P. and Deckwer, W.-D. (1982) Design parameters estimations for bubble column reactors. *AIChE J.*, **28**, 353–379.

[2] Rietema, K. and Ottengraf, S.P.P. (1970) Laminar liquid circulation and street formation in a gas-liquid system. *Trans. I. Chem. Eng.*, **48**, T54–T62.

[3] Groen, J.S. (2004) Scales and structures in bubbly flows, Ph.D. thesis, Technical University of Delft.

[4] See for example, Růžička, M., Drahoš, J., Mena, P.C. and Teixeira, J.A. (2003) Effect of viscosity on homogeneous–heterogeneous flow regime transition in bubble columns. *Chem. Eng. J.*, **96**, 15–22.

[5] Letzel, H.M., Schouten, J.C., Krishna, R. and van den Bleek, C.M. (1999) Gas holdup and mass transfer in bubble column reactors operated at elevated pressure. *Chem. Eng. Sci.*, **54**, 2237–2246.

[6] Cheng, H., Hills, J.H. and Azzopardi, B.J. (1998) A study of the bubble-to-slug transition in vertical gas-liquid flow in columns of different diameters. *Int. J. Multiphase Flow*, **24**, 431–452.

[7] Pioli, L., Bonadonna, C., Azzopardi, B., Phillips, J. and Ripepe, M. (2011) Experimental constraints on the outgassing dynamics of basaltic magmas. *Earth Planet. Sci. Lett.*, in preparation.

[8] Lin, T.J., Tsuchiya, K. and Fan, L.S. (1998) Bubble flow characteristics in bubble columns at elevated pressure and temperature. *AIChE J.*, **44**, 545–560.

[9] Urseanu, M.I., Guit, R.P.M., Stankiewicz, A. *et al.* (2003) Influence of operating pressure on the gas hold-up in bubble columns for high viscous media. *Chem. Eng. Sci.*, **58**, 697–704.

[10] (a) Kelkar, B.G., Godbole, S.P., Honath, M.F. *et al.* (1983) Effect of addition of alcohols on gas holdup and backmixing in bubble columns. *AIChE J.*, **29**, 361–369; (b) Zahradnik, J., Fialová, M., Růžička, M. *et al.* (1997) Duality if the gas-liquid regimes in bubble column reactors. *Chem. Eng. Sci.*, **52**, 3811–3826; (c) Krishna, R., Urseanu, M.I. and Dreher, A.J. (2000) Gas hold-up in bubble columns: influence of alcohol addition versus operation at elevated pressures. *Chem. Eng. Proc.*, **39**, 371–378; (d) Ribeiro Jr, C.P. and Mewes, D. (2007) The influence of electrolytes on gas hold-up and regime transition in bubble columns. *Chem. Eng. Sci.*, **62**, 4501–4509.

[11] Omebere-Iyari, N.K., Azzopardi, B.J., Lucas, D. *et al.* (2008) The characteristics of gas/liquid flow in large risers at high pressures. *Int. J. Multiphase Flow*, **34**, 461–476.

[12] Jamailahmadi, M., Zehtban, M.R., Muller-Steinhagen, H. *et al.* (2001) Study of bubble formation under constant flow conditions. *Trans. I. Chem. Eng.*, **79**, 523–532.

[13] Gaddis, E.S. and Vogelpohl, A. (1986) Bubble formation in quiescent liquids under constant flow conditions. *Chem. Eng. Sci.*, **41**, 97–105.

[14] Davidson, J.F. and Schüler, B.O.G. (1960) Bubble formation at an orifice in a viscous liquid. *Trans I. Chem. Eng.*, **38**, 144–154.

[15] Davidson, J.F. and Schüler, B.O.G. (1960) Bubble formation at an orifice in an inviscid liquid. *Trans I. Chem. Eng.*, **38**, 335–342.

[16] Lin, T.J. and Fan, L.S. (1999) Heat transfer and bubble characteristics from a nozzle in high pressure bubble columns. *Chem. Eng. Sci.*, **54**, 4853–4859.

[17] Idogawa, K., Ikeda, K. and Fukuda, T. (1987) Formation and flow of gas bubbles in a pressurized bubble column with a single orifice or nozzle gas distributor. *Chem. Eng. Commun.*, **59**, 201–212.

[18] Wang, T. and Wang, J. (2007) Numerical simulations of gas–liquid mass transfer in bubble columns with a CFD–PBM coupled model. *Chem. Eng. Sci.*, **62**, 7107–7118.

[19] Prince, M.J. and Blanch, H.W. (1990) Bubble coalescence and break-up in air-sparged bubble columns. *AIChE J.*, **36**, 1485–1499.

[20] Kaji, R., Zhao, D., Licence, P. and Azzopardi, B.J. (2009) Studies of the interaction of ionic liquid and gas in a small diameter bubble column. *Ind. Eng. Chem. Res.*, **48**, 7938–7944.

[21] Sreekumar, D. (2009) Effect of physical properties on gas-liquid flow in a bubble column. M.Sc. dissertation, University of Nottingham.

[22] Tomiyama, A., Celata, G.P., Hokosawa, S. and Yoshida, S. (2002) Terminal velocity of single bubbles in surface tension force dominant regime. *Int. J. Multiphase Flow*, **28**, 1497–1519.

[23] Wellek, R.M., Agrawal, A.K. and Skelland, A.H.P. (1966) Shape of liquid drops moving in liquid media. *AIChE J.*, **12**, 854–862.

[24] Fan, L.-S. and Tsuchiya, K. (1990) *Bubble Wake Dynamics in Liquids and Liquid-Solid Suspensions*, Butterworth-Heinemann, Boston.

[25] Celata, G.P., D'Annibale, F., Di Marco, P. *et al.* (2007) Measurements of rising velocity of a small bubble in a stagnant fluid in one- and two-component systems. *Exp. Thermal Fluid Sci.*, **31**, 609–623.

[26] Žun, I. (2005) The principles of complexity in bubbly flow. *Multiphase Sci. Technol.*, **17**, 169–190.

[27] Harmathy, T.Z. (1960) Velocity of large drops and bubbles in media of infinite or restricted extent. *AIChE J.*, **6**, 281–288.

[28] Mendelson, H.D. (1967) The prediction of bubble terminal velocity from wave theory. *AIChE J.*, **13**, 250–253.

[29] Jamailahmadi, M., Branch, C. and Muller-Steinhagen, H. (1994) Terminal bubble rise velocity in liquids. *Trans. I. Chem. Eng.*, **72**, 112–119.

[30] (a) Hills, J.H. (1975) The rise of a large bubble through a swarm of smaller ones. *Trans. I. Chem. Eng.*, **53**, 224–233; (b) Hills, J.H. and Darton, R.C. (1976) Rising velocity of large bubble in a bubble swarm. *Trans. I. Chem. Eng.*, **54**, 258–264.

[31] Richardson, J.F. and Zaki, W.N. (1954) Sedimentation and Fluidization: Part I. *Trans. I. Chem. Eng.*, **32**, 35–53.

[32] (a) Sarrafi, A., Jamailahmadi, M., Muller-Steinhagen, H. and Smith, J.M. (1999) Gas holdup in homogeneous and heterogeneous gas-liquid bubble column reactors. *Can. J. Chem. Eng.*, **77**, 11–21; (b) Růžička, M., Drahoš, J., Mena, P.C. and Teixeira, J.A. (2003) Effect of viscosity on homogeneous–heterogeneous flow regime transition in bubble columns. *Chem. Eng. J.*, **96**, 15–22; (c) Thorat, B.N. and Joshi, J.B. (2004) Regime transition in bubble columns: experimental and predictions. *Exp. Thermal Fluid Sci.*, **28**, 423–430.

[33] Růžička, M., Zahradnik, J., Drahoš, J. and Thomas, N.H. (2001) Homogeneous–heterogeneous regime transition in bubble columns. *Chem. Eng. Sci.*, **56**, 4609–4626.

[34] Ribeiro Jr, C.P. (2008) On the estimation of the regime transition point in bubble columns. *Chem. Eng. J.*, **140**, 473–483.

[35] Wilkinson, P.M., Spek, A.P. and van Dierendonck, L.L. (1992) Design parameters estimation for scale-up of high-pressure bubble columns. *AIChE J.*, **38**, 544–554.

[36] Reilly, I.G., Scott, D.S., De Bruin, J.W. and MacIntyre, D. (1994) The role of gas phase momentum in determining gas hold up and hydrodynamic flow regimes in bubble column operations. *Can. J. Chem. Eng.*, **72**, 3–12.

[37] Wallis, G.B. (1969) *One-Dimensional Two-Phase Flow*, McGraw-Hill, New York.

[38] Zuber, N. and Findlay, J.A. (1965) Average volumetric concentrations in two-phase flow systems. *J. Heat Trans.*, **87**, 453–468.

[39] Hills, J.H. (1976) The operation of a bubble column at high throughput. *Chem. Eng. J.*, **12**, 89–99.

[40] Clark, N.N. and Flemmer, R. (1986) Effect of varying gas voidage distributions on average hold up in vertical bubbly flow. *Int. J. Multiphase Flow*, **12**, 299–303.

[41] Kataoka, I. and Ishii, M. (1987) Drift flux model for large diameter pipe and a new correlation for pool void fraction. *Int. J. Heat Mass Trans.*, **30**, 1927–1939.

[42] Cumber, P. (2002) Modelling top venting vessels undergoing level swell. *J. Hazard. Mater.*, **A89**, 109–125.

[43] Beattie, D.R.H. and Suguwara, S. (1986) Steam-water void fraction for vertical upflow in a 73.9 mm pipe. *Int. J. Multiphase Flow*, **12**, 641–653.

[44] (a) Nikuradse, J. (1932) Gesetzmassigkeiten der Turbulenten Stromung in glatten Rohren. *VDI-Forschungsheft*, **356;** (b) Nunner, W. (1956) Waermeubergang und Druckfall in rauhen Rohren. *VDI-Forschungsheft*, **455**.

[45] Azzopardi, B.J. (2006) *Gas-Liquid Flows*, Begell House, New York.

[46] Krishna, R., Urseanu, M.I. and Dreher, A.J. (2000) Gas hold-up in bubble columns: influence of alcohol addition versus operation at elevated pressures. *Chem. Eng. Proc.*, **39**, 371–378.

[47] Ribeiro Jr, C.P. and Lage, P.L.L. (2005) Gas-liquid direct-contact evaporation: a review. *Chem. Eng. Technol.*, **28**, 1081–1107.

[48] Kumar, S.B., Moslemian, D. and Dudukovic, M.P. (1997) Gas-holdup measurements in bubble columns using computed tomography. *AIChE J.*, **43**, 1414–1425.

[49] Parasu Veera, U. and Joshi, J.B. (1990) Measurement of gas hold-up profiles by gamma ray tomography: effect of sparger design and height of dispersion in bubble columns. *Trans. I. Chem. Eng.*, **77**, 303–317.

[50] Hills, J.H. (1974) Radial non-uniformity of velocity and voidage in a bubble column at high throughput. *Trans. I. Chem. Eng.*, **52**, 1–9.

[51] Wu, Y., Ong, B.C. and Al-Dahhan, M.H. (2001) Predictions of radial gas holdup profiles in bubble column reactors. *Chem. Eng. Sci.*, **56**, 1207–1210.

[52] Wu, Y. and Al-Dahhan, M.H. (2001) Prediction of axial liquid velocity profile in bubble columns. *Chem. Eng. Sci.*, **56**, 1127–1130.

[53] Mudde, R.F. and Saito, T. (2001) Hydrodynamical similarities between bubble column and bubbly pipe flow. *J. Fluid Mech.*, **437**, 203–228.

[54] Joshi, J.B. and Sharma, M.M. (1979) A circulation cell models for bubble columns. *Trans. I. Chem. Eng.*, **57**, 244–251.

[55] Chen, R.C., Reese, J. and Fan, L.-S. (1994) Flow structure in a three-dimensional bubble column and three-phase fluidised bed. *AIChE J.*, **40**, 1093–1104.

[56] Danckwerts, P.V. (1951) Significance of liquid-film coefficients in gas absorption. *Ind. Eng. Chem.*, **43**, 1460–1467.

[57] (a) Tsuchiya, K., Saito, T., Kajishima, T. and Kosugi, S. (2001) Coupling between mass transfer from dissolving bubbles and formation of bubble-surface wave. *Chem. Eng. Sci.*, **56**, 6411–6417; (b) Tsuchiya, K., Ishida, T., Saito, T. and Kajishima, T. (2003) Dynamics of interfacial transfer in a gas-dispersed system. *Can. J. Chem. Eng.*, **81**, 647–654.

[58] Akita, K. and Yoshida, F. (1973) Gas holdup and volumetric mass transfer coefficient in bubble columns. *Ind. Eng. Chem. Proc. Des. Dev.*, **12**, 76–80.

[59] Jordan, U. and Schumpe, A. (2001) The gas density effect on mass transfer in bubble columns with organic liquids. *Chem. Eng. Sci.*, **56**, 6267–6372.

[60] Garner, F.H., Ellis, S.R.M. and Lacey, J.A. (1954) The size distribution and entrainment of droplets. *Trans. I. Chem. Eng.*, **32**, 222–235.

[61] Günther, A., Wälchli, S. and von Rohr, P.R. (2003) Droplet production from disintegrating bubbles at water surfaces. Single vs. multiple bubbles. *Int. J. Multiphase Flow*, **29**, 795–811.

[62] Verlaan, C.C.J. (1991) Performance of novel mist eliminators, Ph.D. thesis, Delft University of Technology, The Netherlands.

[63] Wilkinson, D. (1999) Optimizing the design of waveplates for gas-liquid separation. *Proc. I. Mech. Eng.*, **213**, 265–274.

[64] Burkholz, A. (1989) *Droplet Separation*, Wiley-VCH Verlag GmbH, Weinheim, Germany.

[65] (a) Wang, W. and Davies, G.A. (1996) CFD studies of separation of mists from gases using vane-type separators. *Chem. Eng. Res Design*, **74**, 232–238; (b) Wang, Y. and James, P.W. (1998) The calculation of wave-plate demister efficiencies using numerical simulation of the flow field and droplet motion. *Chem. Eng. Res Design*, **76**, 980–985.

[66] Gillandt, I., Riehle, C. and Fritsching, U. (1996) Gas-particle flow in a comparison of measurements and simulations. *Forsch. Ing. Wess.*, **62**, 315–321.

[67] Wang, Y. and James, P.W. (1999) Assessment of an eddy-interaction model and its refinement using predictions of droplet deposition in a wave plate demister. *Chem. Eng. Res Design*, **77**, 692–698.

[68] Gosman, A.D. and Ioannides, E. (1981) Aspects of computer simulation of liquid-fuelled combustors. Paper presented at AIAA 19th Aerospace Sciences Meeting, St Louis, MO, USA, Paper AIAA-81-0323.

[69] (a) Houghton, H.G. and Radford, W.H. (1939) Measurements on eliminators and the development of a new type for use at high gas velocities. *Trans. Am. Inst. Chem. Eng.*, **35**, 427–433; (b) Monat, J.P., McNulty, K.J., Michelson, I.S. and Hansen, O.V. (1986) Accurate evaluation of Chevron mist eliminator. *Chem. Eng. Prog.*, **82**, 32–39.

[70] Azzopardi, B.J. and Sanaullah, K. (2002) Re-entrainment in wave plate mist eliminators. *Chem. Eng. Sci.*, **57**, 3557–3563.

[71] Phillips, H.W. and Deakin, A.W. (1991) Measurement of the collection efficiency of various demisters. Paper presented at 5th Annual Conference of the Aerosol Society, Loughborough, March 25–27, 1991, pp. 169–174.

[72] Loffler, F. and Muhr, W. (1972) Die absceidung von festoffteichen und trophen an kreiszylindren infolge von traegheitkraeften. *Chem. Ing. Tech.*, **42**, 247.

[73] Feord, D., Wilcock, E. and Davies, G.A. (1993) A stochastic model to describe the operation of knitted mesh mist eliminators, computation of separation efficiency. *Trans. I. Chem. Eng.*, **71**, 282–294.

[74] Dickson, A.N. and Morrison, D.S. (1974) The performance of wire-mesh demisters. *I. Chem. E. Symp. Ser.*, (38), Paper K2.

[75] El-Dessouky, H.T., Alatiqi, I.M., Ettouney, H.M. and Al-Deffeeri, N.S. (2000) Performance of wire mesh mist eliminator. *Chem. Eng. Proc.*, **39**, 129–139.

[76] Mugele, R.A. and Evans, H.D. (1951) Droplet size distribution in sprays. *Ind. Eng. Chem.*, **43**, 1317–1324.

[77] Fewel, K.J. and Kean, J.A. (1992) Vane separators in gas/liquid separation. Paper presented at the Proceedings of the ASME Energy Sources Technology Conference, Houston, TX, January 26–29 1992.

3

Sparged Stirred Vessels

Hydrodynamics of Gas-Liquid Reactors: Normal Operation and Upset Conditions, First Edition.
B. J. Azzopardi, R. F. Mudde, S. Lo, H. Morvan, Y. Yan and D. Zhao.
© 2011 John Wiley & Sons, Ltd. Published 2011 by John Wiley & Sons, Ltd.

3.1 Introduction

Stirred tanks are used in many industrial applications. They are amongst the most important devices when it comes to mixing of different species, both for single and multiphase flows. Stirred tanks are mechanically powered flow devices with one or more impellers and usually have baffles at the side to prevent too much of a solid-body rotation in the stirred liquid. As a consequence, the flow inside the tank is complex, regardless of whether it is laminar or turbulent. However, the majority of stirred tanks are operated in the turbulent regime. Various types of impellers exist. They can be classified according to the circulation they induce in the stirred tank. Radially pumping impellers, such as the Rushton turbine, generate radially outward flowing jets from their blades (Figure 3.1). These jets hit the vessel wall and flow partly upwards, and also partly downwards creating two big circulation loops: one below the impeller plane and one above it. Pitch blade impellers are part of the family of downward pumping impellers. They create one large-scale circulation with downward flow in the central region and upward flow along the vessel wall.

Under single-phase conditions, the power consumption is related to the rotation speed of the impeller, the properties of the liquid and the geometry. Usually, it is specified as the dimensionless power draw, which is a function of the Reynolds number and the geometry, including the type of impeller.

Figure 3.2 shows various types of impellers, the top row being axial pumping ones. The disc turbine, or Rushton turbine, is radially pumping.

Each of these impellers, or even combinations of them in industrial-scale equipment, are also used in aerated vessels. Obviously, they serve a double purpose: mixing of the liquid phase and dispersing of the gas.

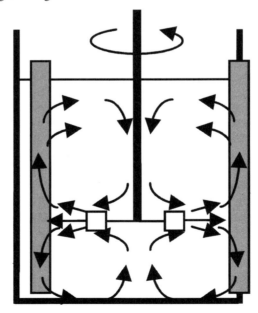

Figure 3.1 *Circulation pattern of a radially pumping impeller*

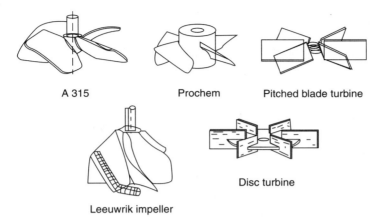

A 315   Prochem   Pitched blade turbine

Disc turbine

Leeuwrik impeller

Figure 3.2   *Different types of impellers. Provided by Andre Bakker with permission from Robert Mudde and the Department of Multi-Scale Physics, Delft University of Technology, The Netherlands*

3.2   Flow Regimes

In aerated stirred tanks, the gas is usually introduced in the bottom section of the tank. Obviously, in this way the gas is automatically distributed via buoyancy over the larger part of the tank, that is, with no impeller action a type of bubble column is mimicked. The stirrer is used to break the gas into small bubbles and to 'mix' them over the entire vessel.

Single-phase flow in a stirred tank can be classified based on the Reynolds number, $Re = \rho_l N D^2/\eta_l$, where ρ_l is the liquid density, N the rotational speed of the impeller, D the impeller diameter and η_l the liquid viscosity. In gas-liquid stirred tanks the gas inflow is also taken into account. This gives rise to a second dimensionless number, the gas flow number, $Fl_g = \phi_g/ND^3$, with ϕ_g being the gas flow rate. In some instances, gravity is important and a Froude number, $Fr = N^2 D/g$, with g being the acceleration of gravity, is used. The flow map is a function of the last two of these numbers. For a low gas flow number, the impeller will be able to disperse the gas in the vessel giving two regimes: loading and recirculation. For higher gas flow numbers, the impeller will have to generate a larger liquid circulation to drag sufficient gas phase to be able to disperse the gas completely. Therefore, if the pumping capacity is insufficient only partial loading is found. The bottom part of the vessel may be operating under almost single-phase conditions. If the impeller speed is further reduced, the liquid flow is too weak to disperse the bubbles and the flow close to the impeller blades can no longer break up the gas jet. This is called flooding.

Figure 3.3 shows the flow map for a standard tank equipped with a Rushton turbine. The lines indicating the border between recirculation and loading, flooding and loading and complete versus incomplete dispersion are given by, for example, Nienow [1] (for low viscosity liquids):

$$\text{Recirculation}-\text{loading} : \left(Fl_g\right)_{recir} = 13\left(\frac{D}{T}\right)^5 Fr^2 \qquad (3.1)$$

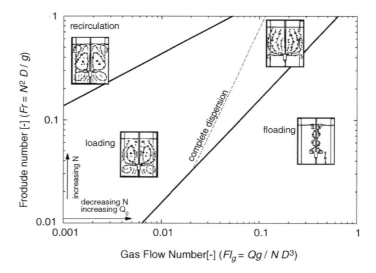

Figure 3.3   *Flow map of a gassed stirred tank equipped with a Rushton turbine*

$$\text{Loading}-\text{flooding} : \left(Fl_g\right)_{flooding} = 30\left(\frac{D}{T}\right)^{3.5} Fr \tag{3.2}$$

$$\text{Complete dispersion} : \left(Fl_g\right)_{comp\ dis} = 0.2\left(\frac{D}{T}\right)^{0.5} Fr^{0.5} \tag{3.3}$$

At the downstream side of the blades, a low-pressure region is present. This is especially apparent for the Rushton type of impellers. Gas gets trapped in these regions and forms a cavity behind the blades, see Figure 3.4. This has consequences for the gas dispersion

Figure 3.4   *Gas cavities behind the blades. Photo: Provided courtesy of Robert Mudde, Delft University of Technology*

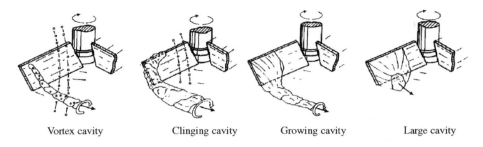

| Vortex cavity | Clinging cavity | Growing cavity | Large cavity |

Figure 3.5 *Cavity formation at increasing gas loading. Provided by Andre Bakker with permission from Robert Mudde and the Department of Multi-Scale Physics, Delft University of Technology, The Netherlands*

characteristics of the impeller as well as on the power draw of the liquid. If the gas flow rate is too high, flooding of the impeller occurs. Dispersion is no longer possible and thus the impeller cannot mix the liquid efficiently.

If the cavities are small, the gas is discharged via the vortices behind the blades. The process is sketched in Figure 3.5 for a pitch blade impeller.

Also under these conditions, the power draw decreases with increasing cavity formation.

The vortex behind the blades can also be exploited to efficiently disperse the gas. Special impellers have been designed, for example the Leeuwrik impeller (see Figure 3.2), which has some of the features of a wing.

3.3   Variations

A conventional stirred tank consists of a vessel equipped with a rotating mixer. The vessel is generally a vertical cylindrical tank. The rotating mixer, entering the vessel from its top, has several components, such as one or multiple impellers, a shaft, shaft seal, gearbox and a motor drive. Wall baffles are generally installed for transitional and turbulent mixing, in order to prevent fluid swirls.

Sometimes non-standard geometry stirred tanks are used in various industries. Non-standard vessels, such as those with a square or rectangular cross-section, or horizontal cylinder vessels, have been found to be more suitable for some specific requirements. A mixer can be positioned at the bottom of a tall tank to avoid using a long shaft and this arrangement provides mechanical stability. The side-entering and angled-entering mixers produce asymmetric flow and are often used in large product storage and blending tanks. In horizontal cylindrical tanks, the mixer can be installed from the side or from the top.

3.4   Spargers

Gas is usually introduced into the stirred vessels through a sparger. Spargers are important components of the stirred tank reactors used in gas-liquid applications. A well-designed and located sparger can enhance the gas-liquid dispersion by maximising contacting interface and eliminating maldistribution. The design of the sparger depends on the ratio of the volumetric gas flow to the liquid volume, and the sizes of the particles, if any, that are present in the reactor. Different types of spargers, such as nozzle, orifice and porous

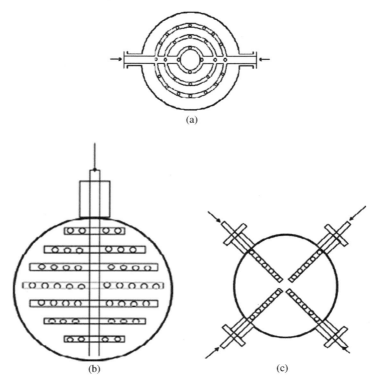

Figure 3.6   *Schematic diagrams of different types of spargers [2]. (a) Multiple ring sparger; (b) spider sparger; and (c) radial pipe sparger (Reprinted from Kulkarni, A. V., Design of a Pipe/ Ring Type of Sparger for a Bubble Column Reactor, Chemical Engineering & Technology, 33, 6, 1015–1022. Copyright (2010) Wiley-VCH Verlag GmbH & Co. KGaA. Reproduced with permission.)*

spargers, are available. The most commonly used sparger is probably the ring sparger made of perforated pipes with equally spaced holes positioned below the impeller. The sparger diameter should be less than the impeller diameter, typically 0.8 times the impeller diameter. More complicated spargers can also be found in industries. Typical examples include the multiple ring sparger (Figure 3.6a), spider sparger (Figure 3.6b) and radial pipe sparger (Figure 3.6c). These configurations produce more uniform gas distribution and small gas bubbles.

Some important parameters need to be considered in the sparger design. The number and diameter of the holes, the size of the sparger and the sparger location with respect to the impeller blades all play important roles in gas sparging performance. Small orifices can generate smaller sized bubbles but have high pressure drops and are easily blocked by solid particles. For an even gas distribution between the sparger holes, the number and size of the holes on the ring sparger should be selected so that the gas velocity through the holes is at least three times the velocity through the pipe that forms the sparger ring. The location of the sparger with respect to the impeller has an important effect with respect to the gas dispersion inside the tank. A sparger of good design should be able to generate high gas hold-up and prevent the impeller from being flooded.

Gas can also be entrained into the liquid by self-inducers. The simplest self-inducer for an agitated vessel is an impeller located near the surface, sometimes with the upper part of the baffles removed, so as to encourage the formation of a surface vortex. This is, however, a sensitive and unstable arrangement. It is better, although probably more expensive, to use a self-inducing impeller system in which gas is drawn down a hollow shaft to the low-pressure region behind the blades of a suitable, often shrouded, impeller such as the Ekato gasjet Praxair AGR or the Frings Friborator.

3.5 Impellers

In process industries, many different operations such as liquid blending, gas dispersion and solid suspension are performed in stirred tank reactors. In complex processes, two or more different operations are sometimes involved simultaneously. Each operation puts different requirements on the agitator. Liquid blending and solid suspension systems require the impeller to give a quick and sufficient liquid axial circulation in order to achieve rapid liquid macro-mixing and to maintain solid particle suspension. For gas-dispersion applications, an impeller system must develop high shear rates to break-up gas bubbles. In biological processes, these high shear rates may be harmful to micro-organisms. For mixing in sensitive complex reaction processes, the product distribution depends on local reagent segregation, which is determined by the micro-mixing rate. Each of these various requirements can only be fulfilled by different impellers. Therefore, it is impossible to design a 'perfect' impeller that is efficient at carrying out every task. Thus, the understanding of the effect of impeller geometry on the agitator performance becomes very important. The popular modern impellers used in aerated stirred vessels are summarised below.

3.5.1 Disc Turbines

Disc turbines, developed from the early multi-blade paddles, have a number of vertical blades mounted on a disc. The disc forces gas into high shear regions around the blades, where the bubbles are broken up, and this gives much better gas dispersion as compared with a simple paddle. In the 1950s, J.H. Rushton carried out a substantial amount of work on disc turbines [3] and standardised the agitators: the impeller diameter, D, is about one third of the tank diameter, the disc is $2/3D$ and the blade is $D/5$ wide and $D/4$ long. The standard six-blade disc turbine has been named the Rushton turbine (RT) in recognition of his great contribution to mixing and agitators. Since then, the Rushton turbine has been the choice of agitation system, especially for gas dispersion. Even now it is still commonly used in the fermentation industry.

Some weaknesses of the Rushton turbine, have, however, been recognised. Firstly, it has a high power number (approximately 5–6) and runs at a relatively low speed to develop a given power; it thus has a higher torque compared with low power number impellers. Secondly, the Rushton turbine is a radial flow impeller developing upper and lower flow loops in vessels. This can lead to strong compartmentalisation in multiple impeller systems

(a) Vortex cavities   (b) A large smooth-sided cavity   (c) A ragged (flooding) cavity

Figure 3.7   *Different cavities behind a Rushton blade*

and hence result in poor top to bottom mixing [4]. Another disadvantage is that an RT has a significant reduction in power draw (usually more than 50%) upon gas sparging compared with ungassed conditions at the same impeller speed. This decrease implies a reduced heat and mass transfer potential.

The original work [5] by Smith and coworkers throws light on the gassed power reduction mechanism and paves the way to understanding how to improve impeller design. When a blade moves forward, fluid in front of the blade is accelerated as it moves over and around the blade. The combination of pressure and centrifugal forces causes the fluid to turn inward toward the back of the blade. High-speed, low-pressure trailing vortices at the rear of the blade are formed, which are spread outward or away from the blade. In the presence of gas dispersion, the low-pressure core of the trailing vortices attracts gas bubbles, which coalesce to form ventilated cavities behind the blades. These stable cavities reduce the form drag and hence the power consumption. For a Rushton turbine, with an increase in the gas flow rates the cavities develop from vortex cavities, large smooth-sided cavities to large ragged cavities (Figure 3.7).

The size and the stability of the cavity were found to depend on blade geometry. Therefore, suitable blade shape and curvature may retard the formation of the ventilated cavities and reduce their size in such a way that, even under gas sparging, the aerated power draw can remain high.

This was confirmed by the invention of the concave hollow blade [6]. This blade shape is a section of a pipe cross-section. Reduced ungassed power number, flatter gassed power curve and improved gas handling capacity were obtained. Later derivatives of this design, such as deeper semi-circular [7] and parabolic [8] shaped hollow blades, were proposed. Such blades are now commercially available as the Chemineer CD-6 (Figure 3.8a) and SCABA 6SGDT impellers, respectively.

The impeller blades in the above mentioned disc turbines have symmetrical geometry with respect to the plane of the disc. Recently a vertically asymmetric disc turbine, the Chemineer BT-6 (Figure 3.8b), has been developed. The shape of the BT-6 blade is a semi-elliptical cross-section with an extended upper part. Myers *et al.* [9] reported that this impeller has a flatter gassed power curve and higher gas handling capacity than any of the other impellers.

Although these new radial flow impellers give good gas dispersion, they have the same drawback as the traditional Rushton turbines – poor top to bottom mixing. This can be resolved by using an axial flow impeller such as a pitched blade turbine or a hydrofoil impeller.

(a)

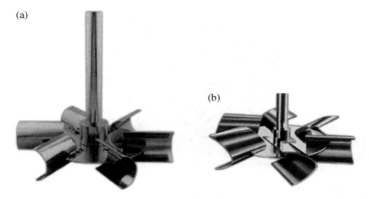

(b)

Figure 3.8 (a) Chemineer CD-6 (Courtesy of Chemineer, Inc. – Reliability and Technology in Mixing.); and (b) Chemineer BT-6 (Courtesy of Chemineer, Inc. – Reliability and Technology in Mixing.)

3.5.2 Pitched Blade Turbines

Pitched blade turbines (PBTs) were introduced in the 1950s. They usually have four flat blades pitched at different angles. PBTs can be operated either pumping down or pumping up, the former being the more traditional mode of operation. PBTs usually have smaller power numbers (about one quarter of those of the Rushton turbines) and are more efficient for solid suspension applications than the standard Rushton turbines. However, when downward pumping pitched blade turbines are used in gas dispersion, two gas loading regimes exist [10]. At low impeller speeds and/or high gas flow rates, some or all of the gas directly loads into the impeller (direct loading), while at high impeller speeds and/or low gas flow rates, gas reaches the impeller only through liquid recirculation (indirect loading). The transition between the two loading regimes often leads to severe hydrodynamic and torque instabilities [11]. Furthermore, extensive research has shown that down-pumping PBTs are easily flooded and therefore have a limited gas handling ability [12].

Pitched blade turbines are not truly axial flow impellers because they produce a flow having a relatively strong radial component of velocity, with a discharge angle of 45–60° to the horizontal [13]. Thus pitched blade turbines are sometimes referred to as mixed flow impellers. In comparison, hydrofoil impellers develop a stronger axial flow [14].

3.5.3 Hydrofoil Impellers

Hydrofoil impellers were developed after the benefits of impellers giving higher flow per unit power consumption were realised. They have profiled and angled blades shaped with a low angle of fluid attack. This blade geometry reduces the drag on the blades and thus decreases impeller power consumption, and at the same time produces a strong bulk recirculation. The low power number hydrofoil impellers can be exploited in retrofitting operations [15]. Compared with the Rushton turbine, a larger diameter hydrofoil impeller can be driven with the same power at the same speed and torque and should produce a better bulk mixing [16].

Figure 3.9   *Lightnin A310 (Courtesy of http://www.lightninmixers.com/)*

Probably the first successful hydrofoil impeller was the Lightnin A310 (Figure 3.9) with a power number 0.3 and flow number, $Fl_g = \phi_g/ND^3$, of 0.56. It has three narrow, cambered and twisted blades. The leading edge is rounded to reduce turbulence. The A310 is very efficient for solid suspensions. For example, with the same impeller diameter, its power draw is only 25 and 66% that of a Rushton turbine and a PBT45, respectively.

Figure 3.10 shows another frequently used hydrofoil impeller known as the Chemineer HE-3 [17]. The HE-3 is basically a PBT with the front corner of the blade bent down at an angle. It is reported that the HE-3 impeller produces more fluid motion per unit of power at constant torque and gives a more uniform flow throughout a vessel than an otherwise similar PBT [17]. The HE-3 has often been used by Chemineer [18] in combination with a CD-6 in dual-impeller systems.

Both the A310 and the HE-3 belong to the class of narrow blade hydrofoil impellers with a low solidity ratio, defined by comparing the projected blade area to the impeller swept area. Oldshue [19] has shown that the solidity ratio has substantial influence on impeller performance. It was illustrated that when a low solidity hydrofoil impeller, for example an A310 (a value of 0.1 is quoted), is operated at a low Reynolds number (say less than 200), its discharge flow pattern becomes more radial and its effectiveness is markedly reduced. Furthermore, with gas dispersion, this impeller tends to flood at relatively low gas flow rate compared with the disc turbines. This has led to the development of the high solidity-ratio hydrofoil impellers, such as the early Prochem Maxflo T [20], Lightnin A315 [19] and more recently the APV-B2 [21], Chemineer Maxflo W and Lightnin A340 [22].

Figure 3.10   *Chemineer HE-3 (Courtesy of Chemineer, Inc. – Reliability and Technology in Mixing.)*

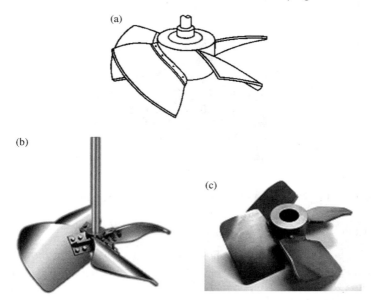

Figure 3.11 *(a) Prochem Maxflo T (Courtesy of Chemineer, Inc. – Reliability and Technology in Mixing.); (b) Prochem Maxflo W (Courtesy of Chemineer, Inc. – Reliability and Technology in Mixing.); and (c) Lightnin A315 (Courtesy of http://www.lightninmixers.com/)*

The Prochem Maxflo T impellers have 3, 4, 5 or 6 trapezoidal cambered blades attached to a large drum-type hub of about 0.4D diameter. The blades are of roughly equal width and length. A four-blade Prochem Maxflo T impeller is shown in Figure 3.11a. An extensive literature suggests that the Prochem Maxflo T agitators have better performance, in both solids suspension and gas dispersion [8, 11, 20, 23], than traditional disc turbines, although these Maxflo T impellers are not well defined in geometry. They have now been superseded by the Prochem Maxflo W impellers shown in Figure 3.11b.

The A315 impellers (Figure 3.11c) have a solidity ratio of about 0.9 and a power number of 0.75. They have very similar aeration characteristics to the Prochem Maxflo T impellers [1]. They both may exhibit large torque fluctuations when operated in the down-pumping mode in gas dispersions.

Lightnin developed a wide blade up-pumping A340 impeller (Figure 3.12). The three blades reduce mass and shaft loading, which is a problem with up-pumping arrangements. It is reported that multiple A340 impeller agitators are ideal for multi-phase applications, such as fermentation, polymerisation and hydrogenation. This agitator is effective in inducing gas drawdown from the surface and in controlling foaming. Improved bulk blending, mass transfer and heat transfer performance is also claimed [21], but the investigation is far from complete.

Another available high solidity ratio impeller is the APV B-2 [24], which is mainly used in the up-pumping mode. Flat gassed power curves and quick liquid mixing can be achieved by single or multiple APV B-2 agitators.

Figure 3.12 *Lightnin A340 (Courtesy of http://www.lightninmixers.com/)*

3.5.4 Multiple Impellers

Multiple impellers on the same shaft should be installed when the ratio of tank height to diameter is more than unity. Many advantages for this configuration include a higher area to volume ratio, which may be important for heat transfer and a better gas utilisation rate due to longer bubble retention times.

Traditional multi-impeller systems, for example fermenters, have used combinations of Rushton-type radial flow impellers because they were considered to be good for gas dispersion. However, the substantial decrease in the gassed power draw reduces the mass transfer potential and their strong compartmentalised flow patterns lead to poor top to bottom liquid mixing, which also limits their use.

In the early 1980s, it was recognised that the purpose of the upper impellers is primarily to provide uniform blending and to maintain gas dispersion. The upper Rushton impellers that produce poor circulation can be replaced with down-pumping axial flow impellers, for example A310 or HE3 hydrofoil impellers. Kaufman *et al.* [22] have shown that through using this combination the zoning formed by radial flow impellers can be greatly reduced and the liquid phase mixing is improved.

Further improvements were made in the late 1980s with the introduction of wide blade hydrofoil impellers (Lightnin A315 and Maxflo T). These impellers produce strong axial flow patterns [25] while providing superior gas dispersion. When narrow blade hydrofoil impellers cannot effectively handle gas, wide blade impellers offer an effective alternative. These wide blade impellers are usually operated in the down-pumping mode.

Since the last decade, increasing attention has been paid to up-pumping multiple-impeller technology [22, 24]. Typical impellers for these applications are the Lightnin A340 and APV B-2. The benefits from using this technology, such as the improved liquid blending and gas dispersion together with increased mass transfer and heat transfer, have been reported.

3.6 Baffles

Wall baffles typically consist of solid surfaces positioned in the path of tangential flows generated by a rotating impeller. The primary aim of using wall baffles is to prevent solid-body rotation during mixing. In the absence of the baffles, the flow created by impeller rotation tends to be two-dimensional and mainly causes a swirling action, that is solid-body rotation. Wall baffles transform tangential flows into vertical flows, provide top-to-bottom mixing without swirl, and minimise air entrainment. Wall baffling has a significant influence on the flow behaviour and resulting mixing quality. Baffles increase the drag and power draw of the impeller and are generally considered to enhance mixing operations.

Baffles are generally used in transitional and turbulent mixing, except in severe fouling systems, which require frequent cleaning of tank internals. For laminar mixing of viscous fluids, baffles are not needed. In square and rectangular tanks, the corners break up the tangential flow pattern and thus provide a baffling effect, and wall baffles may not be needed. Baffles are also not used for side-entering mixers in large product tanks and for angled mixers in small agitated tanks.

A standard baffle configuration consists of four vertical plates having width equal to 8–10% (T/12 to T/10) of the tank diameter. Narrower baffles are sometimes used for high viscosity systems, buoyant particle entrainment (width $= 2\%$ of T), or when a small vortex is desired. A small spacing between the baffles and the tank wall (1.5% of T) is allowed to minimise dead zones particularly in solid-liquid systems.

Other types of baffling (e.g. surface baffles, retractable baffles, twisted baffles and partition baffles) are also used to satisfy specific process needs. Surface baffles can prevent gas entrainment from the vapour head. Retractable baffles are used in systems where rheology changes during the process, and baffles must be removed at low Reynolds numbers. Partition baffles are used for staging of tall vessels. The selection, sizing and location of baffles depend on the process requirements and mixing regime.

3.7 Power Requirements

3.7.1 Single Impellers

Generally speaking, the power draw decreases with increasing gas flow rate for a given stirrer speed. This is caused be the gas cavities formed behind the impeller blades, as mentioned above. For low viscosity liquids, Smith *et al.* [26] divided the power draw into three categories, based on the gas flow number and the occurrence of clinging cavities. They define a specific flow number, $(Fl_g)_{3-3}$, at which behind all six blades of the Rushton turbine cavities are present. This number depends on the Reynolds number and Froude number according to:

$$(Fl_g)_{3-3} = 3.8 \ 10^{-3} \left(\frac{Re^2}{Fr}\right)^{0.07} \left(\frac{T}{D}\right)^{0.5} \qquad (3.4)$$

where

T/D is the tank to impeller diameter ratio.

The power draw is cut into three regimes. For low gas flow number:

$$\left(\frac{P_g}{P_0}\right) = 1 - 16.7 \ Fl_g \ Fr^{0.35} \text{ for } Fl_g < (Fl_g)_{3-3} \qquad (3.5)$$

For intermediate gas flow numbers, that is when small cavities are formed

$$\left(\frac{P_g}{P_0}\right) = C_1 - 0.1\frac{C_2 - C_1}{(Fl_g)_{3-3} - 0.1} + \frac{(C_1 + C_1)}{(Fl_g)_{3-3} - 0.1} \quad \text{for} \quad (Fl_g)_{3-3} < Fl_g < 1.0 \quad (3.6)$$

where

$$C_1 = 0.27 + \frac{0.022}{Fr}$$

$$C_2 = 1 - 17(Fl_g)_{3-3} \, Fr^{0.35}$$

For the large cavities up to the flooding regime these workers proposed:

$$\left(\frac{P_g}{P_0}\right) = 0.27 + \frac{0.022}{Fr} \quad \text{for} \quad 0.1 < Fl_g < (Fl_g)_{flooding} \quad (3.7)$$

For multiple impeller stirred tanks, a rule of thumb is that the total power draw can be calculated as the sum of the power draws of the individual impellers, as if they were on their own. This holds if the distance between two consecutive impellers is large enough, such that the flow field generated by each impeller is not influenced significantly by the others. In practice, this means that the impeller–impeller distance should be at least equal to the vessel diameter. Note that this is most accurate for ungassed systems, where the hydrodynamics are only governed by the action of the impellers. For gassed systems, the action of gravity via the bubble distribution complicates matters. The inflow of gas into one of the impellers will be different to the inflow for which the single impeller data have been derived. For the latter, usually some form of ring sparger aligned with the impeller axis is used. With the multiple impeller, the gas flows in from the zone above another impeller, meaning that the gas comes in much more dispersed over the entire cross-section of the reactor.

Bakker and Van den Akker [27] determined the power draw experimentally. In Figure 3.13 the curve of the power, made dimensionless with the power draw at ungassed

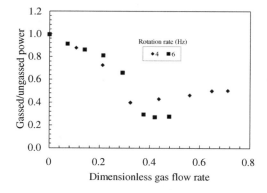

Figure 3.13 *Power consumption (normalised by the ungassed power) as a function of the dimensionless gas flow rate. Pitched blade impeller. Impeller/tank diameter D/T = 0.4, impeller positioned D above base of tank, sparger/impeller separation = 0.8D (Reprinted from Bakker, A. and Van den Akker, H.E.A., Gas-liquid contacting with axial flow impellers, Chem.Eng.Res. Design,* **72**, *573–577. Copyright (1994) with permission from Elsevier.)*

conditions for a pitch blade turbine, is given for two different impeller speeds. Their vessel had a diameter of 440 mm and was equipped with four baffles. The figure shows a steep drop at a gas flow number, Fl_g, of about 0.2–0.3. This drop is caused by a change in flow regime from an indirect to a direct loading. With indirect loading, the downward pumping action of the impeller is sufficient to first drag the bubbles downwards and to the wall region before they circulate back into the impeller zone. In the case of direct loading, the downward flow out of the impeller is insufficient to prevent the bubbles from rising upwards into the impeller zone. The same workers also reported that the type of sparger influences the power consumption. This is again related to the induced flow pattern. A larger sparger and a larger distance between sparger and impeller both require less power for a given gas flow rate to achieve good dispersion.

3.7.2 Multiple Impellers

As mentioned previously, for stirred tanks with multiple impellers, the situation is more complicated. For the ungassed case, if the impeller–impeller distance, C_I, is sufficiently large, that is $C_I/D > 1.5$, the impellers have a weak interaction. Consequently, the power draw of the system is the sum of the power consumption of each of the individual impellers as if they were in a single impeller vessel. For instance, a large-scale vessel equipped with four Rushton turbines, operated in the turbulent regime will have a power draw $Po \sim 4*5.5 = 22$.

The gassed power draw can not so be easily obtained. Even if the impeller–impeller distance is large enough, the bottom impeller (assuming the gassing takes place below the bottom impeller) has a different gas loading than the other ones. Therefore, the structure of the two-phase flow in the bottom impeller region is different to that of the other impellers. Hence, the impeller blade to gas flow interaction is different between the bottom impeller and the others [28, 29]. The decrease in power take of the bottom impeller is significantly more than for the other impellers. Cui *et al.* [29] has reviewed the literature on power consumption of multi-impeller gassed, stirred tanks. The majority of data concerned water as the continuous phase, so the correlations they present do not take the physical properties of the fluids into account. One of the results is that the bottom impeller behaved as a single impeller. The power uptake was correlated with the quantity $\phi_g N^{0.25}/D^2$, which is not a dimensionless quantity. Nevertheless, they found a good correlation for the gassed power draw:

$$\left(\frac{P_g}{P_0}\right) = 1 - 9.9\frac{\phi_g N^{0.25}}{D^2} \quad \text{for} \quad \frac{\phi_g N^{0.25}}{D^2} < 0.055$$

$$\left(\frac{P_g}{P_0}\right) = 0.48 - 0.62\frac{\phi_g N^{0.25}}{D^2} \quad \text{for} \quad \frac{\phi_g N^{0.25}}{D^2} > 0.055$$

$$(3.8)$$

For the other impellers, they correlated the power draw as a function of $\phi_g N$ and found for each of the impellers:

$$\left(\frac{P_g}{P_0}\right) = 1 - K_1\,\phi_g\,N \quad \text{for} \quad \phi_g\,N < 0.013$$

$$\left(\frac{P_g}{P_0}\right) = K_2 - K_3\,\phi_g N \quad \text{for} \quad \phi_g\,N > 0.013$$

$$(3.9)$$

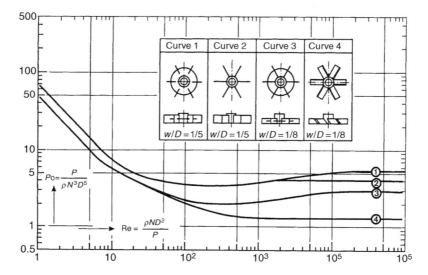

Figure 3.14 *Dimensionless power versus Reynolds number for single-phase flow in a standard stirred tank (w denotes the blade height) [30] Transport Phenomena Data Companion, Janssen and Warmoeskerken*

The constants, K_1, K_2 and K_3 depend on tank geometry. Values of 37.6, 0.375 and 8 were given ($D = 0.256$ m, $T = 0.64$ m, $H = 2.2$ m). Note that the constants in the above equations are not dimensionless.

3.7.3 Single-Phase Power

The above relationships compute only the power input relative to the ungassed case. The latter can be found in the literature relatively easily. In Figure 3.14 the power versus Reynolds number for three different Rushton-type turbines and a pitch blade impeller are given. All have in common the characteristic that the power is inversely proportional to the Reynolds number for laminar flows, whereas they become constant for Re $> 10^4$. The power number is defined as $Po = P_0/\rho N^3 D^5$.

For large Reynolds numbers, this power number tends to a constant, which depends on the type of impeller and the precise geometry of the stirred tank.

The above methods can be illustrated by a specific example of design. A gassed stirred tank needs to be operated in the complete dispersion regime. The gas-liquid volume in the tank is 0.5 m^3. The liquid has physical properties that can be approximated by those of water. A single Rushton turbine is used to mix the phases. The vessel is of standard type, that is $T = H$ and $D = T/3$. For the mixing, it is required that the Reynolds number is about 2×10^5.

a. Calculate the required rotation speed of the impeller.

$$V = \frac{\pi}{4}T^2 H = \frac{\pi}{4}T^3 \rightarrow T = H = \left(\frac{4}{\pi}V\right)^{1/3} = 0.86\,\text{m} \rightarrow D = T/3 = 0.29\,\text{m}$$

$$Re = \frac{\rho N D^2}{\eta} \rightarrow N = \frac{\eta}{\rho D^2} Re = 2.4\,\text{rps}$$

b. Find the maximum gas flow rate into the tank given that the flow regime is complete dispersion.

Using Figure 3.3 or Equation 3.3 for the complete dispersion line:

$$Fr = \frac{N^2 D}{g} = 0.17$$

Thus,

$$\left(Fl_g\right)_{comp\ dis} = 0.2\left(\frac{D}{T}\right)^{0.5} Fr^{0.5} = 0.048$$

from which $\phi_g = Fl_g N D^3 = 2.8$ l/s is obtained.

c. Compute the required power input.

First, compute

$$\left(Fl_g\right)_{3-3} = 3.8\ 10^{-3}\left(\frac{Re^2}{Fr}\right)^{0.07}\left(\frac{T}{D}\right)^{0.5} = 0.041$$

From Section 3.3 it can be concluded that the tank is operated in the intermediate regime and the power draw is given by:

$$\left(\frac{P_g}{P_0}\right) = C_1 - 0.1\frac{C_2 - C_1}{Fl_{g3-3} - 0.1} + \frac{(C_2 - C_1)Fl_g}{Fl_{g3-3} - 0.1}$$

and compute $P_g/P_0 = 0.6$. For a Rushton turbine in a standard stirred tank the power draw in the ungassed case for turbulent flow is given by: $P_0 = 5\rho N^3 D^5 = 140\,\text{W}$ (curve 1 of Figure 3.14). Thus the power draw in the gassed case is $P_g = 84\,\text{W}$.

3.8 Gas Fraction

Reports on the gas fraction are rather scarce. Moreover, the gas fraction is strongly influenced by the physical properties of the liquid phase. Most data in the literature are for air-water systems. These data are only of limited use in more complex systems, for example, in the presence of particles or for different liquids. Differences in properties such as surface tension and viscosity of the gas and liquid phase do influence the

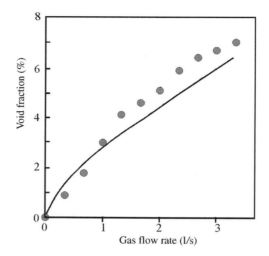

Figure 3.15   *Gas fraction in a 200 l stirred tank. Full line, Calderbank [31] correlation, Equation 3.10; symbols, experimental data taken from Laakkonen et al. [32]*

coalescence and breakup behaviour of the gas bubbles. Consequently, no well-established correlations for the gas fraction can be given. Bubble breakup takes place in the impeller discharge flow, where turbulence intensities are highest. Coalescence is dominating in the more quiescent parts of the reactor. However, the low-pressure area behind the impeller blades (where the cavities are formed) also play an important role as here breakup and coalescence are the dominating processes.

By 1958 Calderbank [31] had already proposed a correlation for the gas fraction, ε_g:

$$\varepsilon_g = \left(\frac{u_{gs}\varepsilon_g}{u_b}\right)^{0.5} + 0.000216\left(\frac{P_g}{V}\right)^{0.4}\left(\frac{\rho_l}{\sigma}\right)^{0.6}\left(\frac{u_{gs}}{u_b}\right)^{0.5} \tag{3.10}$$

where

$u_b = 0.265$ m/s.

This correlation is still in use. Laakkonen *et al.* [32] measured the gas fraction in a 200 l vessel and they also compared their results with the Calderbank correlation. Some of their data are reproduced in Figure 3.15. The Calderbank correlation gives a reasonable prediction of the experimental data.

Vrabel *et al.* [33] investigated the flow in large-scale reactors, up to a volume of 30m^3. Four radially pumping impellers on a single shaft were used to disperse the air and mix the liquid. They found that the gas fraction could be correlated by a single relationship for different types of radially pumping impellers:

$$\varepsilon_g = 0.37\left(\frac{P_g}{V}\right)u_{gs}^{0.55} \tag{3.11}$$

Nienow *et al.* [34] found similar exponents from their experiments in a $14\,m^3$ vessel using Rusthon turbines, the Lightnin A315s or Prochem Maxflo T downwards pumping impellers.

3.9 Mass Transfer

3.9.1 Bubble Size

For isolated droplets and bubbles in a homogeneous turbulent flow, the maximum diameter is often estimated from Hinze's theory [35]. In this approach, the competition between surface tension forces, which try to keep the bubble (or droplet) spherical, and the inertia forces exerted by the eddies that collide with the bubble are balanced. This gives an estimate for the maximum stable bubble size:

$$\frac{\sigma}{d_{bub}} \sim \frac{1}{2}\rho_l v^2 \tag{3.12}$$

where

σ is the surface tension
d is the bubble diameter
ρ_l is the liquid density
v is the fluctuating velocity associated with the eddy that collides with the bubble.

A basic assumption is that breakup can only be achieved by eddies that are of comparable size as the bubbles. The arguments are: a large eddy will only transport the bubble (on the scale of the bubble, only the shear of the eddy is felt, which is mild for the large eddies); a small eddy can only induce some local perturbation of the bubble surface, but has little energy to cause breakage (Figure 3.16).

For eddies in the range of bubble size usually found in stirred tanks, Kolmogorov [36] derived the following relationship between the fluctuating velocity of the eddy, its size, d_e,

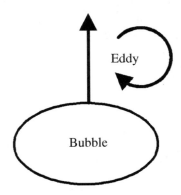

Figure 3.16 *Bubble colliding with an eddy*

and the local turbulent energy dissipation, $\dot{E}$:

$$v^2 = 2(\dot{E}d_e)^{2/3} \tag{3.13}$$

Combining the last two equations gives an estimate of the maximum stable bubble size:

$$d_{bub,max} = C\frac{\sigma^{3/5}}{\rho_l^{3/5}\dot{E}^{2/5}} = C\frac{\sigma^{3/5}}{\rho_l^{1/5}\left(\frac{P}{V}\right)^{2/5}} \tag{3.14}$$

where in the second step the local dissipation is replaced by the global one, that is, $\dot{E} \to P/\rho_l V$. The coefficient C is of order unity. By using the power number, $Po = P_0/\rho N^3 D^5$, the above equation can be rewritten as:

$$\frac{d_{bub,max}}{D} \sim \left(\frac{\rho_l N^2 D^3}{\sigma}\right)^{-3/5} = We^{-3/5} \tag{3.15}$$

The above ideas rely on the assumption that breakup is caused by eddies in the inertial sub-range. Luckily, for most stirred tanks, the turbulence intensities are high and the eddies, which are of the size of the bubbles, fall into this range. Nevertheless, it should be kept in mind, that the local energy dissipation can be an order of magnitude higher than the global one.

The effect of the gas fraction on the prediction of maximum stable bubble size is to increase this size, mainly caused by an increasing probability of coalescence at higher gas fractions. For a stirred tank, a correction has been proposed to account for the effects of finite gas fraction [37]:

$$\frac{d_{max}}{D} = [2.5\varepsilon_g + 0.75]We^{-3/5} \tag{3.16}$$

For air sparged in various alcohols, Hu *et al.* [38] found that the exponent of the Weber number differed significantly from the value -0.6. Their data (for relatively low gas fractions) could be correlated with a similar expression: $d_{32}/D = A\,We^{-B}$ (where d_{32} is the Sauter mean diameter). The exponent B varied from 0.38 to 0.70 depending on the flow regime and on the presence or absence of particles in the flow. The exponent predicted by Kolmogorov's theory for isotropic turbulence did not agree with the findings from these experiments.

3.9.2 Interfacial Area

The interfacial area per unit volume, a, is closely connected to the maximum bubble size. It is usually computed from the gas fraction, ε_g, and estimates of the Sauter mean bubble diameter, d_{32}. The connection between these quantities is given by:

$$d_{32} = 6\frac{\varepsilon_g}{a} \tag{3.17}$$

For non-coalescing media Parathasarathy *et al.* [39] predicted the Sauter diameter from the energy dissipation as found in the impeller swept volume:

$$d_{32} = 2.0 \left(\frac{\sigma^{3/5}}{\rho_l^{1/5} \dot{E}_i^{2/5}} \right) \tag{3.18}$$

Lu *et al.* [40] reported the following correlation:

$$d_{32} = 33.4 \dot{E}^{-0.25} \phi_g^{0.14} \tag{3.19}$$

Note that this equation is dimensionally incorrect: the constant 33.4 carries units and the gas flow rate, ϕ_g, needs to be specified in m^3/s to find the diameter in millimeters.

Barigou and Greaves [41] found that their air-water data could be correlated by:

$$a = 186 \left(\frac{P_g}{V} \right)^{0.3} u_{gs}^{0.51} \tag{3.20}$$

with a scatter of the data within 15% of this relationship. For air-salt water the coefficients changed to 178, 0.36 and 0.45, respectively.

As the power is not distributed homogeneously, the Sauter diameter based on the global power input per unit volume can lead to over-prediction of the bubble size. In the impeller region, the specific power input is much higher than in the bulk of the liquid, the bubbles are broken up into smaller ones in the impeller region more than elsewhere. Therefore, correlations exist that take the power per volume in the impeller region into account. For instance, Alves *et al.* [42]

$$d_{32,imp} = 8.5 \left(1 + 32.5 \frac{\phi_g}{D^2} \right) \left(\frac{P_g}{V_{imp}} \right)^{-0.24} \tag{3.21}$$

where

$$V_{imp} = \pi/4 (D^2 W)$$

and

W denotes the height of the impeller.

Obviously, the above relationship will predict smaller bubbles, as the specific energy input in the impeller region is much higher than if it were based on the entire vessel. In particular, for lower gas fractions or non-coalescing systems this might provide a better estimate of the bubble size.

3.9.3 Mass Transfer

The actual mass transfer is a combination of the bubble-liquid interface, represented by a, and the transfer coefficient, k_l, which represents the resistance to mass transfer in the liquid film surrounding a bubble. The interface and phase concentrations either side of it are illustrated in Figure 3.17.

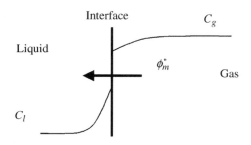

Figure 3.17 *Mass transfer across the gas-liquid interface*

The general formula describing mass transfer is:

$$\phi_m'' = \left(\frac{1}{k_l} + \frac{m}{k_g}\right)^{-1} \left[mc_g - c_l\right] \tag{3.22}$$

where

ϕ_m'' is the mass transfer per unit area
k_l and k_g are the transfer coefficients on the liquid and gas side of the bubble interface, respectively
m is the distribution coefficient defined as $m \equiv \left[c_l/c_g\right]_{eq}$, which is the ratio of the gas to liquid concentration at thermodynamic equilibrium
c_g and c_l are the concentrations of the species in question in the gas and liquid (bulk) phases, respectively.

In many instances, the resistance to mass transfer is almost completely in the liquid phase and the above equation can be reduced to:

$$\phi_m'' = k_l \left[c_l^{sat} - c_l\right] \tag{3.23}$$

with c_l^{sat} the concentration directly at the liquid side of the interface, which is now in equilibrium with the concentration in the gas phase (but not necessarily equal to it). The total mass flow rate per unit volume is thus given by:

$$\phi_m = k_l a \left[c_l^{sat} - c_l\right] \tag{3.24}$$

For isolated bubbles in a simple liquid flow, correlations exist that describe k_l, usually specified in terms of the Sherwood number, $k_l = Sh(D_l/d_p)$, with D_l the diffusion coefficient in the liquid and d_p the diameter of the particle. For instance, the steady-state (liquid side) mass transfer coefficient for a spherical particle is given by:

$$Sh = 2 + 0.66 Re^{0.5} Sc^{0.33} \tag{3.25}$$

for $10 < Re < 10^4$ and $Sc > 0.7$ with the Reynolds number defined as $Re = \rho_l v_{rel} d_p / \eta_l$ (ρ_l and η_l are the liquid density and viscosity and v_{rel} is the relative velocity between the liquid and the particle) and the Schmidt number defined as $Sc = \eta_l / \rho_l D_l$.

For estimates of k_l the Higbie theory [43] is also often used. Here the idea is to follow a package of liquid that moves past the bubble. The contact time is of the order of d_p/v_{rel} and for short times the mass transfer coefficient is estimated from the penetration theory: $k_l = 2\sqrt{(D_l/\pi t)}$. Combining gives:

$$k_l = 2\sqrt{\frac{D_l v_{rel}}{\pi d_p}} \qquad (3.26)$$

However, in a turbulent flow such as that found in a stirred tank, the contact time between fluid elements and the bubbles may also be estimated from turbulent eddy parameters. The velocity is then replaced by an estimate of the fluctuating velocity of the turbulence associated with eddies that are the size of the bubbles. Moreover, the characteristic length scale is then also set at the eddy size. Following Kolmogorov's ideas for the smallest eddies, the ratio of the two gives the required 'contact time':

$$t_e = \frac{\left(v^3 \dot{E}\right)^{1/4}}{\left(v\dot{E}\right)^{1/2}} \qquad (3.27)$$

where

v is the kinematic viscosity of the liquid phase.

The resulting estimate [44] for the mass transfer coefficient is:

$$k_l a = 2\sqrt{\frac{D_l}{\pi}}\left(\frac{\rho \varepsilon}{\eta}\right)^{1/4} \qquad (3.28)$$

Empirical relationships for $k_l a$ are in many cases correlated using the power input per unit volume and the superficial gas velocity. They have the form: $k_l a = C u_{gs}^a \eta_l^b (P/V)^c$, with the constant c depending on impeller type and geometry used and tabulated correlations that are available [42]. A well-known correlation used for both Newtonian and non-Newtonian liquids has been proposed [45]:

$$\frac{k_l a}{\phi_g/V} = 0.22\left(\frac{P_g}{\phi_g \rho_l (v_l g)^{2/3}}\right)^{0.34} \qquad (3.29)$$

where

v_l is the kinematic viscosity of the liquid, which in case of a non-Newtonian liquid is obtained from the apparent viscosity.

Nienow and coworkers [46] experimented with so-called intensified operation: the specific power input and the superficial gas velocity are much higher than at standard operation ($P/V \sim 10$–$100\,\text{kW/m}^3$ and $u_{gs} \sim 10\,\text{cm/s}$). They reported the existence of two regimes: a homogeneous regime at low aeration with small bubbles more or less evenly dispersed over the vessel, and a heterogeneous regime at high aeration rates, fairly similar to the heterogeneous regime in bubble columns, that is a wide distribution of bubble sizes with large bubbles rising through the vessel. The impeller is not able to break these bubbles and disperse them. Nevertheless, these workers could describe their $k_l a$ values by a single correlation:

$$k_l a = 0.0059 \, u_{gs}^{0.36} \left(\frac{P}{V} \right)^{0.607} \tag{3.30}$$

with the specific power in W/m^3. These results hold for both Rushton turbines and a Scaba 6 curved-blade Rushton impeller.

The above methods can be illustrated by a specific example of design. A standard stirred tank (diameter 740 mm) is operated in the loading regime. The liquid has a density of 1100 kg/m^3, a viscosity of 5 mPa s and a surface tension of 0.06 N/m. The gas is air. The power input per unit of volume is 86 J/m^3. The gas flow rate equals 0.3 l/s.

a. Estimate the bubble size.

The maximum bubble diameter is used to estimate the bubble size:

$$d_{32} \sim \frac{\sigma^{0.6}}{\rho_l^{0.2} \left(\frac{P_g}{V} \right)^{0.4}} \sim 8 \text{ mm}$$

b. Compute the gas fraction.

The gas fraction is estimated from the Calderbank relationship. The terminal bubble velocity needs to be calculated for this equation. Assuming the bubbles to be spherical (although in reality they are probably not spherical) with an immobile surface gives: $u_b = 0.29$ m/s. Thus, from Equation 3.10

$$\varepsilon_g = \left(\frac{u_{gs} \, \varepsilon_g}{u_b} \right)^{0.5} + 0.000216 \left(\frac{P_g}{V} \right)^{0.4} \left(\frac{\rho_l}{\sigma} \right)^{0.6} \left(\frac{u_{gs}}{u_b} \right)^{0.5}$$
$$\rightarrow \varepsilon_g = 3\%$$

c. Estimate $k_l a$.

For the mass transfer use Equation 3.29:

$$\frac{k_l a}{\phi_g / V} = 0.22 \left(\frac{P_g}{\phi_g \rho_l (v_l g)^{2/3}} \right)^{0.34} \rightarrow k_l a = 7.4 \cdot 10^{-3} \text{ 1/s}$$

3.10 Mixing Times

The mixing time in single-phase applications depends directly on the circulation velocity. It is generally accepted that in a turbulent regime, the liquid has to pass the impeller three to four times in order to have mixing at the macro-level. In practice, a rule of thumb is used that the dimensionless mixing time, that is the mixing time, t_{mix}, made dimensionless by the rotational speed of the impeller, N, is about $Nt_{mix} = c_{mix} = 3-4$, provided the Reynolds number, $Re = \rho_l ND^2/\eta_l > 10^4$. The precise value of the constant c_{mix} depends on the geometry, including the impeller type. This estimate of mixing takes care of the mixing of the bulk. It does not take into consideration mixing at a molecular level. The turbulence can mix only up to the scale of the smallest eddies, that is to the Kolmogorov length scale of the smallest eddies of size $l_K = (v^3/\dot{E})^{1/4}$. Below this scale, diffusion has to take over, giving a characteristic time scale of $t_{diff,mix} = l_K^2/D_l$. For mixing in gaseous systems, the Kolmogorov scale is the deciding scale. This is caused by the fact that molecular transport of momentum, heat and species all go in the same way in gasses: individual molecules transport their momentum, their mass and their kinetic energy (heat) by transporting themselves. In a macroscopic world this is recognised by the diffusion coefficients for heat, mass and momentum ($a \equiv \lambda/\rho c_p$, D_l, $v \equiv \eta/\rho$ all in m²/s) are all of the same order. Rephrased in dimensionless numbers, the Prandtl ($Pr = v/a$) and Schmidt ($Sc = v/D_l$) numbers are of the order unity. However, for mixing in liquids the situation can be very different, as the Schmidt number is of the order 100–1000. Consequently, the Kolmogorov scales are not the endpoint. Here, the Batchelor scale is governing the chemical species transport at the small scales. The Batchelor length scale, l_B, is connected to the Kolomorov scale: $l_B = l_K Sc^{-1/2}$.

The Batchelor scale is such that it describes the scale at which diffusion starts to dominate the species transport. At the Batchelor scale, the time required for diffusion to even out fluctuations in the species concentration is similar to the time required for viscosity to even out velocity fluctuations at the Kolmogorov scale. The above scales, are merely useful for estimating required dissipation, and hence the power, to be able to mix at a certain micro-level. They do not provide the total mixing time.

Therefore, an alternative way needs to be found to estimate the mixing time, in which it is related to the power input per unit volume. For instance, Prochazka and Landau [47] presented the relationship:

$$t_{mix} = 4.82\left(\frac{P}{V}\right)^{-0.33} \tag{3.31}$$

This provides a macroscopic, empirical correlation for the total process of mixing. Obviously, it is fairly global and will not capture the fine details of mixing in multiphase flows at various gassing rates, or in stirred tanks with a highly non-uniform distribution of turbulence, shear and dissipation.

Questions

3.1 Oxygen needs to be transferred to the liquid phase in a gassed stirred tank (liquid volume 2 m³). The stirred tank is equipped with four Rushton turbines on a single shaft. The distance between the stirrers is equal to the tank diameter, T. The distance from the bottom to the lowest impeller is $T/3$, whereas the highest impeller is $2/3T$

beneath the liquid level. The stirrer speed of each impeller is 2 rps. The inflow of gas is 8 l/s. The density of the liquid is $900\,kg/m^3$ and its viscosity is $3\,mPa\,s$.

a. Compute the tank diameter.
b. Find the flow regime: does this stirred tank operate properly?

0.87 m; flooding.

3.2 The stirred tank from Question 3.1 needs to be operated in the loading regime. The impeller speed can not be increased, so the gas flow rate should be reduced.

    a. What is the maximum gas flow rate such that the tank is still in the loading regime? The reactor will be operated at 90% of this maximum gas flow rate.
    b. Compute the Reynolds number of the flow and estimate the ungassed power consumption.
    c. Estimate the power consumption in the gassed case.

a, 3.5 l/s; b, 55 000 and 289 W; and c, 162 W.

3.3 At the laboratory scale, a gassed stirred tank has been optimised for mixing. The tank diameter and height are both 30 cm. A standard type of stirrer $(D = T/3)$ is used at 300 rpm. The liquid is water and the gas flow rate is rather low (0.01 l/s), so the gas fraction is small. The process needs to be scaled up to a reactor with $1\,m^3$ of liquid, that has a density of $1200\,kg/m^3$ and a viscosity of $7\,mPa\,s$.
    Determine the required impeller speed to keep the mixing at the same micro-scale.

2.1 rps.

3.4 A stirred tank (liquid volume $0.5\,m^3$) is used for a gas-liquid reaction. The gas flow into the tank has a superficial velocity of 80 mm/s. The power input in the reactor is 5 kW. The gas contains a concentration of 10% of a species X that dissolves and reacts in the liquid. This reaction is fast and it is safe to neglect the concentration of X in the liquid phase.
    Calculate the exit concentration of X. The system is operated in steady state.

$0.18c_{in}$.

3.5 A laboratory fermenter of volume 100 l needs to be scaled up by a factor of 100. The scaling strategy recommended is to keep the power input per unit liquid volume constant. Both fermenters need to be operated in the turbulent regime with $Re > 10^4$. The reactors will be geometrically similar. The power number of the impeller used is six. In both reactors the gassed power will be 70% of the ungassed one. The small reactor operates at a gas flow rate of $0.3\,m^3/min$ and a stirrer speed of 120 rpm.

    a. Design the large reactor and calculate the stirrer speed and required power.
    b. What happens to $k_l a$ if both reactors are to be operated at the same superficial gas velocity? Also, what happens to the mass transfer?
    c. What happens to the mixing time?

a, 43 rpm; b, mass transfer is probably not affected too much; and c, the mixing time stays the same in both cases.

3.6 A group of researchers is investigating the efficiency of removing dichlorophenol (DCP) from water effluent by injecting gaseous ozone. In the liquid phase, dissolved ozone reacts with DCP according to the reaction: $2O_3 + DCP \rightarrow$ products.

They run tests in a 5 l, 6 in diameter lab-scale stirred tank. The tank is equipped with a standard impeller (power number five) and four baffles. The aqueous phase (3 l) is put into their small stirred tank and gas containing ozone and oxygen is introduced into the reactor. The solubility, that is the ozone concentration in the water phase divided by that of the gas phase, is 0.24 under the experimental conditions. The impeller speed is 180 rpm and the gas flow rate is set at 1.0 l/min. The inlet concentration of ozone in the gas feed is 0.2 mM. The initial concentration of DCP is 0.5 mM.

The stirred tank can be treated as being ideally mixed and as a first approximation it is assumed that the reaction between ozone and DCP is so fast, that the liquid concentration of ozone stays at zero. The researchers found from the literature that the mass transfer could be described by $k_l a = 0.22 \, u_{gs}^{0.74} (P/V)^{0.4}$.

Estimate the required processing time to remove the DCP.

1650 s.

3.7 Gas is sparged into a stirred tank (volume 1 m^3) at a rate of 5 l/s. The stirrer speed is 2 rps. The power number is four and the blade height is $D/6$. The gassed to ungassed power ratio equals 0.6. The liquid has physical properties equivalent to water.

  a. Estimate the Sauter diameter based on the power input per volume or per swept volume.

  b. Estimate the Kolmogorov length scale in the bulk and in the impeller region.

a, 1.7 mm; and b, 15 μm.

References

[1] Nienow, A.W. (1998) Hydrodynamics of stirred bioreactors. *Appl. Mech. Rev.*, **51**, 3–32.
[2] Kulkarni, A. (2010) Design of a pipe/ring type of sparger for a bubble column reactor. *Chem. Eng. Technol.*, **33**(6), 1015–1022.
[3] Rushton, J.H., Costich, E.W. and Everett, H.J. (1950) Power characteristics on mixing impellers (I and II). *Chem. Eng. Prog.*, **46** (8), 395–404 and **46**(9), 467–476.
[4] Cronin, D.G., Nienow, A.W. and Moody, G.W. (1994) An experimental study of mixing in a proto-fermenter agitated by dual Rushton turbines. *Chem. Eng. Res. Des.*, **72**, 35–40.
[5] van't Riet, K. and Smith, J.M. (1973) The behaviour of gas-liquid mixtures near Rushton turbine blades. *Chem. Eng. Sci.*, **28**, 1031–1037.
[6] van't Riet, K., Boom, J.M. and Smith, J.M. (1976) Power consumption, impeller coalescence and recirculation in aerated vessels. *Chem. Eng. Res. Des.*, **54**, 124–131.
[7] Bakker, A., Myers, K.J. and Smith, J.M. (1994) How to disperse gasses in liquids. *Chem. Eng.*, **101**(12), 98–104.
[8] Nienow, A.W. (1990) Agitators for mycelial fermentations. *Trends Biochem. Technol.*, **8**, 224–233.
[9] Myers, K.J., Thomas, A.J., Bakker, A. and Reeder, M.F. (1999) Performance of a gas dispersion impeller with vertically asymmetric blades. *Chem. Eng. Res. Des.*, **77**, 728–730.
[10] Warmoeskerken, M.M.C.G., Speur, J. and Smith, J.M. (1984) Gas liquid dispersion with pitched blade turbines. *Chem. Eng. Commun.*, **25**, 11–29.

[11] McFarlane, C.M., Zhao, X.M. and Nienow, A.W. (1995) Studies of high solidity ratio hydrofoil impellers for aerated bioreactors 2. Air-water studies. *Biotechnol. Prog.*, **11**, 608–618.

[12] Chapman, C.M., Nienow, A.W., Cook, M. and Middleton, J.C. (1983) Particle-gas-liquid mixing in stirred vessels. *Chem. Eng. Res. Des.*, **61**(Parts I–IV), 71–81, 82–95, 167–181, 182–185.

[13] Ranade, V.V. and Joshi, J.B. (1989) Flow generated by pitched blade turbines I: measurements using Laser Doppler Anemometer. *Chem. Eng. Commun.*, **81**, 197–224.

[14] Jaworski, Z. and Nienow, A.W. (1994) LDA measurements of flow fields with hydrofoil impellers in fluids with different rheological properties. Proceedings of the 8th European Conference on Mixing, Cambridge, September 21–23, 1994, *IChemE Symp. Series*, **136**, 105–112.

[15] Nienow, A.W. (1990) Gas dispersion performance in fermenter operation. *Chem. Eng. Prog.*, **86**, 61–71.

[16] Ruszkowski, S. (1994) A rational method for measuring blending performance and comparison of different impeller types. Proceedings of the 8th European Conference on Mixing, Cambridge, September 21–23, 1994, *IChemE Symp. Series*, **136**, 283–291.

[17] Bakker, A. and Gates, L.E. (1995) Properly choose mechanical agitators for viscous liquids. *Chem. Eng. Prog.*, **91**(12), 25–34.

[18] Myers, K., Fasano, J.B. and Bakker, A. (1994) Gas dispersion using mixed high efficiency/disc impeller systems. Proceedings of the 8th European Conference on Mixing, Cambridge, September 21–23, 1994, *IChemE Symp. Series*, **136**, 65–72.

[19] Oldshue, J.Y. (1989) Fluid mixing in 1989. *Chem. Eng. Prog.*, **85**(5), 33–42.

[20] Gbewonyo, K., DiMasi, D. and Buckland, B.C. (1986) The use of hydrofoil impellers to improve oxygen transfer efficiency in viscous mycelial fermentations. Proceedings of International Conference on Bioreactor Fluid Dynamics, Cambridge, UK (BHRA, Cranfield), 1986, pp. 281–299.

[21] Nienow, A.W. (1996) Gas-liquid mixing studies: a comparison of Rushton turbine with some modern impellers. *Chem. Eng. Res. Des.*, **74**, 417–423.

[22] Kaufman, P., Post, T.A. and Preston, M. (1998) Up-pumping mixing technology – breaking the mould of traditional systems. *Chem. Proc.*, **61**, 10, 63–66.

[23] Nienow, A.W., Buckland, B.C. and Hunt, G. (1994) A fluid dynamic study of the retrofitting of large bioreactors; turbulent flow. *Biotech. Bioeng.*, **44**, 1177–1186.

[24] Hari-Prajitno, D., Mishra, V.P., Takenaka, K. *et al.* (1998) Gas-liquid mixing studies with multiple up- and down-pumping hydrofoil impellers: power characteristics and mixing time. *Can. J. Chem. E.*, **76**, 1056–1068.

[25] Mishra, V.P., Dyster, K.N., Jaworsski, Z. *et al.* (1998) A study of an up- and down-pumping wide blade hydrofoil impeller: Part I. LDA measurements. *Can. J. Chem. E.*, **76**, 577–588.

[26] Smith, J.M., Warmoeskerken, M.M.C.G. and Zeef, E. (1987) *Biotechnology Processes: Scale-up and Mixing* (eds C.S. Ho and J.Y. Oldshue), AIChE, New York, pp. 107–115.

[27] Bakker, A. and Van den Akker, H.E.A. (1994) Gas-liquid contacting with axial flow impellers. *Chem. Eng. Res. Design*, **72**, 573–577.

[28] Warmoeskerken, M.M.C.G. and Smith, M.J. (1988) Impeller loading in multi turbine vessels, in (ed. R. King). *2nd International Conference on Bioreactor Fluid Dynamics, Cambridge, UK, September 21–23*, Elsevier, London.

[29] Cui, Y.Q., Van der Lans, R.G.J.M. and Luyben, K.Ch.A.M. (1996) Local power uptake in gas-liquid systems with single and multiple Rushton turbines. *Chem. Eng. Sci.*, **51**, 2631–2636.

[30] Janssen, L.P.B.M. and Warmoeskerken, M.M.C.G. (2006) *Transport Phenomenon Data Companion*, VVSD, Delft.

[31] Calderbank, P.H. (1958) Physical rate processes in industrial fermentation, part I: the interfacial area in gas–liquid contacting with mechanical agitation. *Trans. IChemE*, **36**, 443–463.

[32] Laakkonen, M., Moilanen, P., Alopaeus, V. and Aittamaa, J. (2007) Modelling local bubble size distributions in agitated vessels. *Chem. Eng. Sci.*, **62**, 721–740.

[33] Vrabel, P., Van der Lans, R.G.J.M., Luyben, K.Ch.A.M. *et al.* (2000) Mixing in large-scale vessels stirred with multiple radial or radial and axial up-pumping impellers: modeling and measurements. *Chem. Eng. Sci.*, **55**, 5881–5896.

[34] Nienow, A.W., Buckland, B.C. and Hunt, G. (1994) A fluid dynamic study of the retrofitting of large bioreactors; turbulent flow. *Biotech. Bioeng.*, **44**, 1177–1186.

[35] Hinze, J.O. (1955) Fundamentals of the hydrodynamic mechanism of splitting in dispersion processes. *AIChE J.*, **1**, 289–295.

[36] Kolmogorov, A.N. (1946) Break up of droplets in turbulent flow. *Dokl. Akad. Mauk USSR*, **66**, 825–828.

[37] Vermeulen, T., Williams, G.M. and Langlois, G.E. (1955) Interfacial area in liquid-liquid and gas-liquid agitation. *Chem. Eng. Progr.*, **51**, 85–94.

[38] Hu, B., Pacek, A.W., Stitt, E.H., and Nienow, A.W. (2005) Bubble sizes in agitated air-alcohol systems with and without particles; turbulent and transitional flow. *Chem. Eng. Sci.*, **60**, 6371–6377.

[39] Parathasarathy, R., Jameson, G.J., and Ahmed, N. (1991) Bubble break up in stirred vessels – predicting the Sauter mean diameter. *Chem. Eng. Res. Design*, **69**, 295–301.

[40] Lu, W.-M., Hsu, R.-C., Chien, W.-C. and Lin, L.-C. (1993) Measurement of local bubble diameters and analysis of gas dispersion in an aerated vessel with disk-turbine impeller. *J. Chem. Eng. Jpn.*, **26**, 551–557.

[41] Barigou, M. and Greaves, M. (1996) Gas holdup and interfacial area distributions in a mechanically agitated gas-liquid contactor. *Chem. Eng. Res. Design*, **74**, 397–405.

[42] Alves, S., Maria, C., Vasconcelos, J. and Serralheiro, A. (2002) Bubble size in aerated stirred tanks. *Chem. Eng. Sci.*, **89**, 109–117.

[43] Higbie, R. (1935) The rate of absorption of a pure gas into a still liquid during short periods of exposure. *Trans. Am. Inst. Chem. Eng.*, **35**, 365–389.

[44] Garcia Ochoa, F. and Gomez, E. (2009) Biorecator scale-up and oxygen transfer rate in microbial processes: An overview. *Biotechnol. Adv.*, **27**, 153–176.

[45] Zlokarnik, M. (1978) Sorption characteristics for gas-liquid contacting in mixing vessels, in *Advances in Biochememical Engineering* Vol. **6** (eds T.K. Ghose, A. Fiechter and N. Blakebrough), Springer, Berlin/Heidelberg, pp. 133–151.

[46] Gezork, K.M., Bujalski, W., Cooke, M. and Nienow, A.W. (2001) Mass transfer and hold-up characteristics in a gassed, stirred vessel at intensified operating conditions. *Chem. Eng. Res. Design*, **79**, 965–972.

[47] Prochazka, J. and Landau, J. (1961) Homogenisation of miscible liquids in the turbulent regime. *Collect. Czech. Chem. Commun.*, **26**, 2961–2973.

4

Thin Film Reactors

4.1 Introduction

For some reactions there can be considerations of heat and mass transfer that would lead to designs where the liquid flows as a film over solids surfaces. Heat transfer might be the driver for very exothermic reactions where it could be advantageous to have all of the liquid as close as possible to the cooling surfaces. Examples of these reactions are the sulfonations and ethoxylations, as used in the manufacture of detergents. For these, control of temperature is important, not just to avoid the runaway of the reaction caused by thermal

Hydrodynamics of Gas-Liquid Reactors: Normal Operation and Upset Conditions, First Edition.
B. J. Azzopardi, R. F. Mudde, S. Lo, H. Morvan, Y. Yan and D. Zhao.
© 2011 John Wiley & Sons, Ltd. Published 2011 by John Wiley & Sons, Ltd.

feedback. Further motivations are the elimination of side reactions that could produce dark coloured by-products which would be unwelcome in the final, consumer, products. Another, very unwelcome, by-product can be dioxins. Obviously, these toxic materials would have to be removed at great expense. Mass transfer considerations would come from reactions such as those termed condensation reactions. In these, molecules such as water are produced as a by-product of the reaction. If these could be removed from the reaction mixture by, for example, evaporation, then this would drive the reaction forwards and considerably shorten reaction times. To effect this, short diffusion paths are sought that point towards thin film reactors. Falling film reactors are considered in Section 4.2.

A limitation of falling film reactors is that the thickness of the film, for a given flow rate, is limited by the driving body force, that is gravity. Other than running the reaction on planets such as Saturn or Uranus, with their higher gravities, an alternative body force has to be sought. Centrifugal force is one possibility. If the mechanical considerations of rotating machinery can be dealt with, the force can be increased through a higher rotation rate. This approach has been applied in what are termed rotating (or spinning) disc reactors and the appropriate aspects of these are considered in Section 4.3.

Carrying out reactions in a pipe is gaining more applications. Although there is significant complexity, when there are two phases flowing simultaneously, there can be advantages. One is that this is one way of converting batch processes into continuous ones. The possibilities of such reactors are considered in Section 4.4, mainly through an application to the production of Nylon-6,6.

Recent advances in manufacturing have allowed the production of units with multiple passages of very small diameter. These are usually referred to as monolith reactors. Their attributes and modelling are considered in Section 4.5.

4.2 Falling Film Reactors

In some reactions, in addition to a requirement for a large specific interfacial area, there is a need for all of the liquid to be close to the cooling surface. This is usually motivated by the need to remove the large heat of reaction that occurs in some instances. It is important to remove the heat of reaction to prevent reactions running away. The rationale for this is given in Chapters 7 and 8.

However, there are also other reasons why good control of heat removal might be required. Examples of such reactions come from the manufacture of detergents. These chemicals tend to be formed by the reaction of organic molecules with sulfur trioxide to yield sulfonated compounds and provide the hydrophilic part of the detergent, which, together with the organic (hydrophobic) end, give it the surfactant action. There are many organic chemicals used. For example, they can be unsaturated aromatic or aliphatic compounds, fatty alcohols or ethoxylated alcohols. In the first group, chemicals such as dodecylbenzene might be involved. These can form dark by-products when the reaction temperature gets too high. Obviously these are unwelcome as the materials are sold to consumers as bright sparkling products. An additional problem can arise with ethoxylated alcohols. Here there might be traces of ethylene oxide present as a result of the previous reaction, the ethoxylation of the alcohol. It might also occur through thermal decomposition, and may recombine to form highly toxic dioxane groups, which would be totally unacceptable in consumer products.

Although bubbly flows, as in bubble columns, can provide a large specific interfacial area, they cannot satisfy the second need, which is for all of the liquid being close to the cooling surface. To meet both requirements, there is a good argument to employ falling film reactors. These are essentially shell and tube heat exchangers with the tube axis mounted vertically and the liquid fed carefully onto the inside of each tube. The liquid then flows down the tube wall under the action of gravity. Gas, often consisting of another of the reagents, usually dispersed in an inert diluent, is also fed into the tubes. This is almost inevitably in a co-current flow, that is downwards. The opposite arrangement, that of the gas flowing upwards in counter-current flow, is not employed as the up-flowing gas can prevent the down flow of liquid through the phenomenon known as flooding. The cooling medium flows on the shell side of the heat exchanger. A characteristic of the films in these units is that they are not smooth but are covered with waves. These wavy films have mass transfer rates several fold greater than those for the equivalent smooth films. In the sub-sections below, the methods of calculating the mean film thickness, the characteristics of waves and methods for calculating mass transfer are laid out.

Before considering boundaries to operation, it is useful to define some dimensional groups. The most important is the Reynolds number. This is usually defined as $\rho_l u L/\eta_l$, where ρ_l is the liquid density, u is the mean film velocity, L is a length scale usually taken as the film thickness and η_l is the liquid viscosity. The first three terms can be combined to give Γ, the mass flow rate per unit perimeter for flow. Although some papers use this definition, others prefer a different version ($4\Gamma/\eta_l$). This is justified on two grounds. Firstly, the length scale for non-circular passages is the hydraulic diameter, which for something wide and thin would be four times the film thickness. Secondly, this alternative definition can also be written in terms of the tube diameter and the superficial velocity of the liquid (the velocity the liquid would have if it alone flowed in the pipe, filling it). Here the second definition will be used and equations have been adapted to take this into account.

A second dimensionless group is used to characterise the differing importance of the physical properties of the liquid, particularly the liquid viscosity and the surface tension, σ. Three different versions have been used. However, they are all related. Bakoploulos [1] used a liquid number, $K_f (= \rho\sigma^3/g\eta_l^4)$, which is, of course, the inverse of the Morton number, Mo, encountered in Chapter 2. In contrast, Meza and Balakotaiah [2] used a Kapitza number, Ka ($=\sigma/\eta_l[\eta_l g/\rho]^{0.33}$), which is equal to $K_f^{0.33}$ or $Mo^{-0.33}$. In many ways the Kapitza number is a more useful version, as it does not have such large ranges of values as do the other two groups.

Now it is important to consider some limitations to the flow to ensure that good operation is achieved. The first of these is how close to the vertical does the surface over which the liquid is flowing have to be? There is only limited information available on this. Butterworth [3] measured the circumferential variation of film flow rate around slightly inclined rods and found that an originally uniform film showed maximum to minimum flow rate ratios at 0.3 m from the inlet, ranging from 6 : 1 to 50 : 1 for angles as small as 2°. Park *et al.* [4] showed that mass transfer could be diminished by 5% for a tilt of 0.2° and by 20% if the tilt reached 0.5°. There is a lower limit to the liquid flow, which is associated with the formation of dry patches. These can be persistent and problematic. They are undesirable because they lower the heat transfer area that is employable. However, more than that, there can be unwanted reactions at the solid-liquid-gas line, which could concentrate certain chemicals, be they liquid or solids. These could cause build-up of the boundary of the dry patch, could cause corrosion or a build-up of solids that might result in tube blockage. When a dry patch initially occurs, it might be

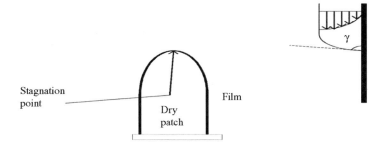

Figure 4.1   *Dry patch*

possible to eliminate it by temporarily turning up the liquid flow rate. However, they are not easy to detect. More persistent dry patches might require shut down, cleaning and then inspection to see that there is no tube damage. Dry patches usually occur at very low liquid flow rates and so knowledge of minimum flow rates is required.

Observation shows that dry patches form at a point and grow downstream to give an area bounded by a parabola, as illustrated in Figure 4.1. The simplest analysis [5] balances the film momentum heading towards the stagnation point with a surface tension force. The momentum force can be written as:

$$F_m = \int_0^\delta \frac{\rho_l u^2}{2} \, dy \qquad (4.1)$$

Using the relationships for laminar flow developed by Nusselt [6], as laid out below, yields

$$F_m = 6.76 \left(\frac{\eta_l^4 g}{\rho_l} \right)^{0.33} \mathrm{Re}^{1.67} \qquad (4.2)$$

where

ρ_l is the liquid density
g is the acceleration due to gravity
η_l is the liquid viscosity.

The surface tension force can be written as:

$$F_s = \sigma[1 - \cos(\theta)] \qquad (4.3)$$

where

σ is the surface tension
θ is the wetting angle.

A dry patch is stable when the two forces are equal. Equating Equations 4.2 and 4.3 and rearranging yields a minimum film Reynolds number:

$$\mathrm{Re}_{min} = 6.76\{Ka[1 - \cos(\theta)]\}^{0.6} \qquad (4.4)$$

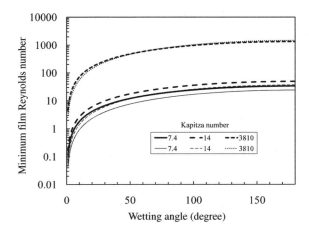

Figure 4.2   *Effect of wetting angle on minimum film Reynolds number for dry patch formation. Thick lines, Equation 4.5; and thin lines, alternative version by Coulon [7]*

A similar result was obtained by Coulon [7] who employed a more empirical relationship for the velocity profile within the film. This gave a pre-constant of 1.0 and a power of 0.67. The minimum Reynolds numbers predicted by these equations as a function of wetting angle are illustrated in Figure 4.2. Water ($Ka = 3810$) and two more viscous liquids are considered. There is little difference between the predictions of the two equations, and the minimum Reynolds number is seen to asymptote to an upper limit with increasing contact angle. This might give a conservative estimate. In reality the minimum might be significantly smaller. For example, Bakopoulos [1] quoted a value of $Re_{min} = 10$ corresponding to a wetted angle of 7°.

The upper limit on liquid flow rate is usually dictated by operational constraints. As will be shown below, film thickness depends on the flow rate of liquid raised to the power of 0.33. Therefore, to maintain the advantages of the thin film flow, not too high a flow rate should be employed.

There is an upper flow rate that should be taken into account when considering the gas flow rate. Although gas flow can make the film thinner, it can also induce the formation of drops. These are accelerated by the gas. Some can be deposited further down the tube. These actions will broaden the residence time distribution of the chemicals in the liquid through increased axial mixing, and so make for poor selectivity of reaction. It should be noted that it is not easy to determine this limit. Observing the presence of the first few drops by eye is almost impossible. A number of workers suggested measuring the flow rate of entrained drops by sampling (or inferring it from measurements of the film flow rate) at a number of gas velocities above the inception point, and then extrapolating back to a zero drop flow rate. However, Andreussi [8] counsels against linear extrapolation suggesting that a more curvilinear fit to the data is more realisitic. To calculate the critical gas velocity for the inception of entrainment of drops from the film, Andreussi started from the work of Taylor [9], who modelled the growth of waves on a liquid of infinite depth. His calculations showed that waves grew with a Weber number ($\rho_g u_g^2/K\sigma$) > 1. In this, ρ_g is the gas density, u_g is the gas velocity, σ is the surface tension and K is a wave number (= $2\pi/\lambda$) with λ being the wavelength. This approach was subsequently adapted to thin film applications [10] and the inequality was written as:

$$\frac{\rho_g u^{*2} \delta}{\sigma} \geq We_c \qquad (4.5)$$

where

δ is the film thickness
u^* is the friction velocity $(= u_{gs}\sqrt{(f_i/2)}$
f_i is the interfacial friction factor obtained as described in Section 4.2.2 below.

The critical gas velocity can be determined by a trial and error approach. For a liquid film Reynolds numbers of > 300, a value of $We_c = 0.0055$ was shown by Andreussi to give a good fit to the data. It should be noted that this approach is based solely on air-water data.

4.2.1   Film Thickness

The relationship between film thickness and liquid flow rate per unit width of wall was first derived by Nusselt [6] in studies on the condensation of steam. It can be developed from the consideration of the force–momentum balance over a small element of film. The film is assumed to have a smooth interface. A coordinate system is considered with directions: x – downwards and y – into the film from the gas-liquid interface. If the flow is assumed fully developed and steady in time and that there are no variations in the third dimension, that is across the flow surface, then the balance over an element of thickness dy, as dy $\rightarrow$ 0, results in:

$$\frac{d\tau}{dy} = \rho_l g \qquad (4.6)$$

where

τ is the shear stress that occurs in the liquid at a distance y from the interface.

Integrating across the complete film yields

$$\tau = \rho_l g y + C \qquad (4.7)$$

For the case of zero gas flow, the shear stress at the interface can be taken as zero so that $C = 0$ and Equation 4.7 reduces to:

$$\tau = \rho_l g y \qquad (4.8)$$

For many fluids the shear stress can be related to the change in velocity across the flow, du/dy, with the constant of proportionality being the viscosity

$$\tau = -\eta_l \frac{du}{dy} \qquad (4.9)$$

Substituting into Equation 4.8 and integrating yields:

$$u = -\left[\frac{\rho_l g}{2\eta_l}\right] y^2 + C' \qquad (4.10)$$

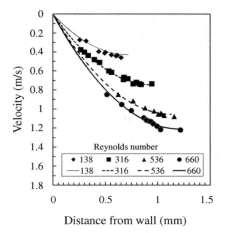

Figure 4.3 *Velocities within a liquid film flowing down a vertical flat plate compared with values predicted by the model of Nusselt [6]*

The constant of integration can be evaluated from the no-slip condition at the wall, that is the assumption that the liquid exactly at the wall does not move so that $u = 0$ at $y = \delta$. Then

$$C' = \left[\frac{\rho_l g}{2\eta_l}\right]\delta^2 \tag{4.11}$$

which, when substituted into Equation 4.10 and rearranged, results in:

$$u = \left[\frac{\rho_l g \delta^2}{2\eta_l}\right]\left[1 - \left(\frac{y}{\delta}\right)^2\right] \tag{4.12}$$

This implies a parabolic velocity profile, a result that has been verified experimentally [11, 12], for example see Figure 4.3 which presents measurements made using stereoscopic, stroboscopic photography.

This equation can provide much information about the flow. For example, the maximum value of velocity is that at the film interface and is given by:

$$u_{max} = \frac{\rho_l g \delta^2}{2\eta_l} \tag{4.13}$$

The total flow rate is obtained from integrating across the film and across the direction perpendicular to the flow, W, to give:

$$\dot{M} = \int_0^W \int_0^\delta \rho_l\, u\, \mathrm{d}y\, \mathrm{d}z = \frac{\rho_l^2\, g \delta^3\, W}{3\eta_l} \quad \text{or} \quad \delta = \left[\frac{3\eta_l \dot{M}}{\rho_l^2 g W}\right]^{0.33} \tag{4.14}$$

From this, the average velocity within the film can be determined from:

$$\bar{u} = \frac{\dot{M}}{W\rho_l\delta} = \frac{\rho_l g \delta^2}{3\eta_l} \tag{4.15}$$

If there is a finite gas flow, then the constant of integration C in Equation 4.7 is not zero but equals τ_i and alternative versions of Equations 4.12 and 4.14 are obtained:

$$u = \left[\frac{\rho_l g \delta^2}{2\eta_l}\right]\left[1 - \left(\frac{y}{\delta}\right)^2\right] + \frac{\tau_i \delta}{\eta_l}\left[1 - \left(\frac{y}{\delta}\right)\right] \tag{4.16}$$

and

$$\dot{M} = \frac{\rho_l^2 g \delta^3 W}{3\eta_l} + \frac{\rho_l \tau_i \delta^2 W}{2\eta_l} \tag{4.17}$$

Film thickness can be determined from Equation 4.14 if the effect of gas flow is negligible or by solving the cubic Equation 4.17 when it is important. The interfacial shear stress, τ_i, is determined by using a correction factor, on the value determined for gas flowing alone, to allow for the effect of the presence of the liquid film. For example, it can be obtained from the method proposed by Henstock and Hanratty [13]. This has been modified by Andreussi [8] and Riazi and Faghri [14]. Both start with a smooth wall gas friction factor and correct it to account for the roughness caused by the waves on the film interface via a function $1 + 1400F\phi$. F is given by:

$$F = \frac{\left[\left(0.707\,Re_l^{0.5}\right)^{2.5} + \left(0.0379\,Re_l^{0.9}\right)^{2.5}\right]^{0.4}}{Re_{gs}^{0.9}} \frac{\eta_l}{\eta_g}\left(\frac{\rho_g}{\rho_l}\right)^{0.5} \tag{4.18}$$

where

$Re_{gs} = \rho_g u_{gs} D_t / \eta_g$

ϕ was correlated with $\tau_i/\rho_g g \delta$, with Andreussi using $\phi = 0.27\,(\tau_i/\rho_g g \delta)^{0.67}$ for $\tau_i/\rho_g g \delta \leq 1.8$ and $0.33\,(\tau_i/\rho_g g \delta)^{0.33}$ for $1.8 < \tau_i/\rho_g g \delta < 30$, and Riazi and Farhgi employing $\phi = 1 - \exp(-\tau_i/\rho_g g \delta)$. Note that the last gives a value of about twice that resulting from Andreussi's equation for $\tau_i/\rho_g g \delta \sim 3$. The interfacial shear was made up of the frictional and mass transfer terms in the definition of Riazi and Farghi. Andreussi only used the frictional part.

An alternative approach also treats the interface as if it were a rough surface [15]. The interfacial shear stress is given by:

$$\tau_i = f_i \frac{\rho_g u_g^2}{2} \tag{4.19}$$

where the interfacial friction factor is given by:

$$\frac{1}{f_i^2} = -4 \log\left[\frac{k\,\delta}{3.7\,D_t} - \frac{5.02}{Re_g}\log\left(\frac{k\delta}{3.7D_t} + \frac{13}{Re_g}\right)\right] \tag{4.20}$$

where

u_g is the average velocity of the gas core $[=u_{gs}D_t^2/(D_t - 2\delta)^2]$
Re_g is the gas core Reynolds number $[= \rho_g u_{gs}D_t^2/(D_t - 2\delta)\eta_g]$
k is a correction factor relating roughness to film thickness.

k can be correlated by:

$$k = 20.55v_{ll} - 0.93 \quad \text{for} \quad v_{ll} > 0.175\text{m/s} \tag{4.21}$$

$$k = 3.59 - 5.1v_{ll} \quad \text{for} \quad v_{ll} < 0.175\text{m/s} \tag{4.22}$$

v_{ll} can be obtained from Equation 4.16 by setting $y = 0$. The system of equations is solved iteratively. An initial value of film thickness is obtained from Equation 4.14, the gas core Reynolds number, liquid interface velocity (from Equation 4.16 assuming $\tau_i = 0$ at the first iteration) and k are calculated. The interfacial stress is obtained using Equations 4.19 and 4.21/4.22 and a new film thickness is determined by solving Equation 4.17 and this process repeated until a constant value of film thickness is the final product.

On considering the accuracy of methods to calculate film thickness, equations such as that derived by Nusselt [6] can be recast in dimensionless terms:

$$\frac{\delta}{L_f} = A\, Re_l^B \tag{4.23}$$

where

$$L_f = (\eta_l^2/g\rho_l^2)^{0.33}$$

Equation 4.14 results in $A = 0.909$ and $B = 0.33$. These constants have been shown to give reasonable predictions for water and good predictions for more viscous liquids [16]. For water at higher Reynolds numbers, the version of Takahama and Kato [17] with $A = 0.228$ and $B = 0.526$ gives better predictions. The statements above pertain to zero gas flow. For gas velocities in the range 0–15 m/s, this parameter was seen to have little effect on the film thickness [16].

It should be noted that if the limitation of having no droplet entrainment occurring is imposed, then the range of possible gas velocities will be such that there will be little effect on film thickness.

4.2.2 Interfacial Waves

The interface of a film flowing under gravity alone or the combined gravity/gas shear is inevitably not smooth unless materials have been added that damp possible waves. From linear stability analysis, it has been shown that the interface is unstable [18, 19]. These results have been confirmed experimentally. However, there has been some confusion because the rate of growth of the amplitude of the waves is sometimes fairly small. This has been taken as an indication that there were no waves. Growth continues until the wave achieves an amplitude of about the mean film thickness, as shown in Figure 4.4, which is the spatial variation of film thickness taken from a film flowing on a flat plate [20]. However, these waves are not themselves stable and as seen in Figure 4.4 they distort to give more widely spaced waves, which are not necessarily uniform around the tube or across the flat plate. It has been observed that increasing the viscosity and co-flowing air velocity does make them circumferentially (or laterally) more regular, as illustrated in Figure 4.5. As can be seen, when the viscosity is only 30 times that of water, the front edge of the waves can be seen to be very steep. This is seen much more clearly in the corresponding plots of film thickness against axial position that are shown in Figure 4.6, which were measured along the centre of the channel. Further examples of such data are shown in Plate B. These are from aqueous solutions of sucrose of viscosities of 0.009, 0.026 and 0.111 Pa s [21]. Please see colour plate B for further information.

To determine the growth behaviour of waves on the film interface, linear stability analysis has been employed [18, 19, 22]. This considers a perturbation of the governing, fourth order differential equation, the Orr–Sommerfeld equation. An alternative approach [2] uses two variables δ and q. The former is the film thickness, the latter the instantaneous flow rate at any axial point in the film at a specific time. Mass and momentum balance are solved both

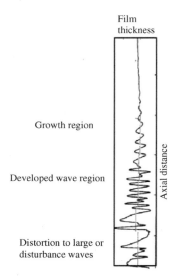

Figure 4.4 *Spatial development of waves on a flat plate (Adapted from Portalski, S. and Clegg, A. J., An experimental study of wave inception on falling liquid films, Chem. Eng. Sci., 27, 1257–1265. Copyright (1972) with permission from Elsevier.)*

linearly and non-linearly. In some instances the momentum balance equation consisted of up to ten terms. These methods give elegant and illuminating solutions. They give good agreement with those experiments where a specific frequency has been imposed. Where the initial disturbances are more random, the experiments can have greater scatter, that is there are often several frequencies present.

Much can be learned about these waves by examining their frequency and celerity (or speed). The former can be obtained by carrying out spectral analysis of the signal of a point measurement probe. The latter is obtained by cross-correlating the signal of probes placed slightly apart axially. From experimental data [16, 23], it is seen in Figures 4.7–4.9 that there is

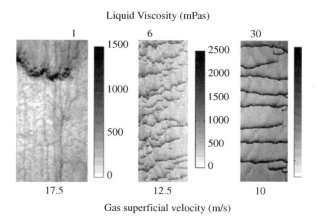

Figure 4.5 *Spatial variation of film thickness for three different liquids. In all cases the liquid flow rate was 0.28 kg/ms*

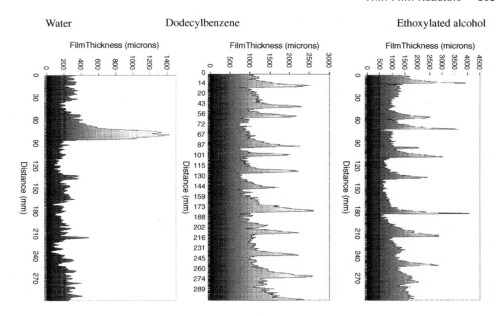

Figure 4.6 *Axial variation of film thickness from cases in Figure 4.5*

no effect of gas flow rate until the gas superficial velocity achieves a value of ~10 m/s. Beyond this, the frequency increases with gas and liquid flow rates. The disturbance wave velocity increases with liquid flow rate. Note that there is a small difference between the plate and pipe data, which might be attributed to the difference between the pipe perimeter and the width of the plate. The velocity is seen to increase in the same way with gas and liquid flow rate.

One question that arises is how the interfacial area for a wavy film differs from that of the equivalent smooth film. Though this information can be deduced from the time varying film thickness determined from a single position in space, it is more directly determined from those experiments (on flat plates) where the spatial distribution of the film thickness is captured. Figure 4.10 shows example results obtained [24, 25]. It is clear that the increase in

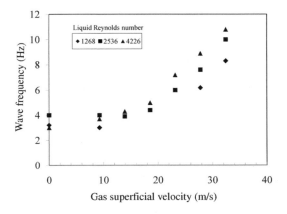

Figure 4.7 *Variation of disturbance wave frequency in a 38 mm diameter pipe; air–water [23]*

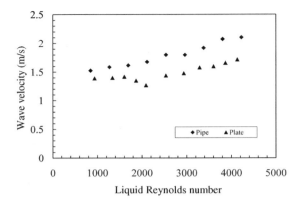

Figure 4.8   *Wave velocity for film flow on a flat plate (width 150 mm) [16] and 38 mm diameter pipe [23]*

interfacial area over the smooth values is less than 1%. The available data show that though there is an increase with liquid flow rate, and that the effect of gas flow rate and liquid viscosity are small. From this it can be seen that the specific interfacial area can be well described by the inverse of the film thickness.

4.2.3   Heat and Mass Transfer

The simplest description of heat transfer associated with falling films is that occurring during condensation. This assumes that the smooth vapour-liquid interface is at the saturation temperature, T_s, whilst the wall is at a temperature, T_w. With the assumption of laminar flow and no sub-cooling and that the heat transfer is by conduction through the film, the heat transfer coefficient can be defined as:

$$\alpha = \frac{\lambda}{\delta} \tag{4.24}$$

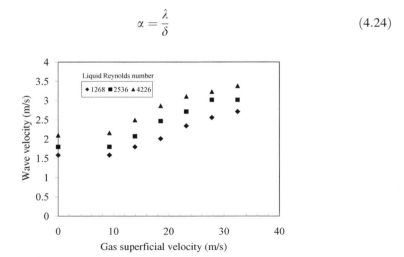

Figure 4.9   *Effect of gas and liquid flow rates on wave velocity; air-water in a 38 mm diameter tube [23]*

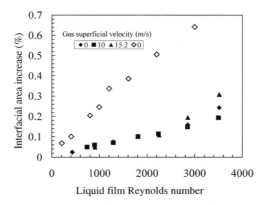

Figure 4.10 *Measured increase in specific interfacial area of falling films on vertical flat plates. Open symbols, Portalski and Clegg [24]; and closed symbols, Sun [25]*

where λ is the thermal conductivity, using Nusselt's relationships for film thickness, Equation 4.14 results in:

$$\alpha = \frac{\lambda}{0.909 \, L_f} \mathrm{Re}^{-0.33} \tag{4.25}$$

which in term of dimensionless groups yields:

$$Nu = \frac{\alpha L_f}{\lambda} = 1.1 \, \mathrm{Re}^{-0.33} \tag{4.26}$$

Using the same arguments and employing the alternative equations for film thickness, a similar equation but with alternative constants is obtained. However, this results in a heat transfer that decreases with increasing Reynolds number. In reality for wavy and particularly turbulent films, the coefficient increases with Reynolds number. For this, correlations such as

$$Nu = \frac{\alpha L_f}{\lambda} = A \, \mathrm{Re}^{B} \, \mathrm{Pr}^{C} \tag{4.27}$$

are more appropriate. The Prandtl number, Pr, is defined as $c_p \eta_l / \lambda$ with c_p being the specific heat capacity. The power on the Reynolds number is positive with values of 0.4 being quoted by a number of workers [26]. This version has $A = 0.0038$ and $C = 0.65$.

Under normal conditions, that is wavy films, mass transfer coefficients have been found to be several times the value determined for a smooth film. It was shown above that there is hardly any increase in specific interfacial area because of the presence of waves. Therefore, there have to be other mechanisms augmenting the mass transfer when waves are present. At lower Reynolds numbers, it is suggested that there are recirculation zones under the waves. Some reports place it directly under the wave, others propose a circulation more towards the tail of the wave. The former have also been identified using computational fluid dynamics calculations. Alternative mechanisms have been proposed. These include disturbances to the boundary layer as it is stretched and compressed by the passage of waves and by a new boundary layer being created after the passage of the wave. The wave could contain a circulation. This might trap material within it. However, that would not in itself augment mass transfer, but a newly created boundary layer might. One suggestion [27] is that a new boundary layer is formed at the surface after a wave. The recirculation in the wave pushes

Table 4.1 *Constants [1] for Equation 4.28*

A	B	C	Range of Re
$0.1025/Ga^{0.05}$	0.73	0.5	< 300
$0.0243/Ga^{0.05}$	0.53	0.5	300–1200
$0.022984/Ga^{0.05}$	0.93	0.5	> 1200

this dissolved component rich layer towards the wall augmenting mass transfer into the film. Obviously, a new layer will have to form after that wave passes. Another mechanism for mass transfer augmentation might be the increased levels of turbulence under the wave crests observed by Karami and Kawaji [28].

If the experimentally determined mass transfer coefficients are examined, it is seen that there are three regions that require different powers to which the film Reynolds number should be raised. A typical form of equation is:

$$Sh = \frac{k_l L_f}{D} = A \, Re^B \, Sc^C \tag{4.28}$$

where

D is the diffusion coefficient
Sc is the Schmidt number ($= \eta_l/\rho_l D$).

Table 4.1 gives the values of the constants, where $Ga = \rho_l^2 L^3 g/\eta_l^2$ with L being the length of the film. Alternative values have been suggested, for example by Lamourelle and Sandall [29] who had $A = 1.76 \ 10^{-5}$, $B = 1.506$, $C = 0.5$.

There is less evidence for the effect of the gas velocity on the liquid side mass transfer. The available data show that a finite gas velocity increases the mass transfer coefficient. Henstock and Hanratty [30] showed that the coefficient was increased to an asymptotic ratio of about 2.8 over the equivalent value for zero gas flow. Figure 4.11 shows that though

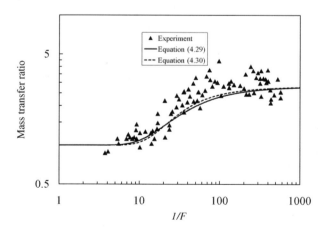

Figure 4.11 *Comparison of equation of Henstock–Hanratty [30] and modified equation for enhancement of mass transfer by gas motion with experimental data (Adapted from Henstock, D. E. and Hanratty, T. J., Gas absorption by a liquid layer flowing on the wall of a pipe, AIChE J.* **25**, *122–131. Copyright (1979) with permission from the American Institute of Chemical Engineers.)*

there is significant scatter, the points in general follow the correlation

$$\frac{k_l}{k_{lo}} = 1 + 1.8 \, e^{-30 \, F} \tag{4.29}$$

where F is defined by Equation 4.18. Note, unlike the original plot, here the ratio is plotted against $1/F$ instead of F. $1/F$ is proportional to the gas velocity raised to the power of 0.9. An alternative version is:

$$\frac{k_l}{k_{lo}} = 2.8 - 1.8 \tanh(21 \, F) \tag{4.30}$$

The modelling techniques mentioned above for calculation of waves have been extended to include mass transfer [31].

The above methods are combined into overall models for falling film reactors incorporating hydrodynamics, mass transfer and reaction kinetics [15, 32].

4.3 Rotating Disc Reactors

There is a limit to the thinness of the film that can be created for a given throughput when driven by gravity. However, if the surface over which the liquid film flows was a horizontal circular disc and it was rotated about a vertical axis, this could produce accelerations in the radial direction with magnitudes that could be many times that of gravity. Consequently, films could be thinner and residence times shorter than what would be achieved in a falling film arrangement. It is realised that the equipment could be much more complex than falling film units, which are essentially shell and tube heat exchangers. If heat transfer is involved, and this is almost inevitable, then a complex arrangement will be required to get the heating or cooling medium into and out of the disc. As with falling film units, careful arrangements will be required to distribute the film uniformly.

It should be noted that there is another way in which rotating discs are employed. Here, the disc, with its axis mounted horizontally, is partially immersed into a pool of liquid. As it is rotated it draws a thin film of liquid from the loop, exposes it to the overlying gas and then returns it to the pool.

A prime requirement in assessing the hydrodynamics of this type of unit is the film thickness–flow rate relationship. In addition, the important feature of interfacial waves must be considered as well as the mass transfer aspects. These are considered in the next three sections.

4.3.1 Film Thickness

In the simplest analysis, the flow is assumed to be purely radial [33]. Assuming a Newtonian liquid, a balance can then be written between viscous and centrifugal forces over a small volume element

$$\eta_l \frac{d^2 u}{dy^2} = -\rho_l \omega^2 r \tag{4.31}$$

where

ω is the rotation rate (radians per second)
r is the radial coordinate.

Here inertial terms are considered negligible. If it is further assumed that there is no slip at the solid surface and the interaction of interface with the gas is negligible, then Equation 4.31 can be integrated across the film to give the characteristic parabolic velocity profile:

$$u = \frac{\rho_l r \omega^2 \delta^2}{2\eta_l} \left[1 - \left(\frac{y}{\delta}\right)^2 \right] \tag{4.32}$$

where

y is the distance from the interface
δ is the local film thickness.

Obviously, the maximum velocity is at the surface, $z = \delta$, whilst the mean velocity is

$$\bar{u} = \frac{\rho_l r \omega^2 \delta^2}{3\eta_l} \tag{4.33}$$

In terms of the volumetric flow rate it is

$$\bar{u} = \frac{1}{3} \left(\frac{3}{2\pi}\right)^{0.67} \left(\frac{\rho_l \omega^2 Q^2}{r\eta_l}\right)^{0.33} \tag{4.34}$$

with Q being the volumetric feed rate. The film thickness is given by:

$$\delta = \left(\frac{3\eta_l Q}{2\pi r^2 \omega^2}\right)^{0.33} \tag{4.35}$$

Note that the radial position, r, is raised to the power of two. Films get thinner both because the centripetal force increases with radius and also because the perimeter, $2\pi r$, over which it flows increases. The way in which the film thickness varies with radius and rotation rate is illustrated in Figure 4.12. These values have been calculated using Equation 4.35 for water

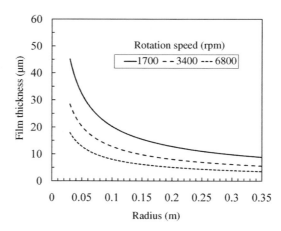

Figure 4.12   *Variation of the thickness of the film on a rotating disc – water; feed rate = 10 ml/s*

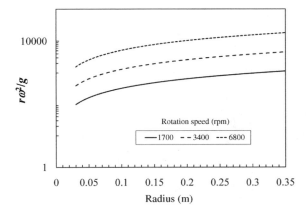

Figure 4.13 *Ratio of acceleration due centrifugal force to that of gravity*

at a flow rate of 10 ml/s. Figure 4.13 shows how the ratio of the centripetal and gravitational accelerations varies with position and rotation rate. Obviously, a minimum rotation rate is required to give films thinner than those produced by gravity.

Equation 4.35 appears to give reasonable values for the film thickness [33, 34]. However, questions have been raised as to whether the flow is purely radial. Rauscher *et al.* [35] developed an asymptotic solution to the more complete Navier–Stokes equations. Simplified versions of these were used by Charwat *et al.* [33] to look at the direction of motion at the film surface. The analysis agreed with their experimental observations and showed that the deviation from the radial direction was greatest close to the centre of the disc but that once a characteristic length scale $[= (9\rho_l Q^2/4\pi^2 \eta_l \omega)^{0.25}]$ became greater than four the deviation was less than 7°.

Obviously, the complete wetting of the disc surface is important for efficient operation. The wetting properties of the disc are very important here and great care needs to be taken to ensure that grease is not put onto the disc surface. Note that even nominally clean fingers can leave residues of grease that can disrupt the film. The occurrence of a circular hydraulic jump has been reported for some operations. However, these occurred at rotation rates where the centrifugal acceleration was well below the value for gravity. A much more important feature is the occurrence of waves of the film interface. This is considered in the next section.

4.3.2   Interfacial Waves

Two main types of waves have been reported for flow on a rotating disc. One type has the form of concentric rings that travel radially. These are caused by instability of the predominantly radial flow and are equivalent to the initial waves of the gravity driven flow in falling film reactors. The second type is termed spiral waves and has the form of spirals unwinding in the direction opposite to the direction of rotation. Unlike concentric waves, which occur at low rotation rates and high liquid flow rates, spiral waves occur at higher rotation rates. Figure 4.14 shows the limits of occurrence of each type of wave. The angle between the wave and the concentric circle is shown in Figure 4.15. The inverse Eckman number is the balance of coriolis and viscous forces $(\rho_l \omega \delta^2/\eta_l)$. The curve shows the results of a linear stability analysis similar to the approach applied to falling films [18, 19].

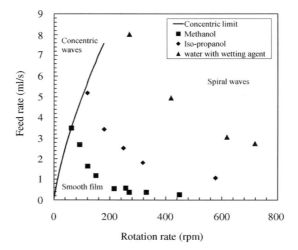

Figure 4.14 *Flow rate–rotation rate plot showing regions of occurrence of different types of waves [33] (Adapted from Charwat, A. F., Kelly, R. E. and Gazley, C., The flow and stability of thin films on a rotating disk, J. Fluids Mech.,* **53***, 227–255. Copyright (1972) with permission from Cambridge University Press.)*

4.3.3 Mass Transfer

Transfer of chemical species across the gas-liquid interface is essential for the effective operation of rotating disc reactors. Figure 4.16 shows typical results for the mass transfer coefficient and shows a strong effect for the rotation rate and a weaker effect for the liquid flow rate [36]. Advanced modelling similar to that employed for falling film reactors which includes calculation of mass transfer has recently been published [37].

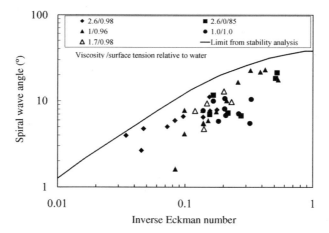

Figure 4.15 *Effect of inverse Eckman number on angle of spiral waves [33] (Adapted from Charwat, A. F., Kelly, R. E. and Gazley, C., The flow and stability of thin films on a rotating disk, J. Fluids Mech.,* **53***, 227–255. Copyright (1972) with permission from Cambridge University Press.)*

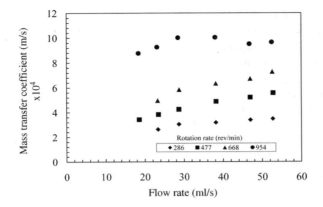

Figure 4.16 *Effect of liquid flow rate and rotation rate on mass transfer in rotating disc reactors [36]*

4.4 Two-Phase Tubular Reactors

In some applications, it might be desirable to ensure that the distribution of residence times is narrow. For these a plug flow reactor would be selected. If in addition, the reaction time required was significant and there was a need to evaporate off the solvent, then a long two-phase tubular reactor might be employed. This arrangement was used in the early production of Nylon-6,6 (see the DuPont patents [38]). The geometry employed was, typically, a 1000 m long pipe made up of four sections of 40–150 mm diameter, formed into a coil to fit into the heater vessel. Before considering the way in which the hydrodynamics of the flow might be modelled, it is instructive to learn about the characteristics of the gas-liquid flow and how some aspects might be advantageous and others disadvantageous.

Because of the very deformable nature of the gas-liquid interface, it is usual to describe the configuration taken by the gas and liquid flowing simultaneously in pipes by use of flow patterns generalised descriptions of the flow. For vertical pipes, four main patterns are usually considered. The first of these is bubbly flow, which consists of a continuous liquid phase with the gas phase dispersed as bubbles within it. The bubbles travel with a complex motion within the flow, may be coalescing and are generally of non-uniform size. In some situations, they congregate mainly at the pipe centre. The concentration of bubbles is not uniform but there are waves of drop concentrations (void waves) that travel along the pipe. The next flow pattern, plug or slug flow occurs when coalescence begins, and the bubble size tends towards that of the channel. Characteristic bullet-shaped bubbles, often called Taylor bubbles, flow up the pipe surrounded by a thin film of liquid. The liquid slug between the Taylor bubbles often contains a dispersion of smaller bubbles. At higher velocities, the Taylor bubbles-liquid slugs in the slug flow break down into an unstable pattern in which there is a churning or oscillatory motion of liquid in the tube. Churn flow with its characteristic oscillations is an important pattern, often covering a fairly wide range of gas flow rates. At the lower end of the range, it may be regarded as a breaking up of the plug flow with occasional bridging across the tube by the liquid phase; whilst at the higher range of gas flow rates it may be considered as a degenerate form of annular flow, with the direction of the film flow changing and very large waves (termed huge waves by some) being formed on the interface. In the latter range the term semi-annular flow has sometimes been used.

Annular flow is characterised by liquid travelling as a film on the channel walls. Part of the liquid can also be carried as drops in the central gas core and there is continuous interchange between film and drops. In fact, for certain fluids and flow rates, the majority of the liquid travels as drops, leading to the term mist flow sometimes being applied to this flow pattern. For horizontal or near horizontal pipes, when gravity acts perpendicularly to the tube axis, separation of the phases can occur. This increases the possible number of flow patterns in horizontal pipes. *Bubbly flow*, like the equivalent pattern in vertical flow, consists of gas bubbles dispersed in a liquid continuum. However, except at very high liquid velocities when the intensity of the turbulence is enough to disperse the bubbles about the cross-section, gravity tends to make bubbles accumulate in the upper part of the pipe.

In *stratified flow* liquid flows in the lower part of the pipe with the gas above it. The interface is smooth. An increase in gas velocity causes waves to form on the interface of stratified flow to yield *wavy flows*. *Plug flow* is characterised by bullet shaped gas bubbles, as seen in vertical flow. However, here they travel along the top of the pipe. *Slug flow*, like plug flow, is intermittent. The gas bubbles are bigger whilst the liquid slugs contain many smaller bubbles. At large levels of aeration, they are called frothy surges or semi-slug, if the surges do not fill the pipe completely. However, this might be more correctly considered as part of wavy flow. A continuous gas core with a complete wall film characterises *annular flow*. As in vertical flow, some of the liquid can be entrained as drops in the gas core. Gravity causes the film to be thicker on the bottom of the pipe but as the gas velocity is increased, the film becomes circumferentially more uniform.

Similar flow patterns have been reported in flow in coiled pipes [39]. Figure 4.17 illustrates the flow pattern boundaries for air-water flow in a 3 m diameter coil of a 25 mm diameter pipe. Altering the coil diameter shifts the flow pattern boundaries.

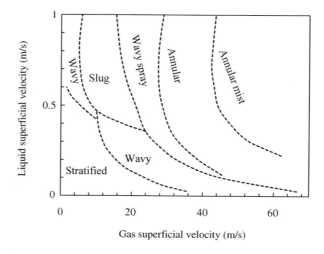

Figure 4.17   *Flow pattern map for gas-liquid flow in coils. Reprinted from Azzopardi, B. J., Gas-Liquid Flows. Copyright (2005) with permission from Begell House, New York*

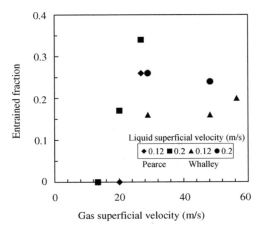

Figure 4.18   *Dependence of entrained fraction in coiled pipe on gas and liquid flow rates (Reprinted from Azzopardi, B. J., Gas-Liquid Flows. Copyright (2005) with permission from Begell House, New York.)*

To ensure a narrow distribution of residence times demands specific flow patterns. Slug flow picks up liquid from ahead of the liquid slug, mixes it with the liquid in the body of the slug and then ejects it from the end of the slug. Annular mist flow has two liquid streams, drops and film.

There is a great deal of interchange between these two streams. Moreover, the drop velocity is much great than that of the film, so that there is significant axial mixing. The ideal flow patterns to minimise axial mixing are stratified and annular flow (with a low fraction of liquid travelling as drops). Figure 4.18 illustrates that a minimum gas velocity is required to create significant entrainment of the film to form drops. Although more elaborate models [40] have been published to calculate the pressure drop in such flow patterns, it has been shown [41] that the Lockhart–Martinelli approach is of a sufficiently sound basis to be employed for this task. The algebraic description by Chisholm [42] of the original graphical methods of Lockhart and Martinelli [43] has been shown to be accurate. For example, Figure 4.19 shows the agreement between the frictional part of this method and data from steam-water flows in coils of three diameters at a pressure of 180 bar. The approach used is a separated flow description, which, from mass and momentum balances, creates pressure drop terms for the frictional, gravitational and accelerational components, Equation 4.36. The two-phase aspects of the frictional part are gathered into a two-phase multiplier, ϕ_l^2, which is specified through Equation 4.37. The Lockhart–Martinelli parameter, X, is the square root of the ratio of the pressure drops that the liquid and gas would have if they were flowing alone in the pipe.

$$-\frac{dp}{dz} = \frac{4 f \dot{m}_l^2}{2 \rho_l D_t} \phi_l^2 + \left[\varepsilon_g \rho_g + (1-\varepsilon_g)\rho_l\right] g \sin \beta + \frac{d}{dz}\left[\dot{m}^2\left(\frac{x_g^2}{\varepsilon_g \rho_g} + \frac{(1-x_g)^2}{(1-\varepsilon_g)\rho_l}\right)\right] \quad (4.36)$$

$$\phi_l^2 = \left[1 + \frac{C}{X} + \frac{1}{X^2}\right] \quad (4.37)$$

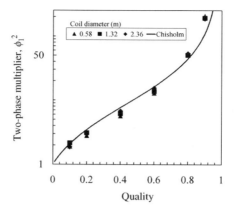

Figure 4.19 *Accuracy of Chisholm's algebraic equation of Lockhart–Martinelli correlation tested against experimental data for coils*

where

$$C = C_{LM} + C_{corr}$$
$$C_{LM} = 5 \text{ for } Re_l \leq 1000. \ Re_g \leq 1000$$
$$C_{LM} = 10 \text{ for } Re_l > 1000. \ Re_g \leq 1000$$
$$C_{LM} = 12 \text{ for } Re_l \leq 1000. \ Re_g > 1000$$
$$C_{LM} = 20 \text{ for } Re_l > 1000. \ Re_g > 1000$$

and C_{corr} is a function of the feed rate and the physical properties of the liquid, which would probably have to be determined from specific experiments. Guidici *et al.* [44] provided such information for the Nylon polymerisation process, which they obtained from fitting plant data.

The void fraction is obtained from:

$$
\begin{aligned}
\varepsilon_g &= 1 - 0.28X^{0.8} & \text{for } 0.01 < X < 0.5 \\
\varepsilon_g &= 1 - 0.175X^{0.32} & \text{for } 0.5 < X < 5 \\
\varepsilon_g &= 1 - 0.143X0.42 & \text{for } 5 < X < 50 \\
\varepsilon_g &= 1 - 1[0.97 + (19/X)] & \text{for } 50 < X < 500
\end{aligned}
\tag{4.38}
$$

The rate of evaporation of water per unit length is driven by the difference between the equilibrium vapour pressure and the system pressure. The constant of proportionality (mass transfer coefficient times specific interfacial area) is found to depend on feed rate and the concentration of the Nylon salt in the feed, as was C_{corr}. The equations were integrated along the pipe [42] and gave results such as those reproduced in Figure 4.20. The left-hand graph shows how liquid is converted into vapour. The right-hand graph shows the velocities calculated and indicates the changes in pipe diameter employed to keep the vapour velocity down to values that would ensure a stratified type of flow.

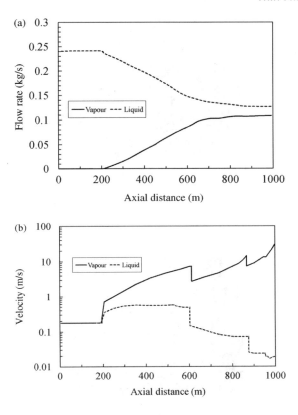

Figure 4.20 *Type of results produced by a model of a Nylon-6,6 reactor by Guidici et al. [44] (Reprinted from Giudici, R. et al., Mathematical modeling of an industrial process of nylon-6,6 polymerization in a two-phase flow tubular reactor, Chem. Eng. Sci.,* **54***, 3243–3249. Copyright (1999) with permission from Elsevier.)*

4.5 Monolith Reactors

Monolith reactors or catalytic monolith reactors have a wide range of applications in industrial processes and as technical devices; these include pollution control in vehicle exhaust, fuel cell reforming and the production of hydrogen from hydrocarbons, and so on. A typical monolith reactor consists of a large number of parallel channels often in a honeycomb type arrangement, as shown in Figure 4.21. Figures 4.22 and 4.23 show typical ceramic monoliths and a metal monolith, respectively. The monolith channels should have a high surface area to volume ratio, and the surface of each channel is covered with a highly porous washcoat containing the catalyst. Reactions take place both in the fluid bulk and, primarily, on the catalyst-coated inner surfaces of the channels. Thus, monolithic reactors are particularly suitable for carrying out fast reactions in applications where space is at a premium.

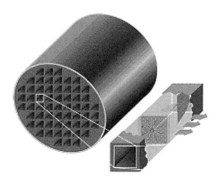

Figure 4.21 Honeycomb monolith reactor

Figure 4.22 Cellular ceramic monoliths [45] (Reprinted from Williams, J. L., Monolith structures, materials, properties and uses, Catalysis Today, **69**, 3–9. Copyright (2001) with permission from Elsevier.)

Figure 4.23 Metal monolith [46] (Reprinted from Burch, R., Breen, J. P. and Meunier, F. C., A review of the selective reduction of NOx with hydrocarbons under lean-burn conditions with non-zeolitic oxide and platinum group metal catalysts, Appl. Catal. B: Environmental, 2002, **39**, 283–303. Copyright (2002) with permission from Elsevier.)

In order to achieve high surface area to volume ratios with a reasonable pressure drop, the channels in monolith reactors are often designed in micro/mini size, the flow in such micro-channels is normally laminar and significant gradients exist in the temperature and concentrations of reactants and products between the bulk fluid and the surface. Substantial heat transfer also takes place along the solid walls of the channels. Micro-channel monolith reactors have advantages such as increased surface area for heat and mass transfer, and have become popular in recent years.

4.5.1 Micro-Channels

Owing to the capabilities exceeding those of traditional macro-scale reactors, micro-channel reactors have increasingly been applied to industrial processes. The high heat and mass transfer rates in micro-channel reactors allow the reaction to be performed under more aggressive conditions with higher yields [47]. The surface to volume ratio can be as high as $1 \times 10^4 \sim 5 \times 10^4 \, \text{m}^2/\text{m}^3$ [48], providing drastically higher heat and mass transfer rates than the traditional chemical reactors. Thus, micro-reactors can remove heat much more efficiently than traditional chemical reactors, and can safely perform highly exothermic or endothermic reactions. In addition, the lesser amounts of catalyst used can help to improve energy efficiency and reduce operational costs. The other benefit of micro-channel reactors is that if the system fails, the amount of accidentally released chemicals is rather small and it could be easily controlled. The integrated sensor and control units could allow the failed reactor to be isolated and replaced while other parallel units continue production [47].

However, there are still some problems for micro-channel monolith reactors. One such problem is clogging occurring in processes containing particles, for example, catalytic reactions. The other is that the processes can cause a rather high pressure drop through the flow channel.

The modelling of monolith reactors, typically micro-channel reactors, remains very challenging. The complex interactions between chemical and physical processes, in addition to the phenomena of mass and heat transfer and fluid flow, need to be fully understood.

4.5.2 Flow Phenomena in Micro-Channels

Fluid transport phenomena in micro-channels have been studied extensively over the past two decades. In parallel with significant experimental observations and studies, much work on theoretical and numerical modelling has also been carried out. A typical gas-liquid two-phase flow pattern termed the Taylor bubble flow can be observed through experiment. Salman *et al.* [52] have developed experimental modelling to demonstrate Taylor bubbles formation and coalescence in a single mini-channel of internal diameter (ID) 0.11, 0.21 and 0.34 mm, respectively. Figure 4.24 shows the bubble formation in a 0.34 mm ID channel with liquid superficial velocity, $u_{ls,} = 7.6$ mm/s and gas superficial velocity, $u_{gs,} = 1.9$ mm/s, respectively.

With respect to modelling, although the flows in most micro-channel reactors can be analysed on the basis of the Navier–Stokes equations, there are a number of papers which argue that the Navier–Stokes equations may have a limitation when describing the transport phenomena in channels with micro-hydraulic diameters or confined micro-boundaries. In

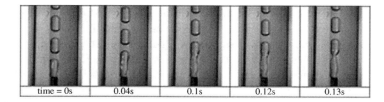

| time = 0s | 0.04s | 0.1s | 0.12s | 0.13s |

Figure 4.24   *Elongated bubble flow in a vertical microchannel [52] (Reprinted from Salman, W., Gavriilidis, A. and Angeli, P., On the formation of Taylor bubbles in small tubes, Chem. Eng. Sci., **61**, 6653–6666. Copyright (2006) with permission from Elsevier.)*

other words, they suggest that the Navier–Stokes equations are incapable of describing the phenomena occurring when the continuum assumption is breaking down. However, it is noted that the mean free path for gases is of the order of 10^{-7} m and that for liquids it is even smaller. Some negligible parameters for flows in conventional channels may become fairly significant and have to be considered for flows in micro-channels [49].

Observations have shown [50] that the molecular effect on the momentum transfer in directions other than the streamwise direction can increase significantly when the length of the flow channels are reduced and the continuum assumption becomes invalid. The variation in fluid properties (e.g. fluid viscosity) can occur by the variation in the temperature of the transport fluid flowing in micro-scale, which causes invalidation of constant properties assumption. To justify the validity of Navier–Stokes equations, a key non-dimensional parameter for gas micro-flows is the Knudsen number (Kn), which is defined as the ratio of the mean free path over a characteristic geometry length or a length over which very large variations of a macroscopic quantity may take place [51]. For two- or multi-phase micro-flow, although the Knudson number for gas micro-flow may not be directly applicable, it has become a very useful reference.

Flow in monolithic reactors with micro-channels is predominantly laminar [49] because of the hydraulic diameter that makes the Reynolds numbers very small [53]. Indeed, for the flows in conventional or normal sized channels, the flow pattern is dominated by the influence of the gravitational force, while for the flows in micro-channels, the flow pattern is mainly a function of the interfacial tension, the wall friction force and the viscosity of the fluid [54]. Surface tension is very important at the scales where the continuum approach is still valid.

Two-phase flow phenomena are more complicated than single-phase ones. Many parameters of two-phase flow affect the flow regimes and flow pattern in flow channels. One of the most significant parameters is the ratio of the flow velocities of each phase, which is critical to the flow regimes. The flow phenomena in two-phase flow are also dependent upon the contact and interaction between the two phases, either continuous-phase contact or dispersed-phase contact [55]. The two-phase flow in micro-channel reactors is typically gas-liquid flow. In continuous-phase contact, two phases are fed separately and form two streams. The streams are also withdrawn separately at the reactor outlet. The main function of such a micro-reactor is to form the interfaces without intermixing between the phases, so the pressures of the two phases have to be carefully controlled to avoid phase intermixing.

In dispersed-phase contact, the dispersion is created by an inlet that induces merging of the gas and liquid streams. The feed is split fed by a multiple feed structure to split the phases in thin lamellae and form dispersion in a mixing section.

4.5.3 Numerical Modelling

Numerical modelling of catalytic monolith reactors has been carried out extensively over the past few years. The models are often based on a single channel to characterise the behaviour of the entire monolith, as every channel within a monolith structure should behave alike (there are of course exceptions). Meanwhile, numerical modelling and simulation of the flow in micro-channels have also been studied very actively due to the trend of increasing applications of micro-channel reactors in chemical processes. In parallel with significant applications of commercial computational fluid dynamic (CFD) software packages, other numerical methods such as the meso-scale lattice Boltzmann method (LBM) and micro-scale molecular dynamics (MD) method have also been applied to the simulation. In general, conventional methods based on the CFD are mainly appropriate for the flow and transport phenomena in macro-scale channels, but often have limitations in dealing with surface tension dominated, micro-scale interfacial transfers and interactions. Normally modifications for boundary conditions have to be considered, because the flow in the micro-scale may be different from that in a macro-scale, and some neglected parameters in macro-scale flow could be important in micro-scale. Also, the domain used in the simulation has to be meshed fine enough in order to obtain the accurate results.

Nevertheless, the volume of fluid (VOF) method and level-set method [56] have become fairly popular in recent years for modelling two-phase flow in micro-channels. The VOF method, which is available in commercial CFD code, such as Fluent and CFX, may be the simplest and easiest to be employed to track the interface and volume fraction of two fluids [57]. The term volume fraction is used as the indicator of a scalar step function for representing the space occupied by one of the fluids. As a technique for tracking interfaces, the VOF method has some advantages, such as that: mass is preserved in a natural way; and no special provision is necessary to perform reconnection or breakup of the interface. In addition, it is relatively simple to implement the VOF algorithms in parallel [58]. Figure 4.25 shows the VOF results of bubbles evolution and coalescence in a micro-channel of 0.2 mm diameter [59]. However, such a simple VOF method can not simulate the mass transfer between the gas-liquid interface and also difficult to deal with the fluids interactions with the channel walls typical of micro-structures.

To date, the meso-scale LBM has demonstrated a significant potential and broad applicability with many computational advantages, including the parallelization of algorithms and the simplicity of programming for multiphase flow [60]. Details of the LBM will be given in Chapter 6.

The modelling shown in Figure 4.26 is concerned with air-water flow in a horizontal rectangular micro-channel of $300 \times 300 \, \mu m^2$ [61]. Two air bubbles with the same diameter

Figure 4.25 *VOF computations of bubble evolution and coalescence in a micro-channel of 0.2 mm diameter (Reprinted from Ji, C. Y. and Yan, Y.Y., A numerical study of bubbly flow in a rectangular microchannel, Proceedings of the 6th International Conference on Nanochannels, Microchannels, and Minichannels, ICNMM2008 (Part A), 293–298. Copyright (2008) with permission from ASME.)*

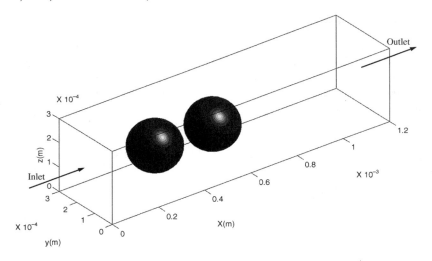

Figure 4.26   *Computational domain and initial/boundary conditions (Reprinted from Y. Y. Yan and Y. Q. Zu, LBM simulation of interfacial behaviour of bubbles flow at low Reynolds number in a square microchannel, WIT Trans. Eng. Sci.* **63***, 495–506. Copyright (2009) with permission from WIT Press.)*

$d = 200\,\mu\text{m}$ are initially placed $300\,\mu\text{m}$ apart in water inside a rectangular channel of the length $L_x = 1200\,\mu\text{m}$, the width and the height $L_y = L_z = 300\,\mu\text{m}$. The channel has an inlet boundary on the left-hand side of the channel and a free outflow boundary on the right-hand side of the channel. The other four sides of the channel are no-slip solid walls. The velocity distribution at the inlet boundary is specified as:

$$\begin{cases} u_x(0, y, z) = 16U(L_y - y)(L_z - z)yz/(L_yL_z)^2; \\ u_y(0, y, z) = 0; \\ u_z(0, y, z) = 0; \end{cases} \tag{4.39}$$

where

U is the maximum value of $u_x(0, y, z)$.

The bubbles flow in the micro-channel at a Reynolds number (Re $= \rho_L U L_x/\mu_L$) of 100, and, meanwhile, capillary numbers ($Ca = \mu_L U/\sigma_{LG}$) of 0.33 are simulated. The evolution with time of bubbles shapes and interaction are shown in Figure 4.27. It can clearly be seen

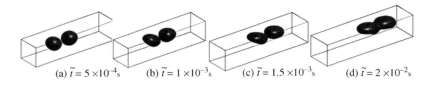

(a) $\tilde{t} = 5 \times 10^{-4}$s   (b) $\tilde{t} = 1 \times 10^{-3}$s   (c) $\tilde{t} = 1.5 \times 10^{-3}$s   (d) $\tilde{t} = 2 \times 10^{-2}$s

Figure 4.27   *Time evolution of bubble shapes at Re= 100 (Reprinted from Y. Y. Yan and Y. Q. Zu, LBM simulation of interfacial behaviour of bubbles flow at low Reynolds number in a square microchannel, WIT Trans. Eng. Sci.* **63***, 495–506. Copyright (2009) with permission from WIT Press.)*

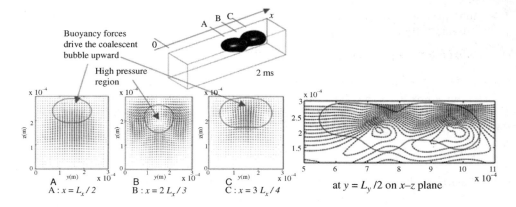

Figure 4.28 *The velocity field at different cross-section (t = 2 ms, Re = 100) (Reprinted from Y. Y. Yan and Y. Q. Zu, LBM simulation of interfacial behaviour of bubbles flow at low Reynolds number in a square microchannel, WIT Trans. Eng. Sci.* **63***, 495–506. Copyright (2009) with permission from WIT Press.)*

that the bubbles move in the x-direction by the thrust force of the surrounding water flow, and at the same time go up in y-direction due to the effect of buoyancy force; and with the progression of time, the two bubbles can finally coalesce into a larger one. Focusing only on the shape evolution of the left bubble at the early stage, it is found that the lower part of the bubble moves more quickly in the x-direction than the upper part, which is mainly caused by the effects of the velocity boundary layer near the solid wall of the channel.

The velocity fields are obtained through the numerical modelling. For example, at $t = 2$ ms, the velocity distribution at different cross-sections of the y–z plane, such as $x = L_x/2$, $2L_x/3$ and $3L_x/4$, respectively, is shown in Figure 4.28, where the solid line, the constant density line, indicates the interface between the two phases. As both pressure and velocity distributions across the interface are normally excellent indicators of numerical stability for the LBM calculations, Figure 4.28 has actually shown that the present LBM can be used to obtain reasonable and stable velocity fields. Similar to the conventional CFD, the numerical instabilities of the LBM for two-phase flow of large density ratios are mainly caused by spurious velocities and/or the large oscillation of the pressure distribution across the phase interface. However, in the present method, the velocity and pressure are both corrected by solving an additional Poisson equation after each collision-stream step. Such corrections are able to ensure the velocity to satisfy the continuity equation and smooth pressure distributions even across the interface, so as to provide the numerical stability.

Questions

4.1 What are the limits of flow rates for a falling film reactor? Consider the minimum liquid flow rate to avoid dry patches and maximum flow rate so as not to create drops. A wetting angle of 15° can be assumed. The reactor will have 50 mm internal diameter tubes. Use air and water physical properties as being at ambient pressure–temperature; 1 tonne $= 10^3$ kg.

0.0165 tonnes per tube per hour; 2.1 m/s.

4.2 A tube falling film reactor (50 mm internal diameter, 6 m long) is fed with 1250 kg/h of an ethoxylated alcohol, which undergoes a sulfonation reaction. If the mean viscosity of the liquid is 50 mPa s, what is the mean film thickness? The liquid density is 875 kg/m^3.

0.95 mm.

4.3 Determine the thickness of the film in a falling film reactor. The unit consists of 120 tubes of 0.076 m internal diameter and 8 m long. The liquid is a dilute aqueous solution whose properties can be assumed to be those of water. This gas is air. The unit is fed at 51.5 tonnes/h. What is the mass transfer coefficient?

0.59 mm; 0.0052 m/s.

4.4 A chemical of viscosity 3 mPa s, density 1060 kg/m^3 is fed at 4 tonnes/h to a 76 tube falling film reactor with 67 mm internal diameter and 4 m length. If the diffusion coefficient for a gaseous component in the liquid is $0.5 \cdot 10^{-9}$ m^2/s, what is the value of $k_l a$?

0.58 1/s.

4.5 What is the mass transfer coefficient for the case in Question 3 if the gas flows at a superficial velocity of 3 m/s?

0.0069 m/s.

4.6 A rotating disc reactor consists of a 300 mm diameter horizontal disc rotating at 2000 rpm (revolutions per minute). It is fed with the liquid considered in Question 4.2. What is the film thickness at the edge of the plate if the liquid flow rate is 1 g/s?

0.48 mm.

4.7 A tubular reactor is used for a condensation polymerisation. It consists of a 0.05 m internal diameter pipe in a downward helical coil with a slope of 10°. Calculate the pressure gradient and specific interfacial area for a feed rate of 3.53 tonnes/h at the point where 5% has been evaporated. The local temperature and pressure are 150 °C and 5 bar (absolute) (1 bar $= 10^5$ Pa), respectively. The gas can be taken to be water. The liquid has a density of 1200 kg/m^3 and a viscosity of 1 Pa s.

0.10 bar/m; 65.5 m^2/m^3.

References

[1] Bakopoulos, A. (1980) Liquid-side controlled mass transfer in wetted wall tubes. *Ger. Chem. Eng.*, **3**, 241–252.

[2] Meza, C.E. and Balakotaiah, V. (2008) Modeling and experimental studies of large amplitude waves on vertically falling films. *Chem. Eng. Sci.*, **63**, 4704–4734.

[3] Butterworth, D. (1967) The laminar flow of liquid down the outside of a rod which is at a small angle from the vertical. *Chem. Eng. Sci.*, **22**, 911–924.

[4] Park, C.D., Nosoko, T., Gima, S. and Ro, S.T. (2004) Wave-augmented mass transfer in a liquid film falling inside a vertical tube. *Int. J. Heat Mass Trans.*, **47**, 2587–2598.

[5] Hartley, D.E. and Murgatroyd, W. (1964) Criteria for the break-up of thin liquid layers flowing isothermally over solid surfaces. *Int. J. Heat Mass Trans.*, **7**, 1003–1015.

[6] Nusselt, W. (1916) Die oberflachenkondensation des wasserdampfes. *VDI Zeitschrift*, **60**, 541–546 and 569–575.

[7] Coulon, H. (1973) Stabilitätsverhältnisse bei Rieselfilmen. *Chemie-Ing. Techn.*, **45**, 362–368.

[8] Andreussi, P. (1980) The onset of droplet entrainment in annular downward flow. *Can J. Chem. Eng.*, **58**, 267–270.

[9] Batchelor, G.K. (ed.) (1962) *The scientific papers of Sir Geoffrey Ingram Taylor*, Cambridge University Press, Cambridge.

[10] Tatterson, D.F., Dallman, J.C. and Hanratty, T.J. (1977) Drop sizes in annular two-phase flow. *AIChE J.*, **23**, 68–76.

[11] Wilkes, J.O. and Nedderman, R.M. (1962) The measurement of velocities in thin films of liquid. *Chem. Eng. Sci.*, **17**, 177–187.

[12] Azzopardi, B.J. (1977) Interaction between a falling liquid film and a gas stream, Ph.D. thesis, University of Exeter.

[13] Henstock, D.E. and Hanratty, T.J. (1976) The interfacial drag and the height of the wall layer in annular flows. *AIChE J.*, **22**, 990–1000.

[14] Riazi, M.-R. and Farghi, A. (1986) Effect of interfacial drag on gas absorption with chemical reaction in a vertical tube. *AIChE J.*, **32**, 696–699.

[15] Talens-Alesson, F.I. (1999) The modelling of falling film chemical reactors. *Chem. Eng. Sci.*, **54**, 1871–1881.

[16] Clark, W. (2000) The interfacial characteristics of falling film reactors, Ph.D. thesis, University of Nottingham.

[17] Takahama, H. and Kato, S. (1980) Longitudinal flow characteristics of vertically falling liquid films without concurrent gas flow. *Int. J. Multiphase Flow*, **6**, 203–215.

[18] Brooke Benjamin, T. (1957) Wave formation in laminar flow down an inclined plane. *J. Fluid Mech.*, **2**, 554–574.

[19] Yih, C.-S. (1963) Stability of liquid flow down an inclined plane. *Phys. Fluids*, **6**, 321–334.

[20] Portalski, S. and Clegg, A.J. (1972) An experimental study of wave inception on falling liquid films. *Chem. Eng. Sci.*, **27**, 1257–1265.

[21] Clark, W.W., Campbell, G.B., Hills, J.H. and Azzopardi, B.J. (2001) Viscous effects on the interfacial structure of falling liquid film/co-current gas systems. Paper presented as the International Conference on Multiphase Flow, New Orleans, May 27 to June 1, 2001.

[22] Anshus, B.E. and Goren, S.L. (1966) A method of getting approximate solutions to the Orr-Sommerfeld equation for flow on a vertical wall. *AIChE J.*, **12**, 1004–1008.

[23] Hewitt, G.F. and Webb, D. (1975) Downward concurrent annular flow. *Int. J. Multiphase Flow*, **2**, 35–49.

[24] Portalski and, S. and Clegg, A.J. (1971) Interfacial area increase in rippled film flow on wetted wall columns. *Chem. Eng. Sci.*, **26**, 771–784.

[25] Sun, H. (2004) Hydrodynamics and mass transfer in Venturi scrubbers, Ph.D. thesis, University of Nottingham.

[26] Chun, K.R. and Seban, R.A. (1971) Heat transfer to evaporating liquid films. *J. Heat Transfer*, **93**, 391–397.

[27] Yoshimura, P.N., Nosoko, T. and Nagata, T. (1996) Enhancement of mass transfer into a falling laminar liquid film by two-dimensional surface waves – some experimental observations and modeling. *Chem. Eng. Sci.*, **51**, 1231–1240.

[28] Karimi, G. and Kawaji, M. (1998) An experimental study of freely falling films in a vertical tube. *Chem. Eng. Sci.*, **53**, 3501–3512.

[29] Lamourelle, A.P. and Sandall, O.C. (1972) Gas absorption into a turbulent liquid. *Chem. Eng. Sci.*, **27**, 1035–1043.

[30] Henstock, D.E. and Hanratty, T.J. (1979) Gas absorption by a liquid layer flowing on the wall of a pipe. *AIChE J.*, **25**, 122–131.

[31] Sisoev, G.M., Matar, O.K. and Lawrence, C.J. (2005) Absorption of gas into a wavy falling film. *Chem. Eng. Sci.*, **60**, 827–838.

[32] (a) Johnson, G.R. and Crynes, B.I. (1974) Modelling of a thin-film sulfur trioxide sulfonation reactor. *Ind. Eng. Chem. Process Des. Dev.*, **13**, 6–14; (b) Davis, E.J., van Ouwerkerk, M. and Ventatesh, S. (1979) An analysis of the falling film gas-liquid reactor. *Chem. Eng. Sci.*, **34**, 539–550; (c) Van Dam, M.H.H., Corriou, J.-P., Midoux, N., Lamine, A.-S. and Roizard, C. (1999) Modeling and measurement of sulfur dioxide absorption rate in a laminar falling film reactor. *Chem. Eng. Sci.*, **54**, 5311–5318; (d) Akanksha, Pant, K.K. and Srivastava, V.K. (2007) Modeling of sulphonation of tridecylbenzene in a falling film reactor. *Math. Comp. Modelling*, **46**, 1332–1344.

[33] Charwat, A.F., Kelly, R.E. and Gazley, C. (1972) The flow and stability of thin films on a rotating disk. *J. Fluids Mech.*, **53**, 227–255.

[34] Lechev, I. and Peev, G. (2003) Film flow on rotating disk. *Chem. Eng. Proc.*, **42**, 925–929.

[35] Rauscher, J.W., Kelly, R.E. and Cole, J.D. (1973) An asymptotic solution for laminar flow of a thin film on a rotating disk. *J. Appl. Mech.*, **40**, 43–47.

[36] Aoune, A. and Ramshaw, C. (1999) Process intensification: heat and mass transfer characteristics of liquid films on rotating discs. *Int. J. Heat Mass Trans.*, **31**, 1432–1445.

[37] Sisoev, G.M., Matar, O.K. and Lawrence, C.J. (2005) Gas absorption in to a wavy film flowing over a spinning disc. *Chem. Eng. Sci.*, **60**, 2051–2060.

[38] Taylor, G.B. (1944) DuPont, US Patent, 2,361,717; Heckert, W.W. (1951) DuPont, US Patent, 2,689,839; Hull, D.R. (1956) DuPont, Canadian Patent No. 527.473.

[39] (a) Banerjee, S., Rhodes, E. and Scott, D.S. (1969) Studies on concurrent gas-liquid flow in helically coiled tubes: Part I. Flow patterns, pressure drop and holdup. *Can. J. Chem. Eng.*, **47**, 445–453; (b) Boyce, B.E., Collier, J.G. and Levy, J. (1969) Hold-up and pressure drop measurements in the two-phase flow of air-water mixtures in helical coils, in *Co-current Gas-Liquid Flow*, Pergamon Press, pp. 203–234.

[40] Azzopardi, B.J. (2006) *Gas-Liquid Flows*, Begell House, New York.

[41] Chen, J.J.J. and Spedding, P.L. (1981) An extension of the Lockhart-Martinelli theory of two pressure drop and hold up. *Int. J. Multiphase Flow*, **7**, 659–675.

[42] Chisholm, D. (1967) A theoretical basis for the Lockhart-Martinelli correlation for two-phase flow. *Int. J. Heat Mass Trans.*, **10**, 1767–1778.

[43] Lockhart, R.W. and Martinelli, R.C. (1949) Proposed correlation of data for isothermal, two-phase, two-component flow in pipes. *Chem. Eng. Prog.*, **45**, 39–48.

[44] Giudici, R., Nascimento, C.A.O., Tresmondi, A. *et al.* (1999) Mathematical modeling of an industrial process of nylon-6,6 polymerization in a two-phase flow tubular reactor. *Chem. Eng. Sci.*, **54**, 3243–3249.

[45] Williams, J.L. (2001) Monolith structures, materials, properties and uses. *Catal. Today*, **69**, 3–9.

[46] Burch, R., Breen, J.P. and Meunier, F.C. (2002) A review of the selective reduction of NOx with hydrocarbons under lean-burn conditions with non-zeolitic oxide and platinum group metal catalysts. *Appl. Catal. B: Environ.*, **39**, 283–303.

[47] Jensen, K.F. (2001) Microreaction engineering – is small better? *Chem. Eng. Sci.*, **56**, 293–303.

[48] Minsker, L.K. and Renken, A. (2005) Microstructured reactors for catalytic reactions. *Catal. Today*, **110**, 2–14.

[49] Papautsky, I., Ameel, T.A. and Frazier, A.B. (2001) A review of laminar single-phase flow in microchannels. Paper presented at ASME International Mechanical Engineering Congress & Exposition, New York, NY, November 2001, pp. 1–9.

[50] Ma, S.W., Gerner, F.M. and Tsuei, Y.G. (1993) Forced convection heat transfer from micro-structures. *J. Heat Transfer*, **115**, 872–880.

[51] Em Karniadakis, G. and Beskok, A. (2002) *Micro Flow*, Springer.

[52] Salman, W., Gavriilidis, A. and Angeli, P. (2006) On the formation of Taylor bubbles in small tubes. *Chem. Eng. Sci.*, **61**, 6653–6666.

[53] Alfadhel, K.A. and Kothare, M.V. (2005) Microfluidic modeling and simulation of flow in membrane microreactors. *Chem. Eng. Sci.*, **60**, 2911–2926.

[54] Waelchli, S. and von Rohr, P.R. (2006) Two-phase flow characteristics in gas-liquid micro-reactors. *Int. J. Multiphase Flow*, **32**, 791–806.

[55] Hessel, V., Angeli, P., Gavriilides, A. and Löwe, H. (2005) Gas-liquid and gas-liquid-solid microstructured reactors: Contacting principles and applications. *Ind. Eng. Chem. Res.*, **44**, 9750–9769.

[56] Sussman, M., Smereka, P. and Osher, S. (1994) A level set approach for computing solutions to incompressible two-phase flow. *J. Comput. Phys.*, **114**, 146–159.

[57] Hirt, C.W. and Nichols, B.D. (1981) Volume of fluids (VOF) method for the dynamics of free boundaries. *J. Comput. Phys.*, **39**, 201–225.

[58] Scardovelli, R. and Zaleski, S. (1999) Direct numerical simulation of free-surface and interfacial flow. *Ann. Rev. Fluid Mech.*, **31**, 567–603.

[59] Ji, C.Y. and Yan, Y.Y. (2008) A numerical study of bubbly flow in a rectangular microchannel. Paper presented at Proceedings of the 6th International Conference on Nanochannels, Microchannels, and Minichannels, ICNMM2008, (Part A), pp. 293–298.

[60] Chen, S. and Doolen, G.D. (1998) Lattice Boltzmann method for fluid flows. *Ann. Rev. Fluid Mech.*, **30**, 329–364.

[61] Yan, Y.Y. and Zu, Y.Q. (2009) LBM simulation of interfacial behaviour of bubbles flow at low Reynolds number in a square microchannel. *WIT Trans. Eng. Sci.*, **63**, 495–506.

5

Macroscale Modelling

Hydrodynamics of Gas-Liquid Reactors: Normal Operation and Upset Conditions, First Edition.
B. J. Azzopardi, R. F. Mudde, S. Lo, H. Morvan, Y. Yan and D. Zhao.
© 2011 John Wiley & Sons, Ltd. Published 2011 by John Wiley & Sons, Ltd.

5.1 Introduction

The approach taken so far has been based on simplified models. The full complexity of the hydrodynamics of multiphase reactors does not allow for analytical solutions, unless serious simplifications are made. However, during the last two decades enormous progress has been made on a computational approach to hydrodynamics, both with single-phase and multi-phase problems. The multiphase flows are even more complex than the single-phase ones. The interaction between the various phases is in principle known. For instance, for gas-liquid systems, the motion of both phases is completely governed by the Navier–Stokes equations. Hence, the fundamental equations are known. However, sub-models governing the interfacial interactions, for example, are required and these flows can only be solved for a limited number of cases, under a number of defining assumptions, even via simulations. Both the occurrence of turbulence and the spatial distribution of the phases with their sharp interfaces make direct simulation of the governing equations a formidable task that quickly gets 'out of control'.

A simple estimate of how huge this task is can be made by estimating the range of length and time scales available in a turbulent single-phase flow. Consider the flow in a river: the water flows at a velocity in the order of $U = 1$ m/s in a river bed with a characteristic length scale of $L = 10$ m. This gives a Reynolds number of 10^7. At the largest scale, the kinetic energy scales with U^2. This energy becomes dissipated in one large eddy turn-over time, which can be estimated as $T = L/U$. Thus, a coarse grain estimate of the dissipation is $\dot{E} \sim U^3/L \sim 0.1$ m^2/s^3. From this we can estimate, using Kolmogorov's ideas, the scale of the smallest eddies. For the length scale we find: $\lambda_K = (v^3/\dot{E})^{1/4} = 50$ μm, where $v = \eta/\rho = 10^{-6}$ m^2/s the kinematic viscosity of water. The lifetime of these small eddies is about $\tau_K = \sqrt{(v/\dot{E})} = 3$ ms. Thus, we find that the biggest to the smallest spatial scale in the flow is 200 000 to 1! Similarly, for the time scale of the biggest eddy to that of the smallest one, we have a ratio of 3000. In a more general way, we can write the above length scale ratio as $\lambda_K/L = \text{Re}^{-3/4}$.

Just imagine that we would like to discretise the flow of that river in a computer. We would need at least three spatial points in the smallest eddies to capture their existence. Furthermore, we need to solve a three-dimensional flow problem. Thus, the number of grid cells also required to cover the largest scales, is at least $3^3 \times \text{Re}^{9/4}$. For the river flow at $\text{Re} = 10^7$, this will give a phenomenal number of data that need to be stored in the computer's memory: 10^{17} computational cells. For each of these we have to remember three velocity components and the pressure, requiring a memory of 10^9 GB! Needless to say, this is way out of reach of present computers and even of those in the near future.

For multiphase reactors it is not much better. The scale may be reduced by a factor of ten, but we need to resolve on a finer grid, as the interfaces between the phases needs to be captured at sufficiently high resolution. Various techniques have been developed to capture the interfaces and they are usually referred to as the volume of fluid (VOF) methods or the level set (LS) methods. They are computationally expensive and only small scale model systems of interest to us here can at present be studied with these techniques. Direct numerical simulation (DNS) is for practical applications out of the question naturally.

Therefore, different strategies have been developed. In the discrete particle approach, usually referred to as the Euler–Lagrange approach, the gas bubbles are treated as point particles that interact with a continuous liquid. The interaction is not calculated directly, as is the case with DNS, but is instead modelled via forces acting on the particles, such as the drag force. The particles exert a force on the liquid, and in many industrial applications this two-way coupling cannot be reduced to something simpler. If the flow of the continuous phase is turbulent, averaged equations can be solved for t. This can be a so-called Reynolds average (resulting in the Reynolds averaged Navier–Stokes equations, or RANS), in which only the average flow structures are solved and the influence of all turbulent eddies is modelled via a turbulence model. The so-called k–$\dot{E}$ model is one of the most popular ways of doing this. Yet this approach is now being replaced in academic work by the large eddy simulation (LES), which allows solving a part of the turbulence energy spectrum and only modelling a fraction of it, corresponding to the smallest more isotropic structures.

However, the Euler–Lagrange approach also has limitations in terms of required computational power (in addition to conceptual limits when the volume fractions of the secondary, Lagrangian, phase is very large). Thus, many industrial applications in multiphase flows are, for the time being, out of reach of these methods. As an alternative we have the so-called Euler–Euler approach, or two-fluid approach. Now, the governing equations are ensemble-averaged right from the start. Here, the idea is that if we cannot solve for the individual bubbles, we can still find averages, such as the local gas fraction and local gas velocity. The averaged equations describe the multiphase flow as interpenetrating fluids, that is, they co-exist in every spot but each with a fraction that sums up to one. The phase interaction is computed based on Lagrangian type sub-models and correlations.

The two-fluid approach is currently the most used method when dealing with industrial multiphase reactors. Its strength is the (relative) computational speed; its weakness is that they cannot provide the full details and have difficulty, for example, with coalescence and breakup of bubbles as collision between bubbles cannot be computed exactly. In this chapter we will deal with the Euler–Euler, or two-fluid approach.

5.2 Eulerian Multiphase Flow Model

5.2.1 Definition

Multiphase flows refer to flows of several fluids in the domain of interest. In general, we associate phases with gases, liquids or solids, and as such some simple examples of multiphase flows are: air bubbles rising in a glass of sparkling water, sand particles carried by wind and rain drops in air. In fact, the definition of 'phase' (fields, strictly speaking) can be generalised and applied to other fluid characteristics, such as size and shape of particles, density, and so on. With this broader definition, multiple phases can be used to represent the entire size distribution of particles in several size groups or 'phases' of a multiphase model, representative of their different topology and behaviour. Mass transfer between the particle phases can, therefore, be used to represent particles moving between different size groups due to breakup and coalescence processes.

In the general Eulerian multiphase model, the phases are treated as interpenetrating continua coexisting in the flow domain. Equations for conservation of mass, momentum and energy are solved for each phase. The share of the flow domain occupied by each phase is given by its volume fraction and each phase has its own velocity, temperature and physical properties. Interactions between phases due to differences in velocity and temperature are taken into account via the inter-phase transfer terms in the transport equations. However, this approach can be relaxed to take advantage of the flow characteristics and account of the level of details required. In a homogeneous formulation, for example, it is assumed that some transported quantities are the same for all phases, in the momentum equation in particular.

The Eulerian multiphase flow model provides a general framework for all types of multiphase flows: both dispersed (e.g. bubble, droplet and particle) and stratified (e.g. free-surface) flows can be modelled. In this chapter we focus on modelling of dispersed flows. In this type of flow, one of the phases is a continuum in which the other phases are dispersed in the form of droplets, particles or bubbles.

5.2.2 Transport Equations

The idea is to represent each phase with its own set of conservation laws of mass, momentum and energy. These are derived from averaging of the underlying equations of motion. One way of doing this is by considering a volume element in the multiphase flow that is large compared with the particle scale (i.e. larger than the particle size and larger than the particle–particle distance), but small compared with the scale of the equipment containing the flow. The governing equations are averaged over this volume. Thus, the fine details of the flow inside the volume are lost and only the average pressure and velocities are left as the unknowns. In addition, the volume fraction of each of the phases is defined. It is the averaged amount (in terms of volume) of each of the phases present in the averaging volume. Note that the volume fraction is, by construction, a smooth quantity that only exists as an average property, that is, it does not have a meaning at sub-scales of the averaging volume in contrast to the velocity, which does have a fine structure that is lost in the averaging. Figure 5.1 gives a graphical impression of the procedure.

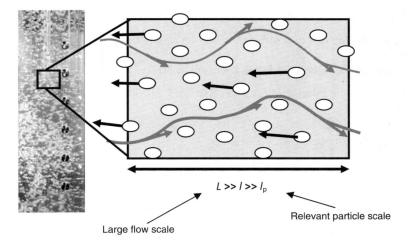

Large flow scale

Relevant particle scale

$L \gg l \gg l_p$

Figure 5.1 *Representative control volume out of a bubbly flow for which the average equations of the Euler–Euler approach are derived*

For a bubbly mixture, the underlying equations are the continuity, Navier–Stokes equations and the energy equation. After averaging, each phase is represented by a set of conservation laws of mass, momentum and energy. Note that volume averaging is just one way of deriving these equations; time averaging or, preferably, ensemble averaging are other options.

5.2.2.1 Continuity Equation
The conservation of mass (or continuity) for phase k is:

$$\frac{\partial}{\partial t}(\varepsilon_k \rho_k) + \nabla \cdot (\varepsilon_k \rho_k u_k) = \sum_{i=1}^{N}(\dot{m}_{ki} - \dot{m}_{ik}) \tag{5.1}$$

where

ε_k is the volume fraction of phase k
ρ_k is the phase density
u_k is the phase velocity
$\dot{m}_{ki}$ and $\dot{m}_{ik}$ are mass transfer rates to and from the phase
N is the total number of phases.

The sum of the volume fractions is clearly equal to unity.

$$\sum_{k} \varepsilon_k = 1 \tag{5.2}$$

5.2.2.2 Momentum Equation
The conservation of momentum for phase k is:

$$\frac{\partial}{\partial t}(\varepsilon_k \rho_k u_k) + \nabla \cdot (\varepsilon_k \rho_k u_k u_k) - \nabla \cdot \left[\varepsilon_k(\tau_k + \tau_k^t)\right] = -\varepsilon_k \nabla p + \varepsilon_k \rho_k g + M \tag{5.3}$$

where

τ_k and τ_k^t are the laminar and turbulence shear stresses, respectively
p is pressure
M is the sum of the interfacial forces (per unit volume).

We will examine the interfacial forces in Section 5.2.3 below.

5.2.2.3   Energy Equation
The conservation of energy for phase k is:

$$\frac{\partial}{\partial t}(\varepsilon_k \rho_k h_k) + \nabla \cdot (\varepsilon_k \rho_k u_k h_k) - \nabla \cdot \left[\varepsilon_k \left(\lambda_k \nabla T_k + \frac{\mu_k^t}{\sigma_h^t} \nabla h_k \right) \right] = Q \tag{5.4}$$

where

h_k is the phase enthalpy
λ_k is the thermal conductivity
Q is the interfacial heat transfer.

5.2.3   Interfacial Forces

The interaction between the various phases follows from the distribution of the pressure and shear stress at the interfaces. However, the details of these are not known and can not be resolved in the Euler–Euler approach, as individual interfaces are not represented in this theoretical framework. Instead, the interaction forces have to be formulated separately and fed back to the momentum equation. There is a reasonably long list of interfacial forces between the phases that have to be considered. These are drag force (F_D), lift force (F_L), virtual mass force (F_M), turbulent drag force (F_T), wall lubrication (F_W), Basset or history force (F_B) and momentum transfer associated with mass transfer. The combined force is then

$$M = F_D + F_L + F_M + F_T + F_W + F_B + \sum_{i=1}^{N}(\dot{m}_{ki} u_i - \dot{m}_{ik} u_k) \tag{5.5}$$

All of these forces need to be evaluated, typically using models and correlations obtained experimentally, some of which are described below. This is an important point to note as many of these relationships can be case or application specific, and based on limited data. This dimension adds another difficulty to exploring multiphase flow systems as, not only the formalism used (RANS versus LES; Eulerian versus Lagrangian), but also the sub-models contain assumptions as to what the system should be.

5.2.3.1   Drag Force
The drag force on a particle in a turbulent flow can be considered in two parts: the mean and turbulent drag forces. The mean drag force on a single particle, F_{Di}, can be calculated from:

$$F_{Di} = C_D A \frac{\rho_c u_r^2}{2} \tag{5.6}$$

where

C_D is the drag coefficient
$u_r = (u_c - u_d)$ is the relative velocity between the two phases
A is the projected area of the particle in the direction of the flow.

Subscript c stands for 'continuous phase' and d for 'dispersed phase'.

In multi-particle systems we simply multiply the single particle drag force by the number of particles in the control volume. The total drag force per unit volume is therefore:

$$F_D = nF_{Di} = \frac{3}{4} C_D \frac{\varepsilon d \rho_c}{d} |u_r| u_r \tag{5.7}$$

where the number of particles per unit volume, n, is given by:

$$n = \frac{6\varepsilon_d}{\pi d^3} \tag{5.8}$$

For convenience in later analyses, we define a drag force coefficient A_D as:

$$F_D = A_D u_r \tag{5.9}$$

where

$$A_D = \frac{3}{4} \frac{C_D \varepsilon_d \rho_d |u_r|}{d} \tag{5.10}$$

5.2.3.1.1 *Commonly Used Equations for Drag Coefficient for Particles.* The correlation for drag coefficient commonly found in the literature is by Schiller and Naumann [1]:

$$C_D = \frac{24}{Re_d} (1 + 0.15\, Re_d^{0.687}) \quad 0 < Re_d \leq 1000 \tag{5.11}$$

$$C_D = 0.44 \quad Re_d > 1000 \tag{5.12}$$

The particle Reynolds number, Re_d, is defined as:

$$Re_d = \frac{\rho_c |u_r| d}{\eta_c} \tag{5.13}$$

Other similar correlations exist for high particle concentrations [2] and to account for particle deformation [3]. Wang's correlation for bubbly flow is outlined in more detail next.

5.2.3.1.2 *Drag Force for Bubbly Flows.* Equation 5.11 given above, for particle flows, can be used to calculate the drag force for spherical bubbles smaller than 1 mm in diameter. Larger bubbles tend to be non-spherical and different drag coefficients must be applied. For simplicity, we can consider bubbles smaller than 1 mm as 'spherical', bubbles between 1 and 15 mm as 'ellipsoidal' and larger bubbles as 'spherical cap' bubbles.

The following empirical correlation by Wang [4] can be used for bubbles rising in water:

$$C_D = \exp\left[a + b \ln Re_d + c (\ln Re_d)^2\right] \tag{5.14}$$

Table 5.1 *Drag coefficients for a single rising bubble in water by Wang [4]*

Re_d	a	b	c
$Re_d \leq 1$	ln 24	−1	0
$1 < Re_d \leq 450$	2.699 467	−0.33 581 596	−0.07 135 617
$450 < Re_d \leq 4000$	−51.77 171	13.1 670 725	−0.8 235 592
$Re_d > 4000$	ln (8/3)	0	0

where the coefficients were derived from curve-fitting measurements taken for a single bubble rising in water. The coefficients are given in Table 5.1.

5.2.3.2 Lift Force

For particles moving in continuous fluid that has a strong velocity gradient, different sides of the particle may experience different flow conditions. Similar to the flow over an airfoil, a lift force is developed in a direction perpendicular to the main flow direction. In the literature (e.g. Auton *et al.* [5]), the lift force is usually given as:

$$F_L = C_L \varepsilon_d \rho_c u_r \times (\nabla \times u_c)$$ (5.15)

where

C_L is the lift coefficient.

For inviscid flow around a sphere, $C_L = 0.5$. The lift force is found to have a major influence on the radial distribution of the two phases, particularly so in bubbly flows in small diameter pipes. According to Tomiyama *et al.* [6] the lift force is a function of bubble diameter d and is given in terms of the Reynolds and Eötvos numbers:

$$C_L = C_{LF} + C_{WK}$$ (5.16)

with:

$$C_{LF} = 0.288 \cdot \tanh (0.121 \cdot Re_d)$$ (5.17)

$$C_{WK} = \begin{cases} 0, & E\ddot{o} < 4 \\ -0.096\,E\ddot{o} + 0.384, & 4 < E\ddot{o} < 10 \\ -0.576 & E\ddot{o} > 10 \end{cases}$$ (5.18)

where

Re_d is the particle Reynolds number defined in Equation 5.13
$E\ddot{o} = \frac{g \cdot (\rho_c - \rho_d) \cdot d^2}{\sigma}$ is the Eötvos number
g is the gravitational acceleration
σ is the surface tension coefficient.

5.2.3.3 Virtual Mass Force

For a particle to accelerate, additional force is required to display (or push away) the surrounding fluid. This additional force is conveniently represented as an additional mass to the particle. Hence, this force is often called the added mass force or the virtual mass force and can be expressed as:

$$F_M = C_M \varepsilon_d \rho_c \left(\frac{Du_d}{Dt} - \frac{Du_c}{Dt} \right)$$ (5.19)

where

C_M is the virtual mass coefficient.

For inviscid flow around an isolated sphere, $C_M = 0.5$.

5.2.3.4   Turbulent Drag Force

Turbulent drag force considers the fluctuating component of the drag force to account for the additional drag due to interaction between the dispersed phase and the surrounding turbulent eddies.

$$F_T = -A_D \frac{v_c^t}{\varepsilon_d \varepsilon_c \sigma_\alpha} \nabla \varepsilon_d$$ (5.20)

where

v_c^t is the continuous phase turbulent kinematic viscosity
σ_α is the turbulent Prandtl number (value of 1 is often used).

5.2.3.5   Basset Force

The Basset (or history) force is another force that arises due to relative acceleration of the dispersed particles. It is associated with viscous momentum transfer to or from the interfaces and represents the effect of acceleration on the drag force. Unfortunately, this force depends on the history of the flow, that is, on the acceleration at previous times (hence, the name history force). This makes it complicated in analytical and numerical work. For a spherical particle, the force is given by:

$$F_B = \frac{3}{2} d_p^2 \sqrt{\pi \eta_l \rho_l} \int_{-\infty}^{t} \left[\frac{d(u_p - u_l)}{dt} \right]_{t=\tau} \frac{dt}{\sqrt{t - \tau}}$$ (5.21)

Note that the history integral is weighted by the inverse of the square root of the elapse time $t - \tau$, indicating a diffusive transport of vorticity away or towards the interface. If the relative acceleration, $|\partial(u_p - u_l)/\partial t|$, is small compared with $(u_p - u_l)^2/d_p$, the Basset force can be neglected.

5.2.3.6   Wall Lubrication Force

When a particle approaches a wall the flow field around the particle is modified and gives rise to a viscous lubrication force, which tends to push the particles away from the wall. This force is modelled as [7]:

$$F_W = \frac{\varepsilon_d \rho_c u_r^2}{d} \max \left[C_1 + C_2 \frac{d}{y_w}, 0 \right] n_w$$ (5.22)

where

y_w is the distance to the nearest wall
n_w is the unit normal pointing away from the wall.

Typical values of the coefficients are $C_1 = -0.1$ and $C_2 = 0.147$ for a sphere.

5.2.4   Turbulence Models

Modelling turbulence in multiphase flows is clearly a highly complex topic and is not well understood. Here we assume that a RANS approach is used and present a simple and commonly used model based on the k–ε model. To calculate the continuous and dispersed phase turbulence stresses, values for k and ε are required. These can be computed using the extended k–ε equations containing extra source terms that arise from the interphase forces present in the momentum equations. The additional terms account for the effect of particles on the turbulence field. The relevant equations are:

$$\frac{\partial}{\partial t}\varepsilon_c\rho_c k + \nabla \cdot \varepsilon_c\rho_c u_c k = \nabla \cdot \left[\frac{\varepsilon_c(\mu_c + \mu_c^t)}{\sigma_k} \nabla k \right] + \varepsilon_c(G - \rho_c\dot{E}) + S_{k2} \tag{5.23}$$

$$\frac{\partial}{\partial t}\varepsilon_c\rho_c\dot{E} + \nabla \cdot \varepsilon_c\rho_c u_c\dot{E} = \nabla \cdot \left[\frac{\varepsilon_c(\mu_c + \mu_c^t)}{\sigma_E} \nabla\dot{E} \right] + \varepsilon_c(C_1 G - C_2\rho_c\dot{E}) + S_{E2} \tag{5.24}$$

where

$$S_{k2} = -A_D \frac{v_c^t}{\varepsilon_c\varepsilon_d\sigma_\alpha}(u_d - u_c)\cdot\nabla\varepsilon_d + 2A_D(C_t - 1)k \tag{5.25}$$

$$S_{E2} = 2A_D(C_t - 1)\dot{E} \tag{5.26}$$

$$G = \eta_c(\nabla u_c + \nabla u_c^T) : \nabla u_c \tag{5.27}$$

In the above equations, C_t is a response coefficient defined as the ratio of the dispersed phase velocity fluctuations to those of the continuous phase:

$$C_t = \frac{u'_d}{u'_c} \tag{5.28}$$

The turbulent stress, τ_c^t, in the continuous phase momentum equation can be modelled using the eddy–viscosity concept:

$$\tau_c^t = \eta_c^t \left(\nabla u_c + \nabla u_c^T - \frac{2}{3}\nabla u_c I \right) - \frac{2}{3}\rho_c kI \tag{5.29}$$

with the turbulent viscosity given by:

$$\eta_c^t = C_\eta\rho_c\frac{k^2}{\dot{E}} \tag{5.30}$$

The dispersed-phase turbulent stress, τ_d^t, is correlated to the continuous-phase turbulent stresses, τ_c^t, via the response coefficient, C_t, such that

$$\tau_d^t = \frac{\rho_d}{\rho_c}C_t^2\tau_c^t \tag{5.31}$$

There is evidence that C_t is a strong function of the volume fraction and that beyond a certain limiting value, which could be as small as 6%, C_t approaches unity. For practical calculations Equation 5.31 can be simplified to:

$$\tau_d^t = \frac{\rho_d}{\rho_c} \tau_c^t \qquad\qquad (5.32)$$

Modelling turbulence for multiphase flow using typical models is often done by mimetism and by extending the existing models. This is another area of uncertainty to take into account and another modelling challenge.

5.2.5   Case Study – Cylindrical Bubble Column

Bubble columns are one of the work horses of chemical engineering and biotechnology. Their simplicity of construction, absence of moving parts and possibilities for large scale make them attractive. The design and fine tuning is, however, hampered by the complex fluid flow inside these reactors. Computational fluid dynamics can be used for a better understanding and design of the bubble columns. As a case study, we will consider the bubble flow in a 200 mm diameter, cylindrical bubble column [8]. The gas (air) is introduced into the liquid phase (water) via the bottom through 314 holes of 1.2 mm in a square pitch of 6 mm. The gas injection takes place uniformly over the entire cross-section of the bottom. The mixture height in the column is 1.0 m.

The Euler–Euler equations are numerically solved, using the drag, lift and virtual mass force between the two phases. The turbulence is taken into account via a modified k–ε model (based on the mixture velocity, see below).

The bubble size is fixed at 5 mm. An unstructured grid of 40×10^3 cells has been used, with a time step of 1 ms in the initial phase of the simulations. After the initial transient phase, a time step of 5 ms is employed. The gas inlet and outlet velocities were fixed at 0.02 m/s.

A snapshot of the gas fraction distribution as well as the liquid flow field are, for a superficial velocity of 0.02 m/s, given in Figure 5.2. The maximum velocity here is about 0.08 m/s. The flow in the simulations is transient and shows the familiar large-scale motion.

The simulations are compared with measured pressure data and the gas fraction at different heights. There is definitely room for improvement. These workers attribute this partly to the use of a single bubble size in the simulations, whereas in the actual experiments a bubble size ranging from 1 to 20 mm has been observed. This is no surprise, given the relatively high superficial gas velocity. Improvement is expected through use of a bubble size distribution in the simulations, that is, invoking population balances, see Section 5.3.

5.2.6   Homogenous and Mixture Modelling

Full Eulerian simulation can be relatively costly, in particular as a result of having to solve for a full set of momentum equations for each phase, or field, and taking into account the necessary coupling between the velocities as part of the resolution of the relative velocity u_r.

Certain flow problems can be simplified by taking advantage of specific flow characteristics. This is true when the flow is well segregated or stratified for example, or when the disperse bubbles can be assumed to have the same velocity as the carrying fluid, in which case a homogenous formulation is suited.

When, in a dispersed flow situation, the velocity of the gas phase can be characterised as being directly connected to the mixture velocity, then the mixture model is a suitable

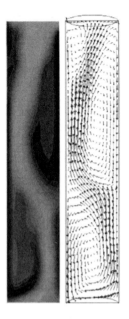

Figure 5.2 *Gas fraction (left) and liquid velocity (right) for gas superficial velocity = 0.02 m/s [8] (Reprinted from Rampure, M.R., Buwa, V.V. and Ranade, V.V., Modelling of gas-liquid/gas-liquid-solid flows in bubble columns: Experiments and CFD simulations, Can. J. Chem. Eng., 81, 692–706. Copyright (2003) with permission from John Wiley & Sons.)*

proposition. It leaves the formulation open to a so-called closure problem, but the latter can be resolved via suitable sub-models.

These two concepts, usually referred to as algebraic slip mixture models (ASMM), are presented next.

5.2.6.1 General Formulation

A homogeneous model is a simplification of the full Eulerian approach highlighted in Section 5.2.2, in which one assumes that a particular transported quantity is shared amongst all the phases or fields (note: with the exception of the volume fractions). Therefore, densities, viscosities and velocities are computed as:

$$\rho_m = \sum_{k=1}^{N} \varepsilon_k \rho_k \tag{5.33}$$

$$\eta_m = \sum_{k=1}^{N} \varepsilon_k \eta_k \tag{5.34}$$

$$u_m^i = \frac{1}{\rho_m} \sum_{k=1}^{N} \varepsilon_k \rho_k u_k^i \tag{5.35}$$

where

u_k^i is the average velocity of phase k.

With the associated continuity equation for the mixture density and the Navier–Stokes equation for each of the mixture velocity components becoming:

$$\frac{\partial}{\partial t}(\rho_m) + \frac{\partial}{\partial x_j}(\rho_m u_m^j) = 0 \tag{5.36}$$

$$\frac{\partial}{\partial t}(\rho_m u_m^i) + \frac{\partial}{\partial x_j}(\rho_m u_m^i u_m^j) = -\frac{\partial}{\partial x_i}p - \frac{\partial}{\partial x_j}\left(\tau^{ij} + \tau_t^{ij} + \sum_{k=1}^{N}\varepsilon_k \tau_{k,D}^{ij}\right) + \rho_m g^i + M^i \tag{5.37}$$

to be compared with Equation 5.3 for example. The third stress term on the right-hand side is showing that the flow is actually a mixture and not a single phase: it accounts for the relative velocities of the phases, see below. If the relative velocity between the phases is identical to zero, this term disappears and Equations 5.36 and 5.37 are equivalent to the single phase Navier–Stokes equations.

If the flow is drag dominated for example, and the body forces are negligible, then the velocities for each of the phases will become equal over a very short spatial length scale, and the above is an accurate representation of the flow. Consider a case when the flow is strongly gravity dominated, for example in the case of a stratified flow, or if very large bubbles do exist in the flow, these being of a scale larger than the typical mesh size. Then in most of the solution domain the volume fractions will either be equal to zero or one, and the above governing equation will again be representative of the flow (except maybe at the interface, although this can be improved through recourse to local reconstruction schemes, and in the event that significant shear exists between the phases).

5.2.6.2 Mixture Model

The mixture model solves the above mixture momentum equation and relies on the concept of relative velocities to ascribe momentum properties to the dispersed phase. This approach is very much suited to cases with relatively low loading or particles of relatively small size (small volume fraction; scales smaller than the mesh resolution).

An important nuance needs to be underlined: The explicit assumption that there exists a velocity difference between the mixture and the disperse gas phase leads to the introduction of an extra term in the governing equations, due to the non-linearity of the Navier–Stokes equations. This is explained below and constitutes a significant departure from Equation 5.37 for example.

Before this, however, some additional terminology is necessary. The slip velocity is defined as the difference between the continuous fluid velocity (c) and the disperse phase velocity (d)

$$u_{s,k}^i = u_{d,k}^i - u_c^i \tag{5.38}$$

The diffusion or drift velocity is defined as:

$$u_D^i = u_d^i - u_m^i \tag{5.39}$$

Using the above definition of the diffusion velocity, the momentum equation for the mixture can be written as:

$$\frac{\partial}{\partial t}(\rho_m u_m^i) + \frac{\partial}{\partial x_j}(\rho_m u_m^j u_m^i) = -\frac{\partial}{\partial x_i}p - \frac{\partial}{\partial x_j}\left(\tau^{ij} + \tau_t^{ij} + \underline{-\sum_{k=1}^{N}\varepsilon_k \rho_k u_{D,k}^i u_{D,k}^j}\right) + \rho_m g^i$$

(5.40)

in terms of mixture variables. One notes the presence of an extra term, underlined in Equation 5.40, the so-called diffusion stress due to the assumed drift velocity between the mixture and the disperse phase.

What is required is essentially a relationship to obtain the diffusion velocities. A key assumption that is made with these methods is that the disperse phase reaches its terminal velocity instantaneously.

Applying the kinematic closure condition:

$$\sum_{k=1}^{N}\varepsilon_k \rho_k u_{D,k}^i = 0$$

(5.41)

the diffusion stress can be written as:

$$\tau_{k,D}^{ij} = -\rho_m \sum_{k=1}^{N}\varepsilon_k u_{s,k}^i u_{s,k}^j + \rho_m \sum_{k=1}^{N}\varepsilon_k \varepsilon_l u_{s,k}^i u_{s,l}^j$$

(5.42)

which, in turn, simplifies to:

$$\tau_{k,D}^{ij} = -\rho_m \varepsilon_d (1-\varepsilon_d) u_{s,d}^i u_{s,d}^j$$

(5.43)

when only one dispersed phase is present.

What is required is the ability to compute the slip velocity.

5.2.6.2.1   Terminal Settling Speed.   In the seminal work of Richardson and Zaki [9] on batch sedimentation, it was found that the settling speed, for a given particle size, is directly related to the terminal settling speed of a single particle in a large body of fluid, V_T, via a power law of the background fluid volume fraction, that is $(1-\varepsilon_d)^n$. This leads to the slip velocity being equal to:

$$u_{s,d} = V_T(1-\varepsilon_d)^{n-1}$$

(5.44)

A range of correlations for the terminal velocity are available for bubbly flow, for example, Peebles and Garber [10] and subsequent publications. Wallis [11] also provides a range of correlations between dimensionless terminal velocities and dimensionless bubble sizes, thereby covering a range of forces acting on the particles (from pure viscous to fully developed turbulent wake effects for the largest particles; see Kolev [12] for more details).

5.2.6.2.2   General Approach to Relative Velocity.   A general group of models have been developed since the mid-1960s to compute the velocity differences, arising from gravity and inertial forces mainly, on the basis of local equilibrium. Depending on the form of the equations used, these models have received various names, although they aim to achieve the same thing: the drift–flux model [13], the mixture model [14], the algebraic slip model or ASM [15], the diffusion model [14] or, more recently, the local equilibrium model [16]. Manninen *et al.* [17] offer a good overview and, in fact, provide further references, including software specific implementations for the most popular commercial codes.

5.3 Poly-Dispersed Flows

5.3.1 Methods of Moments

The evolution of the bubble size distribution as a result of coalescence and breakup can be modelled by means of the population balance equation (PBE). A detailed description of the bubble size distribution will require a large number of population classes, for example from 10 to 20. The computational effort required to solve 10–20 PBEs simultaneously in a CFD (computational fluid dynamics) calculation can be large. A simpler alternative can be derived by assuming that the bubble size distribution conforms to a pre-defined shape, and this shape is retained during the process being investigated. Under these assumptions, the complete bubble size distribution can be represented by a limited number of parameters and the PBE could be reformulated in terms of these parameters. S_γ is conserved on a volumetric basis and is related to the moment M_γ of the distribution:

$$S_\gamma = nM_\gamma = n \int_0^\infty d^\gamma P(d)\, \mathrm{d}(d) \tag{5.45}$$

where

n is the number density of the bubbles.

Note that $n = S_0$, that is the zeroth-moment of the distribution, is the bubble number density. The second-moment of the distribution, S_2, is related to the interfacial area density, a_i:

$$a_i = \int_0^\infty n\pi d^2 P(d)\mathrm{d}(d) = \pi S_2 \tag{5.46}$$

The third-moment of the distribution, S_3, is related to the volume fraction of the bubbles, ε:

$$\varepsilon = \int_0^\infty n\frac{\pi d^3}{6} P(d)\mathrm{d}(d) = \frac{\pi}{6} S_3 \tag{5.47}$$

The Sauter mean diameter (surface area weighted mean diameter) can be obtained:

$$d_{32} = \frac{S_3}{S_2} = \frac{6\varepsilon}{\pi}\frac{1}{S_2} \tag{5.48}$$

Alternatively, the particle number weighted mean diameter can be obtained from:

$$d_{30} = \left(\frac{S_3}{S_0}\right)^{\frac{1}{3}} = \left(\frac{6\varepsilon}{\pi S_0}\right)^{\frac{1}{3}} \tag{5.49}$$

The transport equation for S_γ is given by:

$$\frac{\partial S_\gamma}{\partial t} + \nabla \cdot (S_\gamma \mathbf{u}_G) = s_{br} + s_{cl} \tag{5.50}$$

where s_{br} and s_{cl} are the source terms for breakup and coalescence, respectively.

5.3.1.1 Breakup Model

Breakup will occur only if the bubble is larger than the critical diameter, d_{cr}, that is, the so-called maximum stable bubble diameter. Viscous breakup is found in laminar flows and in turbulent flows for bubbles smaller than the Kolmogorov length scale. Larger bubbles are subjected to inertial breakup. The Kolmogorov length scale, λ_k, is given by:

$$\lambda_k = \left(\frac{v^3}{\dot{E}}\right)^{\frac{1}{4}} \tag{5.51}$$

where

v is the continuous phase kinematic viscosity
$\dot{E}$ is the continuous phase dissipation rate of turbulent kinetic energy.

5.3.1.1.1 Viscous Breakup.   The breakup criterion follows from a balance between disruptive and restoring forces: the viscous stress and Laplace pressure, respectively. This force balance is expressed in terms of the capillary number, Ω:

$$\Omega = \frac{\eta_c d}{2\sigma}\dot{\gamma} \tag{5.52}$$

where

η_c is the dynamic viscosity of the continuous phase
σ is the surface tension coefficient
$\dot{\gamma}$ is the shear rate of the continuous phase.

The breakup criterion is given as $\Omega \geq \Omega_{cr}$. The critical diameter is therefore given by:

$$d_{cr} = \frac{2\sigma\Omega_{cr}}{\eta_c \dot{\gamma}} \tag{5.53}$$

5.3.1.1.2 Inertia Breakup.   Inertia breakup is found in turbulent flows for bubbles larger than the Kolmogorov length scale, L_k. The breakup criterion is formulated in terms of the dimensionless Weber number:

$$We = \frac{\rho_c \dot{E}^{2/3} d^{5/3}}{2\sigma} \tag{5.54}$$

Breakup occurs when $We \geq We_{cr}$, where We_{cr} is a function of Re_{cr}, the Reynolds number based on the critical bubble diameter:

$$d_{cr} = (1 + C_\alpha)\left(\frac{2\sigma We_{cr}}{\rho_c}\right)^{3/5} \dot{E}^{-2/5} \tag{5.55}$$

In the work of Yao and Morel [18], the value of $We_{cr} = 0.31$ was used.

5.3.1.2   *Coalescence Model*

When two bubbles collide they interact for a certain amount of time, forming a dumbbell. During this interaction, the film of the continuous phase between the bubbles will start to drain. If drainage proceeds down to a certain critical film thickness, δ_{cr}, within the provided interaction time, coalescence will take place; otherwise, the bubbles will separate. Coalescence is therefore a function of collision of two bubbles and their interaction time. Similar to the breakup models described above, we have a viscous coalescence regime in laminar flows and an inertia regime in turbulent flows. Full details of the models can be found in Lo and Rao [19]. For simplicity, only the drainage time and interaction time in viscous regimes are considered further here.

5.3.1.2.1   *Film Drainage.*   For the viscous collision, the coalescence probability, P_{cl}, is linked to the ratio of the interaction time, t_i, and the film drainage time, t_d:

$$P_{cl} = \exp\left(-\frac{t_d}{t_i}\right) \tag{5.56}$$

The interaction time scale is given by:

$$t_i = \frac{1}{\dot{\gamma}} \tag{5.57}$$

The drainage time, t_d, depends on the mobility of the interface. For bubbly flows, we can consider three different drainage modes, indicating the extent of the bubble surface mobility.

Drainage mode 2 (partially mobile interface due to heavy concentration of surface contaminant):

$$t_d = \frac{3}{2}\left(\frac{F_i d^2 \sqrt{\rho_d \eta_d}}{32\pi\sigma^2 \delta_{cr}}\right)^{2/3} \tag{5.58}$$

Drainage mode 3 (partially mobile interface due to moderate concentration of surface contaminant):

$$t_d = \frac{\pi\eta_d\sqrt{F_i}}{2\delta_{cr}}\left(\frac{d}{4\pi\sigma}\right)^{3/2} \tag{5.59}$$

Drainage mode 4 (fully mobile interface with no surface contaminant):

$$t_d = \frac{3\eta_c d}{4\sigma}\ln\left(\frac{d}{8\delta_{cr}}\right) \tag{5.60}$$

The interaction force during the collision is given by:

$$F_i = \frac{3\pi}{2}\eta_c\dot{\gamma}d^2 \tag{5.61}$$

and the critical film thickness is obtained from:

$$\delta_{cr} = \left(\frac{A_H d}{24\pi\sigma}\right)^{1/3} \tag{5.62}$$

where

$A_H = 5 \times 10^{-21}$ is the Hamaker constant

It can be seen that the modelling of breakup and coalescence is indeed very complex.

5.3.2   Case Study – Hibiki's Bubble Column

The experimental data sets published by Hibiki *et al.* [20] provide measured data on void fraction, interfacial area density, bubble diameter and gas and liquid velocities for bubbly flows in a vertical pipe. A wide range of gas and liquid flow rates were studied. These data sets therefore provide very valuable information for checking and validating the S_y model for gas-liquid flows.

The test section of Hibiki's experiments is a vertically placed circular pipe. The inner diameter of the pipe is 50.8 mm and the height is 3.06 m. Water and air at atmospheric conditions were used. Measurements were taken in two axial positions: $z/D = 6.0$ and $z/D = 53.5$, radial position varies from $r/R = 0$ to $r/R = 0.95$. Hibiki *et al.* [20] reported that the initial bubble diameter was about 3 mm at the pipe inlet. Dispersed air bubbles and water were injected into the pipe through the bottom plane.

5.3.2.1   Numerical Solution Method

All the numerical simulations are carried out with the commercially available CFD package STAR-CD 3.27. The flow domain is two-dimensional axisymmetric and it is sub-divided by 3000 grid cells (20 in the radial and 150 in the axial directions). In this work, the bubbly flow is treated as steady state. The second-moment S_2 model with breakup and coalescence was solved to capture the evolution of the bubble size. The convergence criteria are all residuals below 1×10^{-6}. A summary of the simulation parameters and physical properties used are presented in Table 5.2.

5.3.2.2   Results and Discussion

5.3.2.2.1   Applicability of the S_y Model.   Figure 5.3 presents the simulated bubble size distribution of test case 1. First of all, it is observed in Figure 5.3 that the bubble size ranges from 1.7 to 4.7 mm, which is in reasonable agreement with the experimental data of Hibiki *et al.* [20], where the bubble size varied from 2 to 5 mm in the test case. It is clearly shown in this figure that the bubble size increases with the increasing distance from the sparger. This

Table 5.2   *Case definition and parameters used*

Case	j_G (m/s)	j_L (m/s)	γ	Drainage mode
1	0.321	0.986	2	3
2	0.321	0.986	0	3
3	0.321	0.986	2	2
4	0.321	0.986	2	4
5	0.471	2.01	2	3
6	0.624	2.01	2	3

$C_D = 1.071$, $C_L = -0.288$, $C_{VM} = 0.5$
$\rho_L = 1000 \text{ kg/m}^3$, $\rho_G = 1.29 \text{ kg/m}^3$
$\sigma = 0.07\ 275 \text{ N/m}$
$\eta_{l,Lam} = 0.001 \text{ kg/(m s)}$, $\eta_{g,Lam} = 1.812 \times 10^{-5} \text{ kg/(m s)}$
$d_{B,ini} = 3 \text{ mm}$

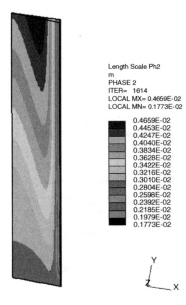

Figure 5.3   *Predicted Sauter mean diameter contour from the S_γ model*

finding is also consistent with the experimental results and is due to the coalescence occurring inside the column.

Furthermore, it is found in Figure 5.3 that bubble size near the wall is smaller. There are two reasons for this. As observed in Figure 5.4, the gas phase volume fraction profile exhibits

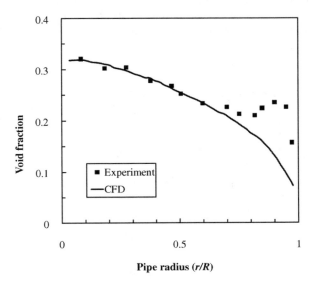

Figure 5.4   *Comparison of the simulated voidage radial distribution with the corresponding experimental data*

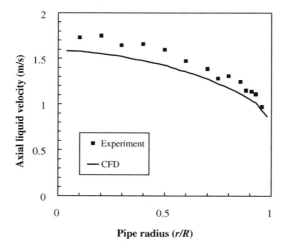

Figure 5.5 *Comparison of the simulated axial velocity profile with the corresponding experimental data*

a core peaking shape, the volume fraction near the wall is smaller, subsequently the collision probability is reduced, which weakens the coalescence; furthermore, as can be easily deduced from Figure 5.5, the gradient of the liquid phase velocity is very high near the wall, which leads to a high shear rate near the wall and consequently, breakup is enhanced in this region. A more quantitative comparison is provided in Figure 5.6, in which the predicted bubble size is compared with the experimental measurements in the radial direction and here, numerical results fit well with the experimental data. Based on Figures 5.3–5.6, it can be concluded that the S_γ model is capable of reasonably and accurately predicting the bubble size and bubble size distribution in gas-liquid bubbly flows.

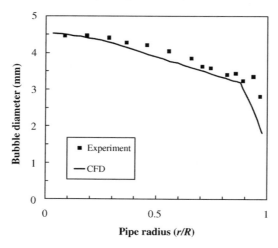

Figure 5.6 *Comparison of the simulated Sauter mean diameter radial distribution with the corresponding experimental data*

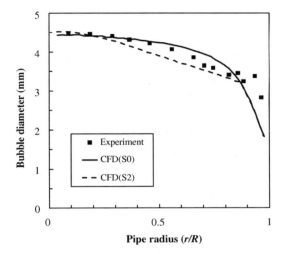

Figure 5.7 *Comparison of the simulated Sauter mean diameter obtained from different moments (γ) with the corresponding experimental data*

It should be pointed out that in all our simulations, good agreement was achieved between the numerical results and the experimental measurements for the velocity profiles and radial void distributions for both phases.

5.3.2.2.2   Sensitivity of the S_γ Model to the Moment γ.   As described in the Section 5.3.1, S_γ is a volumetric conserved γth-moment distribution. S_0 represents the bubble number density, n, and S_2 is the interfacial area density. In principle, bubble size, d_B, and/or interfacial area density, a_i, could be obtained from the zeroth-moment distribution (S_0) as well as those obtained from the second-moment distribution, S_2. Figure 5.7 displays a comparison of the bubble size predicted by both moments. It is seen in Figure 5.7 that the bubble size distribution predicted by the S_0 model slightly differs from that obtained from the S_2 model, which is also found in a comparison of the interfacial area density, a_i, as illustrated in Figure 5.8. Although the difference is very small, it seems that the S_0 model produces a better solution for the bubble size while S_2 provides a better solution for the interfacial area density. Nevertheless, Figures 5.7 and 5.8 suggest that the S_γ model is compatible for all moments in practice.

5.3.2.2.3   Sensitivity of the S_γ Model to the Drainage Mode.   The drainage mode provides the drainage time during the viscous collision. Currently, the selection of the drainage mode is based more on experience than the theoretical. Drainage modes 2, 3 and 4 are valid for bubbles and droplets, which qualitatively provides the mobility of the particle (bubble/droplet) surface. When the liquid is not contaminated, the bubble surface is clear and is fully mobile (drainage mode 4). However, when the liquid is slightly contaminated, the mobility of the bubble surface is reduced (drainage mode 3), and when the liquid is highly contaminated, there are small particles aggregating on the bubble surface, so the mobility of the bubble surface is greatly reduced (drainage mode 2).

With the help of test cases 1, 3 and 4, the effect of the drainage mode on the S_2 model is investigated. As found in Equation 5.56, the drainage time, t_d, determines the viscous

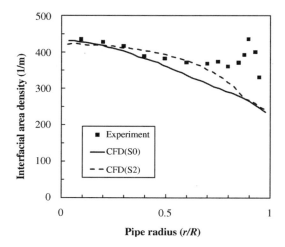

Figure 5.8   *Comparison of the simulated interfacial area density with the corresponding experimental data. Here, S_0 and S_2 models were used to simulate the interfacial area density transport*

collision probability, the drainage time $t_{d,4} < t_{d,3} < t_{d,2}$, therefore, the predicted bubble size with drainage mode 4 is bigger than those obtained from modes 2 and 3, as seen in Figure 5.9. As the interfacial area density is proportional to $1/d_B$, hence, in Figure 5.10, drainage mode 4 under-predicts the interfacial area density. Based on Figures 5.9 and 5.10, it

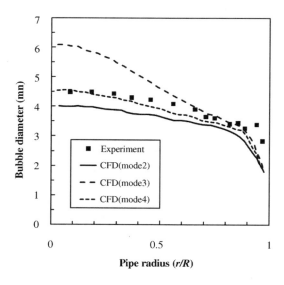

Figure 5.9   *Comparison of the simulated interfacial area density with the corresponding experimental data. Different drainage modes were studied*

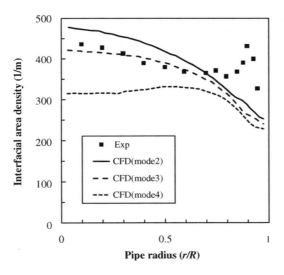

Figure 5.10 *Comparison of the predicted interfacial area density with the corresponding experimental data. Different drainage modes were studied*

could be concluded that for the Hibiki test cases [20], drainage mode 3 produces a better solution. So in the following simulations, drainage mode 3 is adopted.

5.3.2.2.4 Performance of the S_γ Model. Figure 5.11 shows the comparison of the radial bubble size distributions obtained from the numerical simulation with the S_γ model and the experiment measurements. It is seen here that, first of all, the trend of the predicted bubble size distribution agrees with the experimental profile. The difference appears near the wall, where the numerical results under-predicted the measurements for both cases. This is probably because the S_γ model gives a stronger breakup rate and relatively weaker

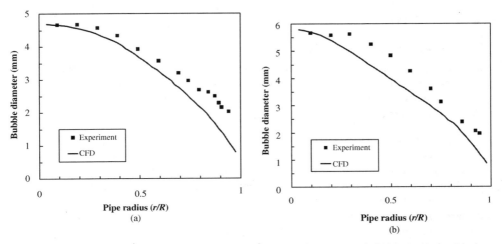

Figure 5.11 *Comparison of the simulated Sauter mean diameter radial distribution with the corresponding experimental data. (a) Case 5, (b) Case 6*

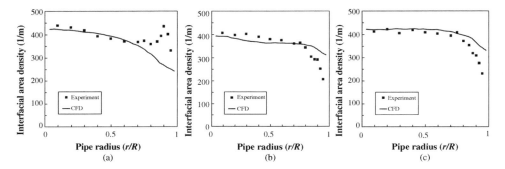

Figure 5.12 *Comparison of the simulated and experimental profiles of the interfacial area density. (a) Case 1, (b) Case 5, (c) Case 6*

coalescence in the near-wall region. Meanwhile, the complex local fluid structure and turbulent intensity also affect the breakup and coalescence models, especially in cases where the gas and liquid superficial velocities are higher. It is also easily observed in this figure that the increase in the gas phase superficial velocity directly leads to an increase in the bubble size in the core region. This is due to the fact that higher gas phase superficial velocity leads to a higher gas phase volume fraction and this in turn increases the coalescence. It should be pointed out that, in order to achieve a good agreement between the numerical results and the experimental measurements, different calibration coefficients for coalescence rate, F_{cl}, were used in case 1, $F_{cl} = 3 \times 10^{-3}$; in case 5, $F_{cl} = 5 \times 10^{-3}$; and in case 6, $F_{cl} = 3 \times 10^{-2}$. This in turn shows that the current model constants are valid for the dilute systems, and volume fraction correction should be accounted for in the near future.

Figure 5.12 shows the interfacial area density radial distributions obtained from simulation and experimental measurements. Similar to the findings in the comparison of the bubble size distribution, the difference between the numerical results and the measurements lies in the near-wall region. The main reason for this is the under-prediction of the bubble size in the near-wall region as the interfacial area density is related to the bubble size $a_i = 6\varepsilon_G/d$. With an under-prediction of the bubble size in cases 5 and 6, the interfacial area density, a_i, near the wall is over-predicted.

It is further found here, that with the increase in the gas phase superficial velocity, the interfacial area density does not dramatically change, as is found in the gas phase volume fraction. The main reason is that although there is an increase of the voidage, the bigger bubbles greatly reduce the interfacial area density, which leads to the results that the shape of the interfacial area density radial distribution differs from that of the void fraction.

5.3.2.3   Summary of Case Study

The numerical results were compared with the available experimental data of Hibiki *et al.* [20]. Good agreement was achieved for the phase axial velocity and radial void fraction for all tested cases. It was found that the second-moment S_y model is capable of predicting a bubble size and distribution with reasonable accuracy, even in a high void fraction. Except in the near-wall region, simulated bubble size and therefore the interfacial area density fit well with the experimental measurements. This can be attributed to the fact that in the near-wall region, the liquid phase shear rate is very high, which leads to a breakup rate that is too high in the current model.

It was observed that the predicted bubble size and interfacial area density obtained from the S_0 and S_2 models are more or less the same, which indicates that the numerical results are independent of the distribution moment, γ. It was further found that the drainage mode greatly affects the bubble size: the increase of the mobility of the bubble surface enhances the coalescence in the current model and leads to an over-prediction of the bubble size in the pipe core. The bubble size increases with an increase in the gas phase superficial velocity, while the interfacial area density varies less, as the interfacial area density is a combined function of the bubble size and local gas hold-up.

5.4 Gassed Stirred Vessels

Stirred vessels are commonly found across industry, performing a wide range of duties, for example: the blending and mixing of various chemicals and compounds to make household products such as detergent, hand cream and toothpaste. Because of the wide variety of applications and duties, mixing vessels come with many different types of impellers and stirrers. Very often the vessels have to cope with complex multiphase flow systems. For example, vapour bubbles could be generated by evaporation of the liquid or by injection of gas bubbles via spargers or dip pipes into a slurry. Adequate and continuous stirring of the multiphase mixture is necessary in order to keep the solid particles in the slurry well suspended and the gas bubbles uniformly distributed. Heat and mass transfers and chemical reactions can take place in the vessel simultaneously, adding further complications to the analyses.

When designing a new mixing vessel, the engineers need to ensure that adequate shaft power is available to perform the range of mixing duties required. More importantly, for the efficient and economical running of the plant, engineers need to know and be able to set the optimum shaft power for a given duty. Some information based on experimental data can be found in the literature. However, it is not easy to find an empirical correlation that is suitable for both the impeller type and the fluid system under consideration. In these cases, computational methods based on CFD, which account for the actual geometric information, the fluid physical properties and operating conditions, can be a very useful tool to have.

5.4.1 Impeller Model

In modelling a stirred vessel, we have the extra complication of the impellers rotating inside a stationary vessel with fixed baffles. Various methods have been developed to model the rotating impellers:

i. Modelling the effects of the impellers by momentum sources.
ii. 'Cut out' the impellers and set measured data, such as velocity and turbulence parameters, on the impeller boundary.
iii. Divide the vessel into two domains. The inner domain contains the impellers and the outer domain contains the baffles and the vessel walls. Equations for the inner domain are transformed and solved in rotating coordinates. Equations for the outer domain are solved in the fixed coordinates. This approach is the multiple reference frame model, which is discussed in Section 5.4.2.

iv. Similar to (iii) above, the inner region is rotated during the computation. The shape and the rotation of the impellers are, therefore, represented exactly. Because the grid of the inner region is made to rotate and slide along the interface with the outer region, this method is often called the 'sliding-grid' method.

5.4.2   Multiple Reference Frame

The sliding-mesh method discussed in (iv) above is computational intensive. An alternative method has been developed in which the flow domain is split into several zones. For the inner zone with the impeller, a rotational reference frame is defined in which the impeller is at rest, and the corresponding momentum equation is solved, to which centrifugal and Coriolis forces are added. For the outer zone consisting of the tank wall and baffle, a fixed inertial reference frame is defined in which the tank walls and baffles are at rest, and the standard momentum equation is solved. At the interface between the two zones, the flow is assumed to be steady and coupling of the two solutions is achieved by matching velocities locally via velocity transformations from one frame to the other. It is a steady-state method (as the tank components are stationary at the corresponding frame of reference), and therefore avoids the large computational demands associated with the sliding-grid method.

The multiple reference frame model is an approximation, as it solves the flow for a fixed position of the impeller with respect to the baffles. However, it can provide a reasonable model of the time-averaged flow for mixing tanks with small impeller to tank diameter ratios, such as $D/T = 1/3$, as the impeller–baffle interactions are then relatively weak. Moreover, it speeds up the computations by an order of magnitude compared with the sliding-grid methods.

5.4.3   Multiple Impellers

In larger equipment a single impeller will be insufficient to properly stir and mix the multiphase flow. Therefore, these reactors are equipped with multiple impellers, for example on a single shaft. Here, the strategy discussed above can be repeated: a different zone is assigned to each impeller. Again, sliding grids or multiple reference frames can be used. This type of simulation has the tendency to use a large number of grid cells: each impeller needs a sufficient number of grid cells and the space in between the impellers also needs to be covered accurately. In order to speed up the simulations, researchers have investigated the possibilities of simulating only a part of the vessel, for example a pie-part of 60° for a six blade Rushton turbine (see Figure 5.13) rather than the total 360°. The side boundaries are treated as periodic boundaries. It was found that the difference with a full 360° simulation was marginal, but, obviously, the speed was much higher. Furthermore, it was found that, due to the weak coupling between the impellers and the baffles, it does not matter much whether four or six baffles are used. This finding allows for taking a pie-piece of 60° or 90° (depending on the number of blades per impeller).

Gunyol and Mudde [21] simulated a large-scale fermenter (liquid volume $22 \, m^3$, diameter 2.09 m and liquid height 6.55 m). The reactor had been experimentally investigated by Vrabel *et al.* [22]. It is equipped with four Rushton turbines on a single shaft. Gunyol used the commercial package Fluent to perform the simulations. A two-phase k–ε model was used to deal with the turbulence. Only 60° were simulated using the multiple reference frame method. The set-up is shown in Figure 5.13. The computational mesh had 325×10^3 cells. The authors report that the drag force was the dominating force.

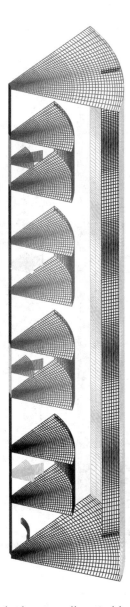

Figure 5.13 *60° of a four-impeller, six-blade gassed stirred tank*

Simulations were performed for various fixed bubble sizes, see Figure 5.14. Obviously, the predicted gas fraction depends strongly on the bubble size chosen. The results for the average gas fraction [ε_g (%)] of 2.6, 3.7 and 6.8% for d_b values of 3, 2 and 1 mm, respectively, were compared with the experimental data of 4.7%. Agreement could only be found for bubble sizes that were too small. Therefore, various drag formulations were tested and it was found that the drag force needed to be corrected for the effects of the highly turbulent flow.

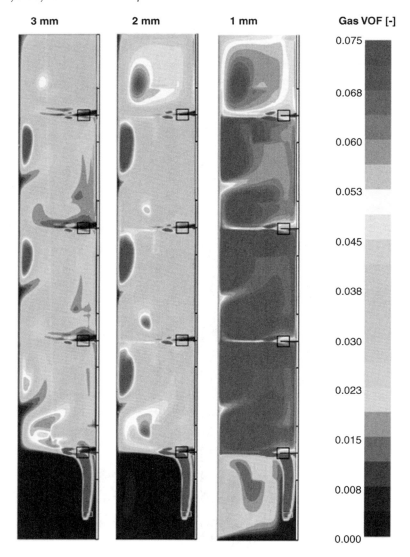

Figure 5.14 *Comparison of gas fractions using different bubble sizes [21] (Reprinted from Gunyol, O. and Mudde, R.F., Computational study of hydrodynamics of a standard stirred tank reactor and a large–scale multi–impeller fermenter, Int. J. Multiscale Comput. Eng., 7, 559–576. Copyright (2009) with permission from Begell House, Inc.)*

The correction of Brucato *et al.* [23] can be used:

$$\frac{C_D - C_{D,0}}{C_{D,0}} = K\left(\frac{d_b}{\lambda_K}\right) \tag{5.63}$$

where

$C_{D,0}$ is the original drag coefficient
d_b is the bubble diameter
λ_K is the Kolmogorov scale.

The correction increases the drag coefficient in line with the reported lower slip velocity of bubbles in turbulent flows at high Reynolds numbers. For the value of K, a value of 6.5×10^{-6}, as proposed by Lane *et al.* [24], was taken. A comparison with the data from Vrabel *et al.* [22] showed that this correction was sufficient at the lower stirrer speeds (70 rpm), but that the gas fraction was under-predicted for higher values (115 rpm). The simulations were performed using a fixed bubble size of 3 mm.

A comparison of the gas fraction for two different drag formulations is given in Figure 5.15. The average gas fraction in the fermenter according to the experiments was

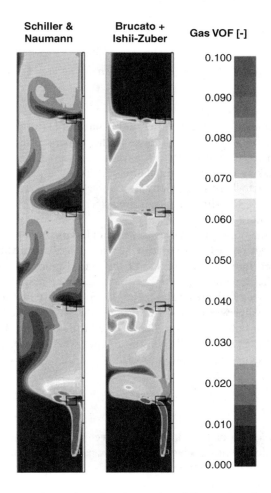

Figure 5.15 *Comparison of gas fractions using different drag force closures [21]*

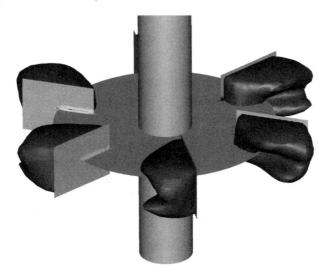

Figure 5.16 *Gas pockets formed behind the impeller blades*

4.0%. The CFD prediction for the Schiller and Naumann [1] drag were 2.2%, with the Brucato correction 4.1% was obtained. However, the higher stirring case showed that the correction is not universal: the correction gave an over-prediction of the gas fraction. Moreover, these workers performed simulations including a population balance for the bubble size. According to these simulations, the bubble size in the reactor could range up to 7–8 mm. At these sizes, the slip velocity is significantly higher than for small bubbles, reducing the gas fraction.

A correct prediction of the gas cavities formed behind the impeller blades is important for finding the right two-phase flow regime in the reactor, as these are the regions where the bubbles are dispersed into the bulk region of the tank. Figure 5.16 shows the clinging cavities present at the impeller, when the aeration rate is increased at high stirring rate. When the impeller speed is decreased at the high aeration rate investigated (i.e. 70 rpm, 0.0263 m³/s), the impeller was not able to disperse the gas into the reactor, resulting in behaviour similar to a bubble column, hence instabilities occurred in the steady-state simulations. This is in agreement with the experiments, where the fermenter is reported to be flooded under these operating conditions [22].

5.5 Summary

CFD can be a powerful tool to help process development and adjustments for better performance. In multiphase flows, several options for performing simulations are available. The choice is usually dictated by the required details and the time allowed. For industrial dispersed flows as found in all types of reactors, at present the Euler–Euler or two-fluid approach is the one used most. It allows assessing the multiphase flow field and transport characteristics in terms of continuous fields. Individual bubbles or particles are not tracked, as that is computationally very expensive. Instead, average fields are used, which represent the velocities (in time and space) and the volume fractions of the various phases. Usually these flows are turbulent and a turbulence closure is required. There are again several options, but the k–ε model, which concentrates on the turbulent kinetic energy and its dissipation is the most widely used. One of the reasons is computational speed: this model introduces only two additional balance equations.

More sophisticated models, such the Reynolds stress model, can deal with anisotropy in the flow, as it solves for the turbulent stress tensor rather than only the kinetic energy. The drawback is obviously a larger complexity and a slower computation. Moreover, the closure of turbulence terms is still a weaker point in multiphase flows. The same holds for the interaction forces at higher volume fractions. This means that aiming for the most advanced modelling will not necessarily render better simulations: the more advanced the model, the more terms need to be closed. Furthermore, some of the 'standard' forces, such as the lift force or the turbulent dispersion force, are not known very accurately. Therefore, one might wonder whether more sophisticated models that use the same forces could really do better.

Nevertheless, CFD is an important step forward and will become more reliable and faster in the near future. It has the main advantage that it is generic compared with engineering approaches and can deal with hostile conditions (high pressure, temperature). Furthermore, it can provide quantities that are experimentally hard to find, especially in industrial equipment. CFD will also help in keeping experimental costs down, as a good simulation can bypass the need for experiments that are particularly costly at the pilot plant scale. However, CFD has not reached the status that a push on the button is sufficient: it requires skilled people to run the simulations and state-of-the-art software on high-end computers. If these are present, CFD allows changes in geometry and comparison of various options. As a tool of the skilled engineer, it offers flexibility and 'desk-top experiments' that are hard to realise in a short time in practise.

Questions

5.1 A bubbly mixture flows through a vertical pipe (4 inches diameter). The bubbles size is 2 mm. The gas has a density of 6 kg/m^3, whereas the liquid density is 900 kg/m^3. The surface tension of the gas-liquid is 55 10^{-3} N/m, the liquid viscosity is 5 mPa s. The liquid superficial velocity is 2 m/s; the gas fraction is 5%. A CFD simulation of the flow needs to be set up.

a. Compute the slip velocity of the bubbles.
b. Discuss whether or not a mixture modelling is possible.
c. If yes, would you opt for a homogeneous model or for the algebraic slip model?

5.2 A 30 cm diameter air-water bubble column operates at a gas fraction of 10%. The bubbles have an average size of 5 mm. The upward velocity of the liquid in the centre of the column is of the order of 0.6 m/s. At a radial position of $r = 0.7\ R$, the average velocity is zero, whereas close to the wall the downward velocity of the liquid is about -0.45 m/s. The magnitude of the turbulent velocity fluctuations is about 0.2 cm/s. The gas fraction in the central region is approximately 15% and it drops towards the wall to about 5%.

a. Estimate the order of magnitude of the forces action on the bubbles: gravity, buoyancy, drag, lift and turbulence diffusion.
b. Which of these forces would you include in your simulation?

References

[1] Schiller, L. and Naumann, A. (1933) A drag coefficient correlation. *VDI Zeits.*, **77**, 318–320.
[2] Wen, C.Y. and Yu, Y.H. (1966) A generalised method for predicting minimum fluidization velocity. *AIChE J.*, **12**, 610–612.
[3] Ishii, M. and Zuber, N. (1979) Drag coefficient and relative velocity in bubbly, droplet or particulate flows. *AIChE J.*, **25**, 843–855.
[4] Wang, D.M. (1994) Modelling of bubbly flow in a sudden pipe expansion. Technical Report II-34, Project BE-4098, Brite/EuRam.
[5] Auton, T.R., Hunt, J.C.R. and Prud'homme, M. (1988) The force exerted on a body in inviscid unsteady non-uniform rotational flow. *J. Fluid Mech.*, **197**, 241–257.
[6] Tomiyama, A., Tarnai, H., Zun, I. and Hosokama, S. (2002) Transverse migration of single bubbles in simple shear flow. *Chem. Eng. Sci.*, **57**, 1849–1858.
[7] Antal, S.P., Lahey, R.T. and Flaherty, J.E. (1991) Analysis of phase distribution in fully developed laminar bubbly two-phase flow. *Int. J. Multiphase Flow*, **17**, 635–652.
[8] Rampure, M.R., Buwa, V.V. and Ranade, V.V. (2003) Modelling of gas-liquid/gas-liquid-solid flows in bubble columns: Experiments and CFD simulations. *Can. J. Chem. Eng.*, **81**, 692–706.
[9] Richardson, J.F. and Zaki, W.N. (1954) Sedimentation and fluidisation: Part 1. *Trans. I. Chem. Eng.*, **32**, 35–53.
[10] Peebles, F.N. and Garber, H.J. (1953) Studies on the motion of gas bubbles in liquid. *Chem. Eng. Proc.*, **49**, 88–97.
[11] Wallis, G.B. (1974) The terminal speed of single drops or bubbles in an infinite medium. *Int. J. Multiphase Flow*, **1**, 491–511.
[12] Kolev, N. (2003) *Multiphase Flow Dynamics, Parts 1 & 2*, Springer, p. 95.
[13] (a) Zuber, N. and Findlay, J.A. (1965) Average volumetric concentration in two-phase flow systems. *J. Heat Transf.*, **87**, 453–468; (b) Ishii, M. (1977) One-dimensional drift-flux model and constitutive equations for relative motion between phases in various two-phase flow regimes. Technical report ANL-77–47, Argonne National Laboratory, USA.
[14] Ishii, M. (1975) *Thermofluid Dynamics Theory of Multiphase Flow*, Eyrolles, Paris.
[15] Pericleous, K.A. and Drake, S.N. (1986) An algebraic slip model of PHOENICS for multiphase applications, in *Numerical Simulation of Fluid Flow and Heat/Mass Transfer Processes* (eds N. C. Markatos, D.G. Tatchell, M. Crossand N. Rhodes), Springer-Verlag, Berlin.
[16] Johansen, S.T., Anderson, N.M. and De Silva, S.R. (1990) A two-phase model for particle local equilibrium applied to air classification of powders. *Powder Technol.*, **63**, 121–132.
[17] Manninen, M., Taivassalo, V. and Kallio, S. (1996) *On the Mixture Model for Multiphase Flow, VTT Publication 288*, VTT, Espoo.
[18] Yao, W. and Morel, C. (2004) Volumetric interfacial area prediction in upward bubbly two-phase flow. *Int. J. Heat Mass Transfer*, **47**, 307–328.
[19] Lo, S. and Rao, P. (2007) Modelling of droplet breakup and coalescence in an oil-water pipeline. Paper presented at the 6th International Conference on Multiphase Flow, ICMF 2007, Leipzig, Germany, July 9–13, 2007.
[20] Hibiki, T., Ishii, M. and Xiao, Z. (2001) Axial interfacial area transport of vertical bubbly flows. *Int. J. Heat Mass Transfer*, **44**, 1869–1888.

[21] Gunyol, O. and Mudde, R.F. (2009) Computational study of hydrodynamics of a standard stirred tank reactor and a large–scale multi–impeller fermenter. *Int. J. Multiscale Comput. Eng.*, **7**, 559–576.

[22] Vrabel, P., van der Lans, R.G.J.M., Cui, Y.Q. and Luyben, K.C.A.M. (1999) Compartment model approach: Mixing in large scale aerated reactors with multiple impellers. *Chem. Eng. Res. Des.*, **77**, 291–302.

[23] Brucato, A., Grisafi, F. and Montante, G. (1998) Particle drag coefficient in turbulent fluids. *Chem. Eng. Sci.*, **45**, 3295–3314.

[24] Lane, G.L., Schwarz, M.P. and Evans, G.M. (2000) Modelling of the interaction between gas and liquid in stirred vessels. Paper presented at Proceedings of the 10th European Conference on Mixing, Delft, The Netherlands, pp. 197–204.

6

Mesoscale Modelling Using the Lattice Boltzmann Method

6.1 Introduction

This chapter introduces numerical modelling using the mesoscale lattice Boltzmann method (LBM). In recent years, along with extensive applications of conventional computational fluids dynamics (CFD) to the study of fluids flow and transport phenomena, the lattice

Hydrodynamics of Gas-Liquid Reactors: Normal Operation and Upset Conditions, First Edition.
B. J. Azzopardi, R. F. Mudde, S. Lo, H. Morvan, Y. Yan and D. Zhao.
© 2011 John Wiley & Sons, Ltd. Published 2011 by John Wiley & Sons, Ltd.

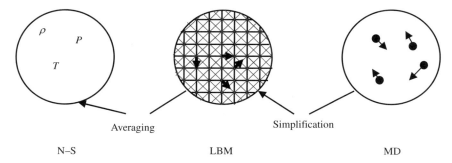

Figure 6.1 *Micro-, meso and macroscale numerical modelling*

Boltzmann method has become an established numerical scheme for simulating single-phase and multiphase fluid flows. The scheme is particularly successful in fluid flow applications involving interfacial dynamics and complex boundaries [1]. The key idea behind the LBM, as shown in Figure 6.1, is to recover the correct macroscopic motion of fluids by incorporating the complicated physics of problems into simplified microscopic models or mesoscale kinetic equations. In this method, kinetic equations of particle velocity distribution functions are first solved; macroscopic quantities are then obtained by evaluating hydrodynamic moments of the distribution function. This intrinsic feature enables the LBM to model phase segregation and interfacial dynamics of multiphase flow, which are difficult to be handled by conventional CFD or the molecular dynamics (MD) method.

The LBM originates from the lattice gas automata (LGA) method, which can be considered as a simplified fictitious molecular dynamics model in which space, time and particle velocities are all discrete. Each lattice node is connected to its neighbours by six lattice velocities, using for example the hexagonal FHP model. There can be either 0 or 1 particles at a lattice node moving in a lattice direction. After a time interval, each particle will move to the neighbouring node in its direction, this process is called the propagation or streaming step. When there is more than one particle arriving at the same node from different directions, they collide and change their directions according to a set of collision rules. Suitable collision rules should conserve the particle number (mass), momentum and energy before and after the collision. However, it was found that LGA suffers from several native defects: lack of Galilean invariance, statistical noise, exponential complexity for three-dimensional lattices and so on. Wolfram [2] presented a complete classification and analysis of the 265 rules for two states, two-neighbour automata.

The main motivation for the transition from LGA to LBM was the desire to remove the statistical noise by replacing the Boolean particle number in a lattice direction with its ensemble average, the so-called density distribution function. Accompanying this replacement, the discrete collision rule is also replaced by a continuous function known as the collision operator. In the LBM development, an important simplification is to approximate the collision operator with the Bhatnagar–Gross–Krook (BGK) relaxation term. This lattice BGK (LBGK) model makes simulations more efficient and allows flexibility of the transport coefficients. Furthermore, it has been shown that the LBM scheme can also be considered as a special discretised form of the continuous Boltzmann equation. Through a Chapman–Enskog analysis, one can recover the governing continuity and Navier–Stokes equations from the

LBM algorithm. In addition, the pressure field is also directly available from the density distributions and hence there is no extra Poisson equation to be solved, as there is in traditional CFD methods.

The numerical simulation of multiphase and multicomponent fluid flows plays an important role in many areas of applied science and engineering, including oil-water flow in porous media, boiling fluids, flows of an immiscible mixture and so on. It is a challenging subject because, in addition to the usual difficulties associated with single-phase motion, it also requires the simulation of some essential physics in multiphase and component flows, such as phase segregation and interfacial dynamics.

Since the foundation of the first lattice Boltzmann model for multiphase flow in 1991, a number of multiphase and multicomponent models based on LBM have been introduced [3]. To date, the LBM has demonstrated a significant potential and broad applicability with many computational advantages, including parallelisation of the algorithm and the simplicity of programming [1]. Based on the two-component lattice gas method, Gunstensen *et al.* [4] proposed a multicomponent model; Shan and Chen [5] have initiated a pseudo-potential model that has the capacity of simulating multiphase and multicomponent immiscible fluid flow of mean-field interactions, and the model was later analysed and evaluated by Shan and coworkers [6] and by Martys and Douglas [7].

Meanwhile, Swift *et al.* [8] proposed the free-energy model, in which the Van der Waals formulation of quasi-local thermodynamics for a two-component fluid in thermodynamic equilibrium state is built. Later, a new LBM model, termed the index model was suggested by He *et al.* [9], which uses an index function to track the interface of a multi-phase flow with large density ratios. To overcome the difficulties of dealing with a large density ratio in two-phase flow, Inamuro *et al.* [10] developed an LBM model based on the projection method to predict the behaviour of incompressible bubbles/particles in bulk liquid. The method calculates two distribution functions of the particle velocity to track the interface and to predict velocities; the corrected velocity field satisfying the continuity equation is obtained by solving the Poisson equation.

Based on the work above, Yan and Zu [11] developed the LBM model for incompressible two-phase (with large density ratio) flow on a partial wetting surface, and a variety of applications of LBM modelling have been attempted at the University ofNottingham since 2006.

In the following sections, we will first introduce LBM simulation applied to single-phase flow and heat transfer over a rotating cylinder, then focus on the application of LBM modelling for two-phase flow.

6.2 Lattice Boltzmann Method and the Advantages

The lattice Boltzmann method (LBM) simplified Boltzmann's original conceptual view by reducing the number of possible particle spatial positions and microscopic momenta from a continuum to just a few and similarly discretising time into distinct steps. It thus offers potentially great advantages over conventional methods for simulating fluids flow and heat transfer typically for micro-electro mechanical systems (MEMS) flows, surface tension dominated flow and two-phase mixing. In the LBM, particle positions are confined to the nodes of the lattice. Variations in momenta that could have been due to a continuum of

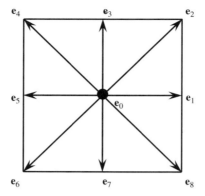

Figure 6.2 *D2Q9 model*

velocity directions and magnitudes and varying particle mass are reduced to, for example, for the simple two-dimensional model, eight directions, three magnitudes and a single particle mass. Figure 6.2 shows the Cartesian lattice where velocities $\mathbf{e}_\alpha$ (where $\alpha = 0, 1, \ldots, 8$) indicate a direction index and $\mathbf{e}_0$ denotes particles at rest. This model is known as D2Q9 as it is two-dimensional and contains nine velocities. This LBM classification scheme was proposed by Qian *et al.* [12] and is in widespread use. Similarly, three-dimensional models known as D3Q15 or D3Q19 are often used. Figure 6.3 shows the Cartesian lattice and velocities $\mathbf{e}_\alpha$ (where $\alpha = 0, 1, \ldots, 14$).

Unlike traditional CFD methods, which are mainly based on the direct numerical approximation to the macroscopic Navier–Stokes equation, the LBM recovers the correct macroscopic motion of the fluid, such as:

$$\rho = \sum_{\alpha=0}^{n} f_\alpha \qquad (6.1)$$

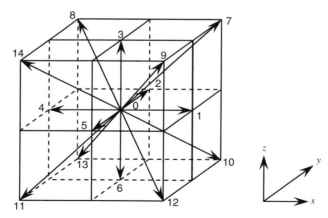

Figure 6.3 *D3Q15 model*

$$\mathbf{u} = \frac{1}{\rho}\sum f_\alpha \mathbf{e}_\alpha \tag{6.2}$$

by incorporating the complicated physics of the problem into simplified microscopic models or mesoscopic kinetic equations. In this method, one solves a kinetic equation for the particle velocity distribution function, such as Equation 6.3:

$$f_\alpha(\mathbf{x} + \mathbf{e}_\alpha \delta t,\; t + \delta t) = f_\alpha(\mathbf{x}, t) - \frac{f_\alpha(\mathbf{x}, t) - f_\alpha^{(eq)}(\mathbf{x}, t)}{\tau} \tag{6.3}$$

where

$f_\alpha(\mathbf{x}, t)$ is the particle distribution function at position $\mathbf{x}$ at time t

$f_\alpha^{(eq)}(\mathbf{x}, t)$ is the corresponding equilibrium particle distribution function

τ is the relaxation time.

Typically, $f_\alpha(\mathbf{x} + \mathbf{e}_\alpha \delta t,\; t + \delta t) = f_\alpha(\mathbf{x}, t)$ is the streaming part and $[f_\alpha(\mathbf{x}, t) - f_\alpha^{(eq)}(\mathbf{x}, t)]/\tau$ is the collision term. Although the streaming and collision can be combined into a single statement as they are in Equation 6.1, collision and streaming steps must be separated if solid boundaries are present, because the bounce back boundary condition is a separate collision.

After solving the particle distribution equations, the macroscopic quantities can then be obtained by evaluating the hydrodynamic moments of the distribution function. This intrinsic feature enables it to model phase segregation and interfacial dynamics in multiphase and component flows, which are difficult to deal with in traditional CFD methods, by incorporating intermolecular interactions at the microscopic or mesoscopic level. Meanwhile, the LBM has demonstrated a significant potential and broad applicability with numerous computational advantages, including the parallelisation of the algorithm and the simplicity of programming [1, 13–15]. In the following sections, we will first introduce the LBM simulation of single-phase flow and heat transfer, then the LBM models for multiphase flow and some examples.

6.3   Numerical Simulation of Single-Phase Flow and Heat Transfer

In this section we introduce numerical simulation of single-phase flow and heat transfer using the D2Q9 LBM model. It is well known that fluid flow around a rotating isothermal cylinder (or between two rotating cylinders) is a common occurrence in a variety of industrial processes. The operations can range from the contact cylinder dryers in the chemical process, food-processing, paper making and the textile industries to the cylindrical cooling devices in the glass and plastics industries. Although the configurations are normally rather simple, as shown in Figure 6.4, the flow past the rotating cylinder and the heat transfer between the fluid and the cylinder are very complex. Factors, such as the viscous wakes including dissipation, diffusion and cancellation of vortices, the effects of cylinder rotation on the production of lift force and moment, the local heat transfer around the cylindrical surface and the evolution of surrounding temperature field with time, greatly increase the difficulties of mathematical formulation and numerical simulation. In the LBM modelling, moving and curved boundaries in particular need to be considered, in addition to heat transfer [16].

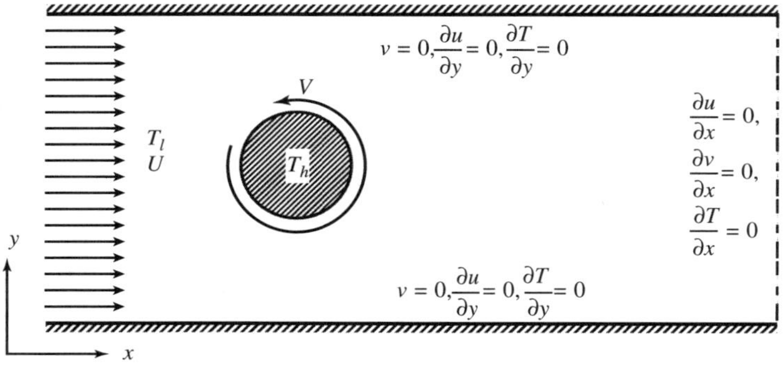

Figure 6.4   *The flow field set-up (Reprinted from Yan, Y. Y., Zu, Y. Q., Numerical simulation of heat transfer and fluid flow past a rotating isothermal cylinder – A LBM approachInt. J. Heat and Mass Transfer, **51** (9–10), 2519–2536. Copyright (2006) with permission from Elsevier.)*

6.3.1   LBM Model

A D2Q9 LBM model (as shown in Figure 6.2) with multiple distribution functions (for velocity and temperature) in the format of Equation 6.3 is employed to simulate incompressible viscous flows. The equilibrium distribution function $f_\alpha^{(eq)}(\mathbf{x}, t)$ in Equation 6.3 is expressed as:

$$f_\alpha^{(eq)}(\mathbf{x}, t) = \omega_\alpha \rho \left[1 + \frac{3}{c^2}(\mathbf{e}_\alpha \cdot \mathbf{u}) + \frac{9}{2c^2}(\mathbf{e}_\alpha \cdot \mathbf{u})^2 - \frac{3}{2c^2}\mathbf{u}^2 \right] \tag{6.4}$$

where

$\mathbf{u}$ and ρ are the macroscopic velocity and density, respectively
c, the streaming speed, is defined as $c \equiv \delta_x/\delta t$
δ_x is the streaming length
ω_α is the weighting coefficient.

In the D2Q9 model,

$$\mathbf{e}_\alpha = \begin{cases} 0 & \alpha = 0 \\ (\cos[(\alpha-1)\pi/4], \sin[(\alpha-1)\pi/4])c & \alpha = 1,3,5,7 \\ \sqrt{2}(\cos[(\alpha-1)\pi/4], \sin[(\alpha-1)\pi/4])c & \alpha = 2,4,6,8 \end{cases} \tag{6.5}$$

$$\omega_\alpha = \begin{cases} 4/9, & \alpha = 0 \\ 1/9, & \alpha = 1,3,5,7 \\ 1/36, & \alpha = 2,4,6,8 \end{cases} \tag{6.6}$$

The lattice Boltzmann equation of the temperature field can be given by:

$$g_\alpha(\mathbf{x} + \mathbf{e}_\alpha \delta t, t + \delta t) - g_\alpha(\mathbf{x}, t) = -\frac{1}{\tau_c}\left[g_\alpha(\mathbf{x}, t) - g_\alpha^{(eq)}(\mathbf{x}, t) \right] \tag{6.7}$$

where

τ_c is the dimensionless relaxation time

$g_\alpha(\mathbf{x}, t)$ is the temperature distribution function in the αth direction

$g_\alpha^{(eq)}(\mathbf{x}, t)$ is the corresponding equilibrium distribution function.

This last equilibrium distribution function can be expressed [17, 18] as:

$$g_\alpha^{(eq)}(\mathbf{x}, t) = \omega_\alpha T \left[1 + \frac{3}{c^2} \mathbf{e}_\alpha \cdot \mathbf{u}\right] \tag{6.8}$$

where T is the fluid temperature and can be evaluated from:

$$T = \sum_\alpha g_\alpha \tag{6.9}$$

In addition, it has been proved that the following macroscopic equation of temperature can be obtained from the Chapman–Enskog analysis as [17]

$$\frac{\partial T}{\partial t} + \nabla \cdot (\mathbf{u}T) = \gamma \nabla^2 T \tag{6.10}$$

where

γ is the diffusivity coefficient, which is represented as $\gamma = (\tau_c - 0.5)/c_s^2 \delta_t$.

In the LBM approach, both Equations 6.3 and 6.7 are computed in two steps, namely, the collision and streaming steps. In the collision steps,

$$\tilde{f}_\alpha(\mathbf{x}, t) = f_\alpha(\mathbf{x}, t) - \frac{1}{\tau_v}\left[f_\alpha(\mathbf{x}, t) - f_\alpha^{(eq)}(\mathbf{x}, t)\right] \tag{6.11a}$$

$$\tilde{g}_\alpha(\mathbf{x}, t) = g_\alpha(\mathbf{x}, t) - \frac{1}{\tau_c}\left[g_\alpha(\mathbf{x}, t) - g_\alpha^{(eq)}(\mathbf{x}, t)\right] \tag{6.11b}$$

In the streaming steps:

$$f_\alpha(\mathbf{x} + \mathbf{e}_\alpha \delta t, t + \delta t) = \tilde{f}_\alpha(\mathbf{x}, t) \tag{6.12a}$$

$$g_\alpha(\mathbf{x} + \mathbf{e}_\alpha \delta t, t + \delta t) = \tilde{g}_\alpha(\mathbf{x}, t) \tag{6.12b}$$

where

$\tilde{f}_\alpha$ and $\tilde{g}_\alpha$ denote the post-collision states of the distribution function of density and distribution function of temperature, respectively.

Obviously, the collision steps are completely local and the streaming processes take little computational effort at each time step, at which the distribution functions of a lattice are only affected by its neighbouring ones. This type of inherent spatial locality of the updating rules makes the LBM perfect for parallel computation.

6.3.2 Treatment for a Curved Boundary

Figure 6.5 shows an arbitrary curved wall (the dashed line) separating a solid region from fluid; where the black solid circles (●) denote intersections of the boundary with various lattice links ($\mathbf{x}_w$), the open circles (○) represent the boundary nodes in the fluid region ($\mathbf{x}_f$), and the grey solid circles (●) indicate those in the solid region ($\mathbf{x}_b$). Obviously, both $\tilde{f}_\alpha(\mathbf{x}_b, t)$ and $\tilde{g}_\alpha(\mathbf{x}_b, t)$ are needed to perform the streaming steps on fluid nodes $\mathbf{x}_f$.

The fraction of an intersected link in the fluid region, Δ, is defined as:

$$\Delta = \frac{|\mathbf{x}_f - \mathbf{x}_w|}{|\mathbf{x}_f - \mathbf{x}_b|} \tag{6.13}$$

It is well known that a bounce-back boundary condition satisfies the no-slip velocity boundary condition with a second-order accuracy if $\Delta = {}^1\!/_2$, so that a thermal boundary condition can be implemented in a similar way to achieve the second-order accuracy. This type of method can certainly be used to treat simple boundaries of straight lines that are in parallel with the lattice grid. However, for a curved boundary, simply placing the boundary at $\Delta = {}^1\!/_2$ will factitiously change the geometry of the boundary and degrade the accuracy of the velocity and temperature fields. Therefore, a new method is introduced to deal with both velocity and temperature boundaries with second-order accuracy. For treating a velocity field with curved boundaries, the method is based on the work by Mei *et al.* [19]; for handling temperature fields with the curved boundaries, an extrapolation method of second-order accuracy has been developed by Yan and Zu [16]. The total force acting on the solid wall by the fluid can be obtained by summing the contribution over all boundary nodes $\mathbf{x}_b$ belonging to the body, namely:

$$\mathbf{F} = \sum_{\text{all } xb} \sum_{\alpha \neq 0} \mathbf{e}_\alpha [\tilde{f}_\alpha(\mathbf{x}_b, t) + \tilde{f}_{\bar{\alpha}}(\mathbf{x}_b + \mathbf{e}_{\bar{\alpha}}\delta_t, t)][1 - \phi(\mathbf{x}_b + \mathbf{e}_{\bar{\alpha}}\delta_t)] \tag{6.14}$$

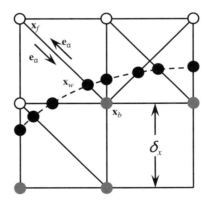

Figure 6.5 *Layout of regularly spaced lattices and curved wall boundary (Reprinted from Yan, Y. Y., Zu, Y. Q., Numerical simulation of heat transfer and fluid flow past a rotating isothermal cylinder – A LBM approachInt. J. Heat and Mass Transfer, **51** (9–10), 2519–2536. Copyright (2006) with permission from Elsevier.)*

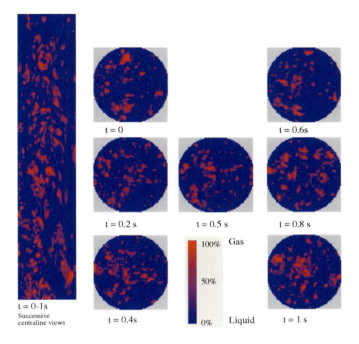

Steam/water – 189 mm diameter column – 46 bar – gas superficial velocity = 0.09 m/s

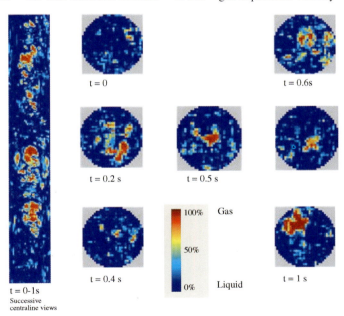

Air/silicone oil – 127 mm diameter column – 1 bar – gas superficial velocity = 0.09 m/s

Plate A *Cross-sectionally resolved phase distribution at different times together with sequence of distributions along a diameter. Measured using wire mesh sensor. Top view, conductance type; and bottom view, capacitance version.*

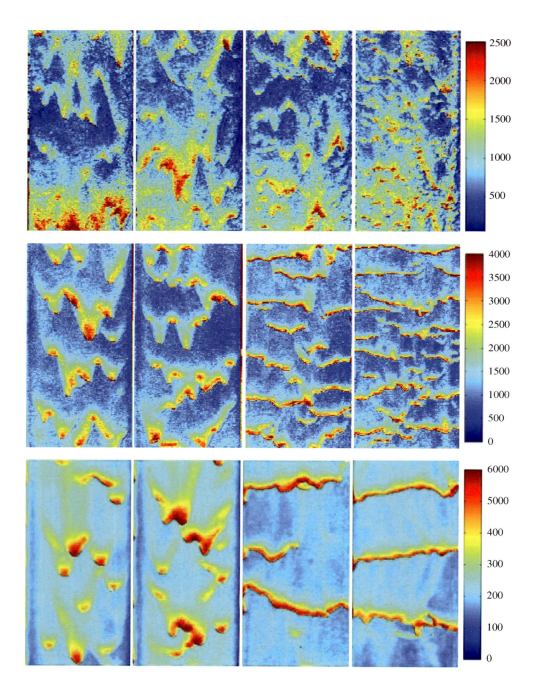

Plate B *Spatial variation of film thickness for liquid flowing down a vertical plate. Light absorption photographs. Co-current gas velocities {across the page}: 0, 4.9, 10, 12.5 m/s. Liquid viscosities (down the page): 0.009, 0.027, 0.111 Pa s.*

where

ϕ is the scalar array

$\phi(i,j) = 0$ refers to the lattice location (i, j) being occupied by fluid

$\phi(i,j)$ is set equal to 1 for those lattice nodes within the solid body.

6.3.3 Numerical Simulation and Results

Based on the flow field set-up shown in Figure 6.4, an initial field of flow is given by $u(x, y) = U$, $v(x, y) = 0$ with uniform temperature T_l, where u and v are, respectively, the x- and y-components of **u**, and the cylinder is stationary with temperature T_h. At the next moment, the cylinder starts to rotate with an angular velocity Ω and the surface temperature remains constant as T_h.

The flow and heat transfer are simulated at Re = 200, 218, 500 and 1000, respectively; where the Reynolds number is defined as Re = $2UR/v$. In order to regard the fluid as incompressible, the flow velocity must be much less than the speed of sound. The velocities are given based on the limit of incompressible flow, $Ma = |u|/c_s < < 1$. The sound speed in LBM is normally defined as $c_s = c/\sqrt{3}$, and the stream speed $c = \delta x/\delta t$ is based on the lattice definition.

Therefore, the inflow velocity U is set at 0.01 for Re = 200 and 218 and at 0.05 for Re = 500 and 1000, respectively. Parameter k is introduced to define the rate of the peripheral velocity $V = \Omega R$ to the inflow velocity U, that is $k = V/U$. A Prandtl number of 0.1, 0.5, 0.71 and 1.0, respectively, is applied for each combination of k and Re; and for all cases of the simulation, $T_h = 40$, $T_l = 20$, $\rho = 6$ and $R = 15\delta_x$ are used. The parameters T_h, T_l and ρ are all the sum of distribution functions that are non-dimensional in the LBM scheme based on the lattice definition. Once U is determined, the kinetic viscosity v and thermal diffusivity γ are determined through Re and Pr; the peripheral velocity V is determined by parameter k.

To compare the current results with those from available previous studies, the following normalisations are conducted:

$$u^* = u/U, \quad v^* = v/U, \quad x^* = x/R, \quad y^* = y/R,$$
$$t^* = Ut/R, \quad T^* = \frac{T - T_l}{T_h - T_l} \tag{6.15}$$

The drag and lift coefficients are defined as:

$$C_D = \frac{F_D}{\rho U^2 R}, \quad C_L = \frac{F_L}{\rho U^2 R} \tag{6.16}$$

where drag force F_D and lift force F_L are, respectively, the x-component and y-components of **F** given by Equation 6.14.

The heat transfer from the cylindrical surface is estimated in terms of a Nusselt number. Once the temperature field is determined, the following Nusselt numbers are defined, respectively, as:

$$\text{Local}: \quad Nu = -\frac{2R}{(T_h - T_l)} \left(\frac{\partial T}{\partial \mathbf{n}}\right)_{wall} \tag{6.17a}$$

$$\text{Surface-averaged}: \quad \langle Nu \rangle = \frac{1}{2\pi} \int_{-\pi}^{\pi} Nu \, d\theta \qquad (6.17b)$$

$$\text{Period-averaged}: \quad \overline{Nu} = \frac{1}{t_p} \int_{t_p} Nu \, dt \qquad (6.17c)$$

$$\text{Period- and surface-averaged}: \quad \overline{Nu} = \frac{1}{t_p} \int_{t_p} \langle Nu \rangle \, dt \qquad (6.17d)$$

where

n is the outer-normal vector of the cylindrical wall
angle θ equals zero at the rearmost point of the cylinder and increases in an anticlockwise direction.

The period-averaged quantities can only be calculated after the flow reaches the periodic state. The frequency of flow evolution, f, is obtained by a Fourier frequency analysis of the periodical variation of v at point $(9R, 0)$; and then the period is given by $t_p = 1/f$. Accordingly, the dimensionless frequency is defined by the following Strouhal number:

$$St = Rf/U \qquad (6.18)$$

The modelling was well validated against experiments. Figure 6.6 shows a comparison [20] between the evolution of the velocity field obtained by LBM simulation and experimental data for Re = 200, and $k = 0.5$ (the ratio of rotation velocity over fluid incoming velocity).

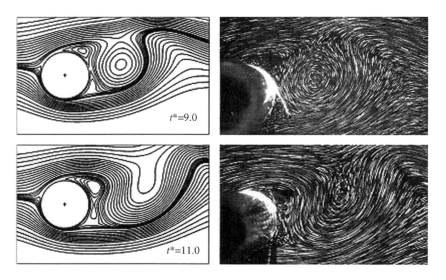

Figure 6.6 *Comparison between LBM and experimental simulations (left, obtained by LBM; right, by experiment) (Reprinted from Yan, Y. Y., Zu, Y. Q., Numerical simulation of heat transfer and fluid flow past a rotating isothermal cylinder – A LBM approachInt. J. Heat and Mass Transfer, 51 (9–10), 2519–2536. Copyright (2006) with permission from Elsevier.)*

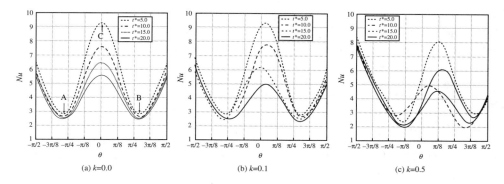

Figure 6.7 *Distributions of local Nusselt numbers for Re = 200, Pr = 0.5, θ ∈ [−π/2, π/2] (Reprinted from Yan, Y. Y. Zu, Y. Q., Numerical simulation of heat transfer and fluid flow past a rotating isothermal cylinder – A LBM approachInt. J. Heat and Mass Transfer,* **51** *(9–10), 2519–2536. Copyright (2006) with permission from Elsevier.)*

Figures 6.7 and 6.8 show distributions of the local *Nu* number at Re = 200, *Pr* = 0.5, θ ∈ [−π/2, π/2] for a ratio of rotational velocity to inflow velocity of k = 0.0, 0.1 and 0.5, respectively.

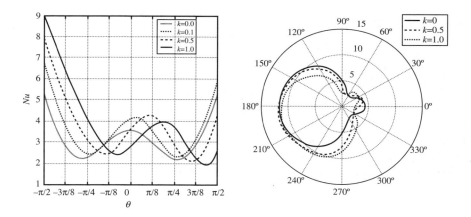

Figure 6.8 *Period-averaged Nusselt number along the cylinder surface at Re = 200, Pr = 0.5*

6.4 Numerical Simulation of Two-Phase Flow

In this section, the LBM model for simulating two-phase flow is discussed, with an introduction of a few examples.

6.4.1 Two-Phase Lattice Boltzmann Model

Since the first LBM model for two-phase flow was proposed in 1991, several LBM models have been successfully developed. These can mainly be summarised as follows.

The *chromodynamics model* [21] has been applied successfully to simulate some multiphase and multicomponent fluid flows, such as two- and three-dimensional simulations of multiphase flows in porous media, the spreading of a droplet on a fluid-fluid interface and blood flow. However, there are a few weak sides to this model. Firstly, a certain degree of anisotropy of the surface tension produces a spurious dependence on the interface orientation, which hence causes unphysical currents near the interfaces. Furthermore, this model is relatively heavy from the computational viewpoint, because of the time-consuming step of colour redistribution. The chromodynamic model was later modified by Dortona *et al.* [22]. In the modification, the step of colour redistribution is replaced by an evolution equation for $f_\alpha^k(\mathbf{x}, t)$, which increases the computational efficiency. In addition, the chromodynamics model was extended by Kono *et al.* [23] to simulate multiphase fluid flows coupled with phase transition by taking account of mass exchange processes in the lattice Boltzmann equation.

Following the chromodynamic models, the *pseudo-potential model*, which has the capability of simulating multiphase and multicomponent immiscible fluids with different masses of constant temperature, was introduced firstly by Shan and Chen [5], and then later analysed carefully by them and other workers [6, 7]. In this model, the additional collision operator is expressed as:

$$(\Omega_\alpha^k)_2 = \mathbf{e}_\alpha \cdot \mathbf{F}^k \tag{6.19}$$

where the effective force $\mathbf{F}^k$ comes from the following pairwise interaction potential,

$$V_{kk'}(\mathbf{x}, \mathbf{x}') = G_{kk'}(\mathbf{x}, \mathbf{x}')\psi^k(\mathbf{x})\psi^{k'}(\mathbf{x}') \tag{6.20}$$

$G_{kk'}(\mathbf{x}, \mathbf{x}')$ is a Green's function,

$$G_{kk'}(\mathbf{x}, \mathbf{x}') = \begin{cases} 0, & |\mathbf{x}\text{-}\mathbf{x}'| > |\mathbf{e}_\alpha| \\ g_{kk'}, & |\mathbf{x}\text{-}\mathbf{x}'| = |\mathbf{e}_\alpha| \end{cases} \tag{6.21}$$

The magnitude of $g_{kk'}$ controls the strength of the interaction potential between phase (or component) k and k', while its sign determines whether it is attractive or repulsive. The quantity $\psi^k(\mathbf{x})$ plays a role as the effective number density for phase k, given by:

$$\psi = \psi(\rho) = \rho_0(1 - e^{-\rho/\rho_0}) \tag{6.22}$$

where

ρ_0 is a reference density marking the border between different phases of the fluid.

The force associated with this pseudo-potential can be written as:

$$\mathbf{F}^k(\mathbf{x}) = -\sum_{k'}\sum_\alpha V_{kk'}(\mathbf{x}, \mathbf{x} + \mathbf{e}_\alpha)\mathbf{e}_\alpha = -\psi^k(\mathbf{x})\sum_{k'} g_{kk'}\sum_\alpha \psi^{k'}(\mathbf{x} + \mathbf{e}_\alpha)\mathbf{e}_\alpha \tag{6.23}$$

In this model, a non-ideal equation of state (EOS), $p = p(\rho)$, can be obtained by letting ψ depend on the fluid density ρ. Phase separation takes place spontaneously whenever the interaction strength exceeds the critical value. This is an important improvement in numerical efficiency compared with previous chromodynamic models. In addition, this model also improves the isotropy of surface tension. Owing to these attractive features, the model is employed frequently to simulate various multiphase and multicomponent flow problems, including porous media flows [24, 25], bubble rising, growing and detachment from a wall [26, 27], droplet deformation, fragmentation and collision [28–30] and multicomponent viscoelastic fluids [31].

In the work of Hou *et al.* [32], two lattice Boltzmann models for multiphase flows, the chromodynamic model and the pseudo-potential model are studied numerically to compare their abilities to simulate the physics of multiphase flows. The test problem is the simulation of a static bubble. Isotropy, strength of surface tension, thickness of the interface, spurious currents, Laplace's law and steadiness of the bubble are examined. The results show that the pseudo-potential model is a major improvement over the chromodynamic model. However, the model has several drawbacks. It leads to inconsistent thermodynamics unless a particular EOS is chosen and, as with the chromodynamic model, it reaches equilibrium distribution, which has unphysical velocity fluctuation within the interfacial region [8]. Another problem of this model is that the local conservation of momentum is not satisfied although it is conserved globally when boundary effects are excluded.

The *free-energy model*, first presented by Swift *et al.* [8], is the third important model of the LBM for multiphase flow. The model builds on the Van der Waals formulation of quasi-local thermodynamics for a two-component fluid in thermodynamic equilibrium at a fixed temperature. The free-energy function is represented as:

$$\Psi(\mathbf{x}) = \int \left\{ \frac{k}{2} |\nabla \rho(\mathbf{x})|^2 + \psi(\rho(\mathbf{x})) \right\} d\mathbf{x} \qquad (6.24)$$

where k is a parameter, the first term gives the contribution from any density gradients and the second describes the bulk free energy density. The non-local pressure is defined by:

$$p(\mathbf{x}) = \rho \frac{\delta \Psi}{\delta \rho} - \Psi(\mathbf{x}) = p_0 - k\rho \nabla^2 \rho - \frac{k}{2} |\nabla \rho|^2 \qquad (6.25)$$

where $p_0 = \rho \psi'(\rho) - \psi(\rho)$ is the EOS of the fluid. For the Van der Waals EOS,

$$\psi = \rho T \ln \left(\frac{\rho}{1 - \rho b} \right) - a\rho^2 \qquad (6.26)$$

where

a and b are free parameters.

To obtain the full pressure tensor in a non-uniform fluid, non-diagonal terms must be added. Based on the Cahn–Hillard theory the pressure tensor can be written as:

$$P_{ij}(\mathbf{x}) = p(\mathbf{x})\delta_{ij} + k \frac{\partial \rho}{\partial x_i} \frac{\partial \rho}{\partial x_j} \qquad (6.27)$$

For a single component non-ideal fluid, the equation of the particle velocity distribution function can be written as:

$$f_\alpha(\mathbf{x} + \mathbf{e}_\alpha, t + 1) - f_\alpha(\mathbf{x}, t) = \Omega_\alpha(\mathbf{x}, t) \qquad (6.28)$$

Ω_α is the collision operator which has the form

$$\Omega_\alpha = -\frac{f_\alpha(\mathbf{x}, t) - f_\alpha^{(eq)}(\mathbf{x}, t)}{\tau} \tag{6.29}$$

where the equilibrium distribution function, $f_\alpha^{(eq)}(\mathbf{x}, t)$, is defined as:

$$f_\alpha^{(eq)}(\mathbf{x}, t) = A + Be_{\alpha i}u_i + Cu^2 + Du_iu_je_{\alpha i}e_{\alpha j} + F_ie_{\alpha i} + G_{ij}e_{\alpha i}e_{\alpha j} \tag{6.30a}$$

$$f_0^{(eq)}(\mathbf{x}, t) = A_0 + C_0u^2 \tag{6.30b}$$

The equilibrium distribution function should satisfy the following three macroscopic constraints, namely the conservations of mass, momentum and momentum flux tensor,

$$\sum_\alpha f_\alpha^{(eq)} = \rho \tag{6.31a}$$

$$\sum_\alpha f_\alpha^{(eq)} e_{\alpha i} = \rho u_i \tag{6.31b}$$

$$\sum_\alpha f_\alpha^{(eq)} e_{\alpha i}e_{\alpha j} = P_{ij} + \rho u_iu_j \tag{6.31c}$$

The coefficients in the expansions (6.30a) and (6.30b) can be determined by combining the constraints (6.31a)(6.32b) and (6.31c) with the equilibrium thermodynamic definitions (6.24),(6.25) and (6.27). Furthermore, an external chemical potential μ_{ex} at the surfaces of a confined system can be used to supplement the usual bounce-back boundary conditions by modifying Equation 6.31b:

$$\sum_\alpha f_\alpha^{(eq)} e_{\alpha i} = \rho u_i - \tau\rho\frac{\partial\mu_{ex}}{\partial x_i} \tag{6.32}$$

As a result, a coefficient of Equation 6.30a, F_i, is changed to:

$$F_i = -\frac{\tau\rho}{3}\frac{\partial\mu_{ex}}{\partial x_i} \tag{6.33}$$

By allowing a non-zero μ_{ex} only at the walls with a different intensity for each of the phases, the affinity of the wall to the fluids can be varied in a simple and physically appealing way.

For binary fluids, the main difference from the one-component case is that, as there are now two independent densities, two sets of lattice Boltzmann distribution functions, f_α and g_α, are now needed to correctly mirror the dynamics of the conserved quantities. These are taken to evolve according to the usual single relaxation-time lattice Boltzmann equation (Swift *et al.* [8]):

$$f_\alpha(\mathbf{x} + \mathbf{e}_\alpha, t + 1) - f_\alpha(\mathbf{x}, t) = -\frac{f_\alpha(\mathbf{x}, t) - f_\alpha^{(eq)}(\mathbf{x}, t)}{\tau_1} \tag{6.34a}$$

$$g_\alpha(\mathbf{x}+\mathbf{e}_\alpha, t+1)-g_\alpha(\mathbf{x}, t) = -\frac{g_\alpha(\mathbf{x}, t)-g_\alpha^{(eq)}(\mathbf{x}, t)}{\tau_2} \tag{6.34b}$$

Let ρ_1 and ρ_2 be the densities of the two fluid components, respectively, $\Delta\rho = \rho_1-\rho_2$ be the density difference and $\rho = \rho_1+\rho_2$ be the total fluid density. Then, f_α and g_α are the distributions of ρ and $\Delta\rho$, respectively, $f^{(eq)}$ and $g^{(eq)}$ are the equilibrium distribution functions. They must satisfy the following conservations of mass, momentum and momentum flux tensor,

$$\sum_\alpha f_\alpha^{(eq)} = \sum_\alpha f_\alpha = \rho \tag{6.35a}$$

$$\sum_\alpha g_\alpha^{(eq)} = \sum_\alpha g_\alpha = \Delta\rho \tag{6.35b}$$

$$\sum_k f_\alpha^{(eq)}\mathbf{e}_\alpha = \sum_\alpha f_\alpha\mathbf{e}_\alpha = \rho\mathbf{u} \tag{6.35c}$$

$$\sum_k g_\alpha^{(eq)}\mathbf{e}_\alpha = \sum_\alpha g_\alpha\mathbf{e}_\alpha = \Delta\rho\mathbf{u} \tag{6.35d}$$

$$\sum_\alpha f_\alpha^{(eq)} e_{\alpha i} e_{\alpha j} = P_{ij} + \rho u_i u_j \tag{6.35e}$$

$$\sum_\alpha g_\alpha^{(eq)} e_{\alpha i} e_{\alpha j} = \Gamma\Delta\mu\delta_{ij} + \Delta\rho u_i u_j \tag{6.35f}$$

where

$\Delta\mu$ is the chemical potential difference between two components
Γ is a mobility.

The free energy for two ideal gasses with repulsive interaction energy can be defined as:

$$\Psi = \int \left\{ \frac{k}{2}|\nabla\rho|^2 + \frac{k}{2}|\nabla\Delta\rho|^2 + \psi(\rho, \Delta\rho) \right\} d\mathbf{x} \tag{6.36}$$

The bulk free-energy density ψ at a temperature T is given by:

$$\psi = \frac{\lambda}{4}\rho\left[1-\left(\frac{\Delta\rho}{\rho}\right)^2\right]-\rho T + \frac{(\rho+\Delta\rho)T}{2}\ln\left(\frac{\rho+\Delta\rho}{2}\right) + \frac{(\rho-\Delta\rho)T}{2}\ln\left(\frac{\rho-\Delta\rho}{2}\right) \tag{6.37}$$

where λ controls the strength of the interaction. $\Delta\mu$ and P_{ij} in Equations 6.35e and 6.35f can be written as:

$$\Delta\mu = -\frac{\lambda}{2}\frac{\Delta\rho}{\rho} + \frac{T}{2}\ln\left(\frac{1+\Delta\rho/\rho}{1-\Delta\rho/\rho}\right) - k\nabla^2(\Delta\rho) \tag{6.38}$$

$$P_{ij}(\mathbf{x}) = p(\mathbf{x})\delta_{ij} + k\frac{\partial\rho}{\partial x_i}\frac{\partial\rho}{\partial x_j} + k\frac{\partial\Delta\rho}{\partial x_i}\frac{\partial\Delta\rho}{\partial x_j} \tag{6.39}$$

where

$$p(\mathbf{x}) = \rho T - k(\rho \nabla^2 \rho + \Delta \rho \nabla^2 \Delta \rho) - \frac{k}{2}(|\nabla \rho|^2 + |\nabla \Delta \rho|^2) \qquad (6.39a)$$

The main new features of this model are the direct introduction of a non-ideal pressure tensor and external chemical potential instead of the introduction of the additional collision operator. By doing so, one can obtain an isothermal model of phase separation, which correctly describes bulk and interfacial dynamics at low temperatures. The model also provides a convenient, physically motivated way of tuning boundary conditions, giving a new method to situations when flow and phase separation are affected by fluid–substrate interactions. Moreover, unphysical velocity oscillations at surfaces and interfaces are substantially reduced. Unlike the previous two LB multiphase models, which can only suit numerical simulation for isothermal multiphase flow, it can be extended to non-isothermal situations where heat transfer is important. It has demonstrated its ability to simulate spinodal phase transitions or decompositions [33–35], bubble flows [36], droplet motion [37, 38], porous media flows [39] and other complex multiphase and multicomponent flows [40]. However, in this model, the interfaces of simulation are not sharp enough, such that the thermodynamic consistency and the Galilean invariance can not be satisfied.

To overcome the Galilean invariance problem Holdych *et al.* [41] introduced a correction term into the model and this solution can remove the non-Galilean terms (at least to low order) from the momentum equation. This was then further developed by Palmer and Rector [42] to model the heat transfer of two-phase systems, through the introduction of two distribution functions, one for the momentum and the other for the energy transfer. The parallel implementation of the free-energy model by a modular LB code, termed LUDWIG, has been described in detail by Desplat *et al.* [43].

Recently, Inamuro *et al.* [10] proposed a method that can simulate multiphase flows with large density differences based on the free-energy model. The method requires two particle distribution functions: one for the order parameter that represents the phase of the fluids, and the other for the predicted velocity of the two fluids without a pressure gradient. The method is then supplemented by the relationship between the velocity and the pressure correction, which is determined by solving an approximate pressure Poisson equation. In addition, some mixed methods that combine the free energy LB method and finite difference scheme have been proposed to simulate complex fluid flows.

The *index function model* is a new LB model for incompressible multiphase flow, which can be derived by systematically removing the unphysical approximations, including spurious current around interfaces, thermodynamic inconsistencies and the lack of Galilean invariance [44], and was presented by He *et al.* [9]. In this model an equation for the index function was introduced to track the interfaces between different fluids and compute the density and viscosity of the flow field. The velocity and pressure field can be given by the equation of the distribution function of pressure. The distribution functions of the index function and pressure satisfy, respectively, the following Equations 6.40 and 6.41:

$$\frac{Df}{Dt} = -\frac{f - f^{(eq)}}{\lambda} - \frac{(\xi - \mathbf{u}) \cdot \nabla \psi(\Phi)}{RT} \Gamma(\mathbf{u}) \qquad (6.40a)$$

$$\frac{Dg}{Dt} = -\frac{g-g^{(eq)}}{\lambda} + (\xi-\mathbf{u})\{\Gamma(\mathbf{u})(\mathbf{F}+\mathbf{G})-[\Gamma(\mathbf{u})-\Gamma(0)]\nabla\psi(\rho)\} \qquad (6.40b)$$

where

ξ is the microscopic velocity
R is the gas constant
T is the constant temperature
λ is a relaxation time
function $\psi(\rho)$ is related to pressure by $\psi(\rho) = p-\rho RT$
$\mathbf{G}$ is the body force
$\mathbf{F} = k\rho\nabla\nabla^2\rho$ is the surface tension in which k determines the strength of the surface tension
f and g are the distribution of the index function Φ and pressure
$f^{(eq)}$ and $g^{(eq)}$ are the respective equilibrium distribution functions.

These equilibrium distribution functions are given by:

$$f^{(eq)} = \Gamma(\mathbf{u})\Phi \qquad (6.41a)$$

$$g^{(eq)} = \Gamma(0)p + [\Gamma(\mathbf{u})-\Gamma(0)]\rho RT \qquad (6.41b)$$

where

$\Gamma(\mathbf{u})$ is a function of macroscopic velocity $\mathbf{u}$ given by

$$\Gamma(\mathbf{u}) = \frac{1}{(2\pi RT)^{D/2}}\exp\left[-\frac{(\xi-\mathbf{u})^2}{2RT}\right] \qquad (6.41c)$$

where

D is the dimension of the space.

As the original index function model does not have the capability for simulating multiphase or multicomponent fluid flows with a high density ratio, Equation 6.40 was suggested by Lee *et al.* [45] to carry out stable discretisation. The discretisation, which is a collection of consistent discretisation strategies, comprises the low Mach number approximation, the use of stress and potential forms of surface tension forces, the incompressible transformation and the consistent discretisation of the intermolecular forcing terms.

6.4.2 Vortices Merging in a Two-Phase Spatially Growing Mixing Layer

Mixing layers are commonly observed in flow fields of many engineering applications, such as in combustion chambers, pre-mixers for gas turbine combustors, chemical lasers, propulsion systems, flow reactors and so on. Controlling the formation and evolution of the coherent structure in a mixing layer can improve efficiencies of combustion, chemical reaction and so on. Certain flow features of mixing layers, such as the instability of flow, evolutions and interactions of vortices layers, have made them very attractive for experimental and computational studies. In this case study, the index function method [9] was

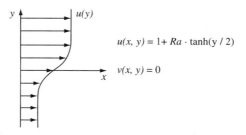

Figure 6.9 *The initial velocity field (Courtesy of WIT Press from Yan, Y. Y. and Zu, Y. Q., Lattice Boltzmann simulation of vortices merging in a two-phase mixing layer, WIT Transactions on Engineering Sciences, **53**, 877–896, 2006.)*

employed to study vortices merging in a two-dimensional (based on D2Q9), two-phase spatial growing mixing layer [46].

With an initial velocity field, which consists of a hyperbolic tangent profile as defined in Figure 6.9, phase distributions of three vortices merging with values of different surface tension were obtained. Figure 6.10 shows the phase distribution of zero surface tension compared with one of the surface tension coefficients at $k = 0.05$ and 0.1. Figure 6.11 shows interface distributions and the corresponding vortices contours with different values of surface tension.

6.4.3 Viscous Fingering Phenomena of Immiscible Two-Fluid Displacement

Applying the pseudo-potential D2Q9 model [5], two types of viscous fingering phenomena of immiscible two-fluid displacement are simulated. The first is the displacement in a channel; the other is the displacement in porous media.

The wettability of reservoir rocks plays an important role in oil recovery in the oil producing industry. It is recognised that the wettability can affect the properties of fluid-rock

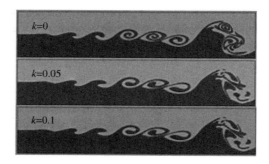

Figure 6.10 *Phase distribution of three vortices merging (Courtesy of WIT Press from Yan, Y. Y. and Zu, Y. Q., Lattice Boltzmann simulation of vortices merging in a two-phase mixing layer, WIT Transactions on Engineering Sciences, **53**, 877–896, 2006.)*

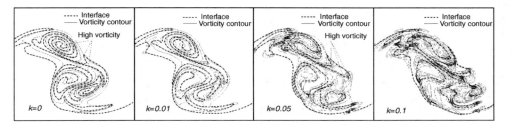

Figure 6.11 *Interface distributions and corresponding vorticity contours (Courtesy of WIT Press from Yan, Y. Y. and Zu, Y. Q., Lattice Boltzmann simulation of vortices merging in a two-phase mixing layer, WIT Transactions on Engineering Sciences,* **53**, *877–896, 2006.)*

interactions, such as residual oil saturation, relative permeability and capillary pressure. Practically, wettability alteration can be achieved by injecting surfactants, altering salinity and increasing temperature. In the oil industry, when supercritical carbon dioxide (SCO_2) is injected into a reservoir, the wettability of reservoir rocks can be changed due to a series of reactions of SCO_2, crude oil and rock materials. In the LBM modelling [47], the effect of wettability of rocks on viscous fingering phenomena, which can account for the break-through of the displacing fluid, has been studied. Additionally, the effect of gravity was considered, because even at the pore-scale level, it still has an impact on fluid distribution and its patterns. The formulation of fluid-solid interaction was considered and applied. The effects of capillary number, bond number, viscosity ratio and the channel surface wettability on the fingering phenomenon were evaluated through a series of numerical simulations. Figure 6.12 shows the effect of gravity on the offsets of finger width and finger length under different conditions of wettability in terms of the strength of fluid-solid interaction parameter Gw1 [47].

The effect of capillary number, Bond number and the viscosity ratio of two immiscible fluids on viscous fingering phenomena were simulated and discussed. Figure 6.13 shows the interface positions at the final stage of the displacement process with different Bond numbers; the interface front has a tendency to move downwards under the action of gravity.

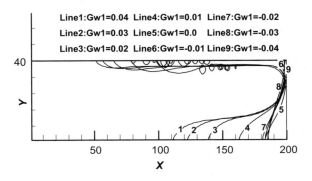

Figure 6.12 *The effect of gravity on finger patterns*

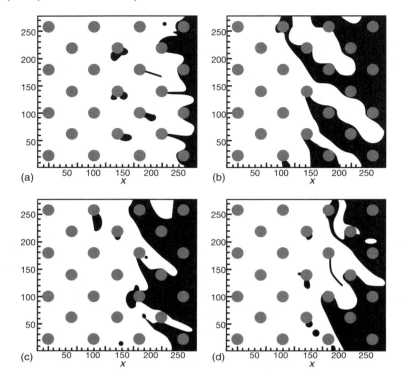

Figure 6.13 *Final finger patterns: (a) Bo = 2.79; (b) Bo = 5.58; (c) Bo = 8.37; and (d) Bo = 11.16*

6.4.4 Bubbles/Drops Flow Behaviour

6.4.4.1 LBM Method

In modelling bubbly flows and coalescence, a D3Q15 free-energy model of the LBM, the projection method [10], can be modified and employed to simulate two-fluid/phase flow with large density and/or viscosity ratio [11, 16]. The D3Q15 LBM model is shown in Figure 6.3, and the particle velocity, $\mathbf{e}_\alpha$, is given by:

$$[\mathbf{e}_0, \mathbf{e}_1, \mathbf{e}_2, \mathbf{e}_3, \mathbf{e}_4, \mathbf{e}_5, \mathbf{e}_6, \mathbf{e}_7, \mathbf{e}_8, \mathbf{e}_9, \mathbf{e}_{10}, \mathbf{e}_{11}, \mathbf{e}_{12}, \mathbf{e}_{13}, \mathbf{e}_{14}] =$$
$$\begin{bmatrix} 0 & 1 & 0 & 0 & -1 & 0 & 0 & 1 & -1 & 1 & 1 & -1 & 1 & -1 & -1 \\ 0 & 0 & 1 & 0 & 0 & -1 & 0 & 1 & 1 & -1 & 1 & -1 & -1 & 1 & -1 \\ 0 & 0 & 0 & 1 & 0 & 0 & -1 & 1 & 1 & 1 & -1 & -1 & -1 & -1 & 1 \end{bmatrix} \qquad (6.42)$$

In this method, two particle velocity distribution functions, f_α and g_α, are introduced.

$$f_\alpha(\mathbf{x} + \mathbf{e}_\alpha \delta_t, t + \delta_t) = f_\alpha^{(eq)}(\mathbf{x}, t) \qquad (6.43a)$$

$$g_\alpha(\mathbf{x} + \mathbf{e}_\alpha \delta_t, t + \delta_t) = g_\alpha^{(eq)}(\mathbf{x}, t) \qquad (6.43b)$$

where

f_α is used to calculate the order parameter
ϕ distinguishes the two phases
g_α is used to calculate the predicted velocity.

The equilibrium states of f_α and g_α are given by:

$$f_\alpha^{(eq)}(\mathbf{x},t) = H_\alpha\phi + F_\alpha\left[p_0 - k\phi\nabla^2\phi - \frac{k}{6}|\nabla\phi|^2\right] + 3\omega_\alpha\phi(\mathbf{e'}_\alpha\cdot\mathbf{u}) + \omega_\alpha k\mathbf{e'}_\alpha\cdot\mathbf{G}(\phi)\cdot\mathbf{e}_\alpha$$

(6.44a)

$$g_\alpha^{(eq)}(\mathbf{x},t) = \omega_\alpha\left[1 + 3(\mathbf{e'}_\alpha\cdot\mathbf{u}) + \frac{9}{2}(\mathbf{e'}_\alpha\cdot\mathbf{u})^2 - \frac{3}{2}\mathbf{u}^2 + \frac{3}{4}\mathbf{e'}_\alpha\cdot(\nabla\mathbf{u}+\mathbf{u}\nabla)\cdot\mathbf{e}_\alpha\right]$$
$$+ \omega_\alpha\frac{k}{\rho}\mathbf{e'}_\alpha\cdot\mathbf{G}(\phi)\cdot\mathbf{e}_\alpha - \frac{2}{3}F_\alpha\frac{k}{\rho}|\nabla\phi|^2 + 3\omega_\alpha\frac{1}{\rho}\nabla\cdot[\mu(\nabla\mathbf{u}+\mathbf{u}\nabla)]\cdot\mathbf{e}_\alpha$$

(6.44b)

where

$$\omega_\alpha = \begin{cases} 2/9, & \alpha = 0 \\ 1/9, & \alpha = 1,\ldots,6 \\ 1/72, & \alpha = 7,\ldots,14 \end{cases}, \quad F_\alpha = \begin{cases} -7/3, & \alpha = 0 \\ 1/3, & \alpha = 1,\ldots,6 \\ 1/24, & \alpha = 7,\ldots,14 \end{cases}, \quad H_\alpha = \begin{cases} 1, & \alpha = 0 \\ \\ 0, & \alpha = 1,\ldots,14 \end{cases}$$

(6.45)

and

$$\mathbf{G}(\phi) = \frac{9}{2}(\nabla\phi)(\phi\nabla) - \frac{3}{2}|\nabla\phi|^2\mathbf{I}$$

(6.46)

where

k is a constant parameter for determining the width of the interface and the strength of
surface tension
$\mathbf{I}$ is the unit tensor of second-order.

Given that $\psi(\phi)$ is the bulk free-energy density, then

$$p_0 = \phi\frac{\partial\psi}{\partial\phi} - \psi$$

(6.47)

The macroscopic quantities $\mathbf{u}^*$, ϕ, ρ and μ can be evaluated as:

$$\phi = \sum_\alpha f_\alpha, \quad \mathbf{u}^* = \sum_\alpha \mathbf{e}_\alpha g_\alpha$$

(6.48a)

$$\rho = \begin{cases} \rho_g, & \phi < \phi_g \\ \frac{\phi - \phi_g}{\phi_l - \phi_g}(\rho_l - \rho_g) + \rho_g, & \phi_g \le \phi \le \phi_l \\ \rho_l, & \phi > \phi_l \end{cases}$$

(6.48b)

$$\eta = \frac{\rho - \rho_g}{\rho_l - \rho_g}(\eta_l - \eta_g) + \eta_g \tag{6.48c}$$

where

ϕ_l and ϕ_g are, respectively, the maximum and minimum order parameter for marking bulk liquid and gas
ρ_l and ρ_g are, respectively, the density of the liquid and gas phases
η_l and η_l are, respectively, the dynamic viscosity of the liquid and gas phases.

To enable the method to treat two-phase fluids interacting with confined solid surfaces with wetting boundary potentials, for the current isothermal system a simple form of representation of the free-energy density $\psi(\phi)$, as suggested in Jamet *et al.* [48], rather than the Van der Waals free energy used in the traditional model, is applied in the present simulation, namely:

$$\psi(\phi) = \beta(\phi - \phi_g)^2(\phi - \phi_{LI})^2 + \mu_b\phi - p_b \tag{6.49}$$

where

β is a constant relating to interfacial thickness
μ_b and p_b are the bulk chemical potential and bulk pressure, respectively.

By substitution of Equation 6.49, Equation 6.47 becomes

$$p_0 = \beta(\phi - \phi_l)(\phi - \phi_g)(3\phi^2 - \phi\phi_l - \phi\phi_g - \phi_l\phi_g) + p_b \tag{6.50}$$

In a plane interface under equilibrium conditions, the density profile across the interface is in equilibrium and can be represented as:

$$\phi(\xi) = \frac{\phi_l + \phi_g}{2} + \frac{\phi_l - \phi_g}{2}\tanh\left(\frac{2\xi}{D}\right) \tag{6.51}$$

where

ξ is the coordinate normal to the interface.

The interface thickness D is given by:

$$D = \frac{4}{\phi_l - \phi_g}\sqrt{\frac{k}{2\beta}} \tag{6.52}$$

The fluid-fluid (liquid-gas) surface tension force, σ, is expressed as [49]

$$\sigma = \frac{(\phi_l - \phi_g)^3}{6}\sqrt{2\,k\beta} \tag{6.53}$$

6.4.4.2 Correction of Pressure

It should be pointed out that the predicted velocity $\mathbf{u}^*$ is not divergence free. To obtain the velocity field that satisfies the continuity equation ($\nabla \cdot \mathbf{u} = 0$), $\mathbf{u}^*$ is corrected by:

$$\mathbf{u} - \mathbf{u}^* = -\frac{\nabla p}{\rho} \tag{6.54}$$

$$\nabla \cdot \mathbf{u}^* = \nabla \cdot \left(\frac{\nabla p}{\rho}\right) \tag{6.55}$$

where

p is the pressure of the two-phase fluid.

Equation 6.55 can be approximated by the LBM framework equation:

$$h_\alpha(\mathbf{x} + \mathbf{e}_\alpha, n + 1) = h_\alpha(\mathbf{x}, n) - \frac{1}{\tau}[h_\alpha(\mathbf{x}, n) - \omega_\alpha p(\mathbf{x}, n)] - \frac{\omega_\alpha}{3}\frac{1}{\rho}\nabla \cdot \mathbf{u}^* \tag{6.56}$$

where

n is the number of iterations
$\tau = 0.5 + 1/\rho$ is the relaxation time.

The pressure at step $n + 1$ is given by:

$$p(\mathbf{x}, n + 1) = \sum_\alpha h_\alpha(\mathbf{x}, n + 1) \tag{6.57}$$

The convergent pressure p is determined when

$$\forall \mathbf{x} \in V, \ |p(\mathbf{x}, n + 1) - p(\mathbf{x}, n + 1)| < \varepsilon \tag{6.58}$$

where

V denotes the whole computational domain.

Substituting the newly obtained pressure p into and solving Equation 6.54 gives the corrected velocity field $\mathbf{u}$.

6.4.4.3 Boundary Treatment

No-slip boundary conditions can be implemented by simply specifying a zero velocity on the solid boundaries, that is the boundary velocities $\mathbf{u}$ and $\mathbf{u}^*$ in Equations 6.44 and are given by:

$$\mathbf{u}_w = 0, \mathbf{u}_w^* = 0 \tag{6.59}$$

There is always a thin liquid layer in the vicinity of the solid boundary surface due to the intermolecular forces between the liquid and solid substrate. Therefore it is assumed in the present model that a thin liquid occupies one layer of the lattice spacing, the order parameter on the boundary used in Equation 6.44 can then be determined by:

$$\phi_w = \phi_l \tag{6.60}$$

Considering the surface wetting condition, according to Young's law, the wetting potential, Ω can be given by [11]

$$\Omega = \frac{4\lambda}{(\phi_l - \phi_g)^2 \sqrt{2k_f \beta}} \tag{6.61}$$

where

$$\lambda = \pm\sqrt{2k_f \psi(\phi_w)}$$

The wetting angle can be determined by:

$$\cos\theta_w = \frac{(1+\Omega)^{3/2} - (1-\Omega)^{3/2}}{2} \tag{6.62}$$

For a given wetting angle at $0 < \theta_w < \pi$, Ω can be obtained from Equation 6.19 as:

$$\Omega = 2\,\mathrm{sgn}\left(\frac{\pi}{2} - \theta_w\right)\left\{\cos\left(\frac{\gamma}{3}\right)\left[1 - \cos\left(\frac{\gamma}{3}\right)\right]\right\}^{1/2} \tag{6.63}$$

where

$$\gamma = \arccos\left(\sin^2\theta_w\right)$$
$\mathrm{sgn}\,(\xi)$ gives the sign of ξ.

It is noted from Equation 6.63 that the required wetting potential Ω can be obtained by choosing a desired contact angle θ_w and then calculating λ by Equation 6.61 with a newly obtained Ω. The finite-difference of the order parameters on the boundary are given by the following form:

$$\left.\frac{\partial\phi}{\partial z}\right|_{z=0} = -\frac{\lambda}{k} \tag{6.64}$$

$$\left.\frac{\partial^2\phi}{\partial\zeta^2}\right|_{\zeta=0} \approx \frac{1}{2}\left(-3\left.\frac{\partial\phi}{\partial\zeta}\right|_{\zeta=0} + 4\left.\frac{\partial\phi}{\partial\zeta}\right|_{\zeta=1} - \left.\frac{\partial\phi}{\partial\zeta}\right|_{\zeta=2}\right) \tag{6.65}$$

where

ζ is the direction perpendicular to the wall.

In this scheme, the first term on the right-hand side of Equation 6.65 is determined by a right-handed finite-difference; the second term is calculated by a standard centred finite-difference formula. Finally, it is found empirically that the best choice for the third term is a left-handed finite-difference formula taken back into the wall, namely,

$$\left.\frac{\partial\phi}{\partial\zeta}\right|_{z=2} \approx \frac{1}{2}\left(3\phi|_{\zeta=2} - 4\phi|_{\zeta=1} + \phi|_{\zeta=0}\right) \tag{6.66}$$

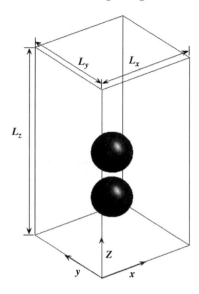

Figure 6.14   *The domain of rising bubbles coalescence*

6.4.4.4   Results of Two Rising Bubbles Coalescence

The method is firstly applied to bubbles coalescence; the coalescence of two rising bubbles is simulated and two cases are calculated. The computational domain is shown in Figure 6.14. In an initial study, two bubbles with the same diameter D are placed $5D/4$ apart in a liquid inside a rectangular domain and are released at time $t = 0$. Calculations are carried out for the liquid and gas phases with different density ratios, ρ_l/ρ_g, and viscosity ratios, η_l/η_g. Dimensionless parameters, Morton number, $Mo = g\eta_l^4(\rho_l-\rho_g)/(\rho_l^2\sigma^3)$, and Eötvös number, $E\ddot{o} = g(\rho_l-\rho_g)D^2/\sigma^3$, are applied for the simulated phenomena. Periodic boundary conditions are imposed on all sides of the computational domain, which is divided into a $64 \times 64 \times 128$ cubic lattice. The diameter of each initial bubble occupies 24 lattice spaces, that is, $D = 24\delta_x$. The behaviour of the two bubbles evolves with time, typically how the lower bubble catches up and then finally coalesces with the upper bubble is studied. Velocity vectors of both inside and around the bubbles during the evolution are also studied.

Figure 6.15 shows time evolution of two bubbles coalescence and velocity vectors at section $y = L_y/2$. The two gas bubbles rising in an unbounded liquid with $\rho_l/\rho_g = 50$, $\eta_l/\eta_g = 50$, dimensionless Morton number $Mo = 1 \times 10^{-5}$, and Eötvös number $E\ddot{o} = 10$ are simulated; t^* refers to dimensionless time ($t^* = tU/D$) and here U is the averaged terminal velocity of the gas phase.

Figure 6.16 shows two bubbles coalescence when gas bubbles rise in an unbounded liquid at $\rho_l/\rho_g = 1000$, $\eta_l/\eta_g = 50$, Morton number $Mo = 1$ and Eötvös number $E\ddot{o} = 15$. The upper figure shows the time evolution of bubble shapes and the lower figure shows the velocity vectors at section $y = L_y/2$.

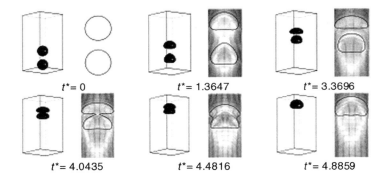

Figure 6.15 *Coalescence of two rising bubbles in liquid ($\rho_l/\rho_g = 50, \eta_l/\eta_g = 50, M_o = 1 \times 10^{-5}$, Eö = 10)*

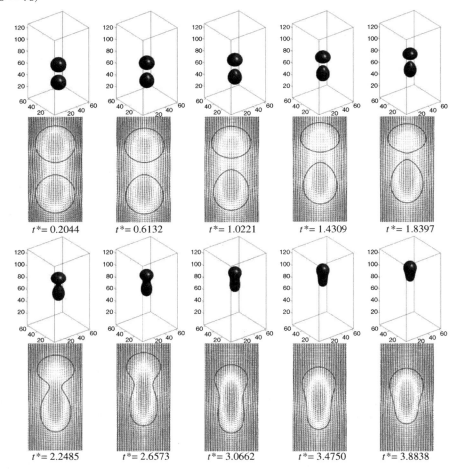

Figure 6.16 *Time evolution of bubble shapes and velocity vectors at section $y = L_y/2$ of coalescence of two rising bubbles in liquid ($\rho_l/\rho_g = 1000, \eta_l/\eta_g = 50, M_o = 1$, Eö =15)*

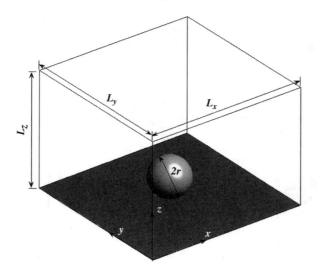

Figure 6.17 *Computational domain of a droplet on partial wetting surface (Reprinted from Yan, Y. Y., Zu, Y. Q., A lattice Boltzmann method for incompressible two-phase flows on partial wetting surface with large density ratio, Journal of Computational Physics,* **227**(1), 763–775. *Copyright (2006) with permission from Elsevier.)*

6.4.4.5   Results of Droplet Spreading on Partial Wetting Surface

The motion of water droplets at normal temperature surrounded by air on a partial wetting wall is considered. The gravitational force is taken into account by adding the term $-3\omega_\alpha e_{\alpha 3}(1-\rho_g/\rho_l)g$ to the right-hand side of Equation 6.43, where g is the dimensionless gravitational acceleration. Naturally, the density ratio of the two fluids are set at $\tilde{\rho}_l/\tilde{\rho}_g = 1 \times 10^3$, and the dynamic viscosity ratio is set at $\tilde{\eta}_l/\tilde{\eta}_g = 50$. The initial surface tension between water and air is $\tilde{\sigma}_{lg} = 1 \times 10^{-3} \mathrm{kgs}^{-2}$ and the gravitational acceleration is set at $\tilde{g} = 9.8\,\mathrm{ms}^{-2}$. To relate the physical parameters to the simulation parameters, a length scale of $L_0 = 1 \times 10^{-4}\,$m, time scale of $T_0 = 1 \times 10^{-6}\,$s and mass scale of $M_0 = 1 \times 10^{-12}\,$kg are chosen. The simulations are within a cuboid computational domain, as shown in Figure 6.17, with a no-slip boundary at the lower surface, that is, the flat partial wetting wall and free outflow/inflow boundaries at the other five surfaces, ε in Equation 6.58, is set as $\varepsilon = 1 \times 10^{-6}$.

Figure 6.18 shows the droplet spreading on a hydrophilic wall, and as time progresses, finally reaches an equilibrium shape; while the contact angle approximates to the initially predicted value, that is, $\theta_w = \pi/4$. Figure 6.19 shows the velocity field on the cross-section of $x = L_x/2$ at $t = 0.006$ s, where the solid line is the interface between two phases is given, to show that the present method can obtain a stable and reasonable velocity distribution.

Figure 6.20 shows evolutions with time of the droplet on a flat moderate wall. The contacting area increases with time due to the effect of gravity, the droplet finally reaches an equilibrium shape with contact angle $\theta_w = \pi/2$. In addition, a water droplet in air spreading on a partial wetting wall is also simulated [11]. Figure 6.21 shows how a small hemispherical

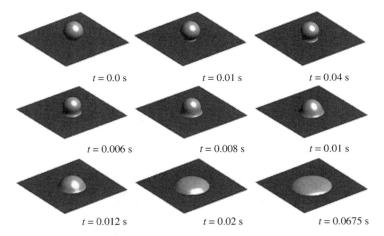

Figure 6.18 *Snapshots of droplet spreading on a uniform hydrophilic surface, $\theta_w = \pi/4$ (Reprinted from Yan, Y. Y., Zu, Y. Q., A lattice Boltzmann method for incompressible two-phase flows on partial wetting surface with large density ratio, Journal of Computational Physics, 227(1), 763–775. Copyright (2006) with permission from Elsevier.)*

water droplet evolves with time on a heterogeneous surface. A narrow hydrophobic strip with width of $l = 6 \times 10^{-4}$ m is located at the centreline of the surface where $\theta_w = 5\pi/6$, and the other areas are occupied by the hydrophilic surface with $\theta_w = \pi/6$. The initial droplet that has a radius $r = 1.5 \times 10^{-3}$ m is set at the centre of the wetting surface.

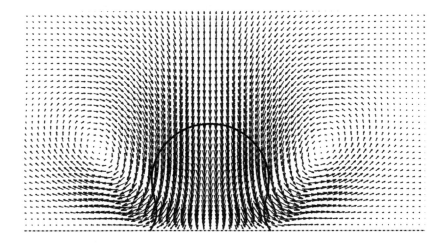

Figure 6.19 *Velocity distribution on the cross-section of $x = L_x/2$, $\theta_w = \pi/2$, $t = 0.006\,s$ (Reprinted from Yan, Y. Y., Zu, Y. Q., A lattice Boltzmann method for incompressible two-phase flows on partial wetting surface with large density ratio, Journal of Computational Physics, 227(1), 763–775. Copyright (2006) with permission from Elsevier.)*

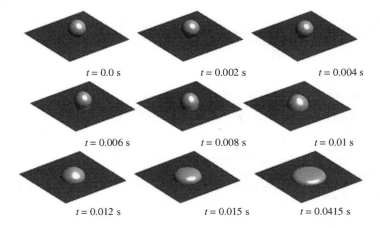

Figure 6.20 *Snapshots of a droplet spreading on a uniform moderate surface, $\theta_w = \pi/2$ (Reprinted from Yan, Y. Y., Zu, Y. Q., A lattice Boltzmann method for incompressible two-phase flows on partial wetting surface with large density ratio, Journal of Computational Physics, 227(1), 763–775. Copyright (2006) with permission from Elsevier.)*

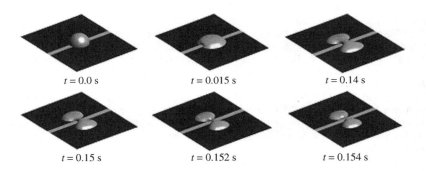

Figure 6.21 *Snapshots of droplet spreading and its break-up on a heterogeneous surface (Reprinted from Yan, Y. Y., Zu, Y. Q., A lattice Boltzmann method for incompressible two-phase flows on partial wetting surface with large density ratio, Journal of Computational Physics, 227(1), 763–775. Copyright (2006) with permission from Elsevier.)*

References

[1] Chen, S. and Doolen, G.D. (1998) Lattice Boltzmann method for fluid flows. *Ann. Rev. Fluid Mech.*, **30**, 329–364.

[2] (a) Wolfram, S. (ed.) (1986) *Theory and Applications of Cellular Automata*, World Scientific Publishing Co., Hong Kong (distributed by Taylor and Francis, Philadelphia) pp. 333–342;(b) Wolfram, S. (2002) *A New Kind of Science*, Wolfram Media, Champaign, IL.

[3] Succi, S. (2001) *The Lattice Boltzmann Equation for Fluid Dynamics and Beyond*, Oxford University Press.

[4] Gunstensen, A.K., Rothman, D.H., Zaleski, S. and Zanetti, G. (1991) Lattice Boltzmann model of immiscible fluids. *Phys. Rev. A*, **43**, 4320–4327.

[5] Shan, X.W. and Chen, H.D. (1993) Lattice Boltzmann model for simulating flows with multiple phases and components. *Phys. Rev. E.*, **47**, 1815–1819.

[6] (a) Shan, X.W. and Chen, H.D. (1994) Simulation of nonideal gases and liquid-gas phase-transitions by the lattice Boltzmann-equation. *Phys. Rev. E*, **49**, 2941–2948; (b) Shan, X.W. and Doolen, G. (1995) Multicomponent lattice-Boltzmann model with interparticle interaction. *J. Stat. Phys.*, **81**, 379–393; (c) Shan, X.W. and Doolen, G. (1996) Diffusion in a multicomponent lattice Boltzmann equation model. *Phys. Rev. E*, **54**, 3614–3620.

[7] Martys, N.S. and Douglas, J.F. (2001) Critical properties and phase separation in lattice Boltzmann fluid mixtures. *Phys. Rev. E*, **6303**, 1/031205-18/031205.

[8] (a) Swift, M.R., Osborn, W.R. and Yeomans, J.M. (1995) Lattice Boltzmann simulation of nonideal fluids. *Phys. Rev. Lett.*, **75**, 830–833; (b) Swift, M.R., Orlandini, E., Osborn, W.R. and Yeomans, J.M. (1996) Lattice Boltzmann simulations of liquid-gas and binary fluid systems. *Phys. Rev. E*, **54**, 5041–5052.

[9] He, X.Y., Chen, S.Y. and Zhang, R.Y. (1999) A lattice Boltzmann scheme for incompressible multiphase flow and its application in simulation of Rayleigh-Taylor instability. *J. Comp. Phys.*, **152**, 642–663.

[10] Inamuro, T., Ogata, T., Tajima, S. and Konishi, N. (2004) A lattice Boltzmann method for incompressible two-phase flows with large density differences. *J. Comp. Phys.*, **198**, 628–644.

[11] Yan, Y.Y. and Zu, Y.Q. (2007) A lattice Boltzmann method for incompressible two-phase flow with large density ratio on partial wetting surface. *J. Comp. Phys.*, **227**, 763–775.

[12] Qian, Y.H., d'Humieres, S. and Lallemand, P. (1992) Lattice BGK models for Navier-Stokes equation. *Europhys Lett.*, **16**, 479–484.

[13] Kandhai, D., Koponen, A., Hoekstra, A.G. *et al.* (1998) Lattice-Boltzmann hydrodynamics on parallel systems. *Comp. Phys. Commun.*, **111**, 14–26.

[14] Pan, C.X., Prins, J.F. and Miller, C.T. (2004) A high-performance lattice Boltzmann implementation to model flow in porous media. *Comp. Phys. Commun.*, **158**, 89–105.

[15] Wang, J.Y., Zhang, X.X., Bengough, A.G. and Crawford, J.W. (2005) Domain-decomposition method for parallel lattice Boltzmann simulation of incompressible flow in porous media. *Phys. Rev. E*, **72**, 01/016706-11/0167062005.

[16] Yan, Y.Y. and Zu, Y.Q. (2008) Numerical simulation of heat transfer and fluid flow past a rotating isothermal cylinder – a LBM approach. *Int. J. Heat Mass Transfer.*, **51**, 2519–2536.

[17] Yang, Z.L., Dinh, T.N., Nourgaliev, R.R. and Sehgal, B.R. (2001) Numerical investigation of bubble growth and detachment by the lattice-Boltzmann method. *Int. J. Heat Mass Transfer*, **44**, 195–206.

[18] Theodoropoulos, C., Sankaranarayanan, K., Sundaresan, S. and Kevrekidis, I.G. (2004) Coarse bifurcation studies of bubble flow lattice Boltzmann simulations. *Chem. Eng. Sci.*, **59**, 2357–2362.

[19] Mei, R.W., Yu, D.Z., Shyy, W. and Luo, L.S. (2002) Force evaluation in the lattice Boltzmann method involving curved geometry. *Phys. Rev. E*, **65**, 041203.

[20] Coutanceau, M. and Menard, C. (1985) Influence of rotation on the near-wake development behind an impulsively started circular cylinder. *J. Fluid Mech.*, **158**, 399–466.

[21] Gunstensen, A.K., Rothman, D.H., Zaleski, S. and Zanetti, G. (1991) Lattice Boltzmann model of immiscible fluids. *Phys. Rev. A*, **43**, 4320–4327.

[22] Dortona, U., Salin, D., Cieplak, M. and Banavar, J.R. (1994) Interfacial phenomena in Boltzmann cellular-automata. *Europhys. Lett.*, **28**, 317–322.

[23] Kono, K., Ishizuka, T., Tsuba, H. and Kurosawa, A. (2000) Application of lattice Boltzmann model to multiphase flows with phase transition. *Comp. Phys. Commun.*, **129**, 110–120.

[24] Spaid, M.A.A. and Phelan, F.R. (1998) Modeling void formation dynamics in fibrous porous media with the lattice Boltzmann method. *Compos. Part A-Appl. S.*, **29**, 749–755.

[25] (a) Sukop, M.C. and Or, D. (2003) Invasion percolation of single component, multiphase fluids with lattice Boltzmann models. *Physica B: Cond. Mat.*, **338**, 298–303; (b) Zhang, J.F. and Kwok, D.Y. (2005) A 2D lattice Boltzmann study on electrohydrodynamic drop deformation with the leaky dielectric theory. *J. Comp. Phys.*, **206**, 150–161; (c) Zhang, J.F. and Kwok, D.Y. (2005) On

the validity of the Cassie equation via a mean-field free-energy lattice Boltzmann approach. *J. Colloid Interf. Sci.*, **282**, 434–438.

[26] (a) Sankaranarayanan, K., Shan, X., Kevrekidis, I.G. and Sundaresan, S. (1999) Bubble flow simulations with the lattice Boltzmann method. *Chem. Eng. Sci.*, **54**, 4817–4823; (b) Sankaranarayanan, K., Shan, X., Kevrekidis, I.G. and Sundaresan, S. (2002) Analysis of drag and virtual mass forces in bubbly suspensions using an implicit formulation of the lattice Boltzmann method. *J. Fluid Mech.*, **452**, 61–96.

[27] (a) Yang, Z.L., Dinh, T.N., Nourgaliev, R.R. and Sehgal, B.R. (2000) Numerical investigation of bubble coalescence characteristics under nucleate boiling condition by a lattice-Boltzmann model. *Int. J. Thermal Sci.*, **39**, 1–17; (b) Yang, Z.L., Dinh, T.N., Nourgaliev, R.R. and Sehgal, B. R. (2001) Numerical investigation of boiling regime transition mechanism by a Lattice-Boltzmann model. *Nucl. Eng. Des.*, **204**, 143–153; (c) Yang, Z.L., Palm, B. and Sehgal, B.R. (2002) Numerical simulation of bubbly two-phase flow in a narrow channel. *Int. J. Heat Mass Transfer*, **45**, 631–639.

[28] Raiskinmaki, P., Koponen, A., Merikoski, J. and Timonen, J. (2000) Spreading dynamics of three-dimensional droplets by the lattice-Boltzmann method. *Comp. Mat. Sci.*, **18**, 7–12.

[29] Sehgal, B.R., Nourgaliev, R.R. and Dinh, T.N. (1999) Numerical simulation of droplet deformation and break-up by lattice-Boltzmann method. *Prog. Nucl. Energy*, **34**, 471–488.

[30] (a) Zhang, J.F. and Kwok, D.Y. (2005) A 2D lattice Boltzmann study on electrohydrodynamic drop deformation with the leaky dielectric theory. *J. Comp. Phys.*, **206**, 150–161; (b) Zhang, J.F. and Kwok, D.Y. (2005) On the validity of the Cassie equation via a mean-field free-energy lattice Boltzmann approach. *J. Coll. Interf. Sci.*, **282**, 434–438.

[31] Onishi, J., Chen, Y. and Ohashi, H. (2006) Dynamic simulation of multi-component viscoelastic fluids using the lattice Boltzmann method. *Physica A: Stat. Mech. Applic.*, **362**, 84–92.

[32] (a) Hou, S.L., Shan, X.W., Zou, Q.S., *et al.* (1997) Evaluation of two lattice Boltzmann models for multiphase flows. *J. Comp. Phys.*, **138**, 695–713; (b) Hou, T.Y., Lowengrub, J.S. and Shelley, M.J. (1997) The long-time motion of vortex sheets with surface tension. *Phys. Fluids*, **9**, 1933–1954.

[33] Orlandini, E., Gonnella, G. and Yeomans, J.M. (1997) Lattice Boltzmann study of spinodal decomposition in structured fluids. *Physica A*, **240**, 277–285.

[34] Osborn, W.R., Orlandini, E., Swift, M.R. *et al.* (1995) Lattice Boltzmann study of hydrodynamic spinodal decomposition. *Phys. Rev. Lett.*, **75**, 4031–4034.

[35] Wagner, A.J. and Yeomans, J.M. (1999) Phase separation under shear in two-dimensional binary fluids. *Phys. Rev. E*, **59**, 4366–4373.

[36] (a) Takada, N. and Tsutahara, M. (1998) Evolution of viscous flow around a suddenly rotating circular cylinder in the lattice Boltzmann method. *Comp Fluids*, **27**, 807–828; (b) Takada, N., Misawa, M., Tomiyama, A. and Hosokawa, S. (2001) Simulation of bubble motion under gravity by lattice Boltzmann method. *J. Nucl. Sci. Technol.*, **38**, 330–341.

[37] (a) Inamuro, T., Konishi, N. and Ogino, F. (2000) A Galilean invariant model of the lattice Boltzmann method for multiphase fluid flows using free-energy approach. *Comp. Phys. Commun.*, **129**, 32–45; (b) Inamuro, T., Tajima, S. and Ogino, F. (2004) Lattice Boltzmann simulation of droplet collision dynamics. *Int. J. Heat Mass Transfer*, **47**, 4649–4657.

[38] (a) Dupuis, A. and Yeomans, J.M. (2004) Lattice Boltzmann modelling of droplets on chemically heterogeneous surfaces, Future Generation. *Comput. Syst.*, **20**, 993–1001; (b) Dupuis, A. and Yeomans, J.M. (2005) Modeling droplets on superhydrophobic surfaces: Equilibrium states and transitions. *Langmuir*, **21**, 2624–2629.

[39] Angelopoulos, A.D., Paunov, V.N., Burganos, V.N. and Payatakes, A.C. (1998) Lattice Boltzmann simulation of nonideal vapor-liquid flow in porous media. *Phys. Rev. E*, **57**, 3237–3245.

[40] Buick, J.M., Cosgrove, J.A. and Greated, C.A. (2004) Gravity-capillary internal wave simulation using a binary fluid lattice Boltzmann model. *Appl. Math. Modelling*, **28**, 183–195.

[41] Holdych, D.J., Rovas, D., Georgiadis, J.G. and Buckius, R.O. (1998) An improved hydrodynamics formulation for multiphase flow lattice-Boltzmann models. *Int. J. Mod. Phys. C*, **9**, 1393–1404.

[42] Palmer, B.J. and Rector, D.R. (2000) Lattice-Boltzmann algorithm for simulating thermal two-phase flow. *Phys. Rev. E*, **61**, 5295–5306.

[43] Desplat, J.C., Pagonabarraga, I. and Bladon, P. (2001) LUDWIG: A parallel Lattice-Boltzmann code for complex fluids. *Comp. Phys. Commun.*, **134**, 273–290.

[44] Cahn, J.W. (1977) Critical-point wetting. *J. Chem. Phys.*, **66**, 3667–3672.

[45] Lee, T. and Lin, C.L. (2005) A stable discretization of the lattice Boltzmann equation for simulation of incompressible two-phase flows at high density ratio. *J. Comp. Phys.*, **206**, 16–47.

[46] Yan, Y.Y. and Zu, Y.Q. (2006) Lattice Boltzmann simulation of vortices merging in a two-phase mixing layer. *WIT Trans. Eng. Sci.*, **53**, 877–896.

[47] (a) Dong, B., Yan, Y.Y., Li, W.Z. and Song, Y. (2010) Lattice Boltzmann simulation of viscous fingering phenomenon of immiscible fluids displacement in a channel. Computers & Fluids, **39**(5), 768–779. (b) Dong, B., Yan, Y.Y., Li, W.Z. (2011) LBM simulation of viscous fingering phenomenon in immiscible displacement of two fluids in porous media. Transport in Porous Media, DOI: 10.1007/s11242–011–9740-y.

[48] Jamet, D., Lebaigue, O., Coutris, N. and Delhaye, J.M. (2001) The second gradient theory: a tool for the direct numerical simulation of liquid-vapor flows with phase-change. *Nucl. Eng. Des.*, **204**, 155–166.

[49] Rowlinson, J.S. and Widom, B. (1989) *Molecular Theory of Capillarity*, Clarendon Press, Oxford.

Part Two

7

Upset Conditions

7.1 Introduction

Deviations from the normal operating conditions of a gas-liquid reactor can occur for a number of reasons. There can be fluctuation in feed rate or stirrer speeds, which cause an oscillation. However, the focus here is on more drastic changes. These are ones that can cause much more harm. The overpressurisation of a storage vessel or a batch type chemical reactor, due to external heating or a runaway reaction, is one of the accidental events that may occur in the chemical industry. The consequences of such an event can be benign (but still costly in terms of lost production) when the products are safely vented to a dump tank or similar device. Alternatively, they can be disastrous if the overpressure causes rupture of the vessel. There can also be terrible consequences for the environment and surrounding people if the products are released to the atmosphere, as for example in the case of the accidents at Seveso and Bhopal.

It has been noted that with the increasing emphasis on consumer orientated chemical products that the use of batch and semi-batch reactions has increased. These systems have a complex chemistry, multi-product use of equipment and high density siting of equipment. They usually have larger inventory than continuous processes and limited zones into which it is safe to discharge.

The importance of venting to these sectors of the industry can be seen in the results of the work by Barton and Nolan [1] who studied case histories of industrial incidents in batch

Hydrodynamics of Gas-Liquid Reactors: Normal Operation and Upset Conditions, First Edition.
B. J. Azzopardi, R. F. Mudde, S. Lo, H. Morvan, Y. Yan and D. Zhao.
© 2011 John Wiley & Sons, Ltd. Published 2011 by John Wiley & Sons, Ltd.

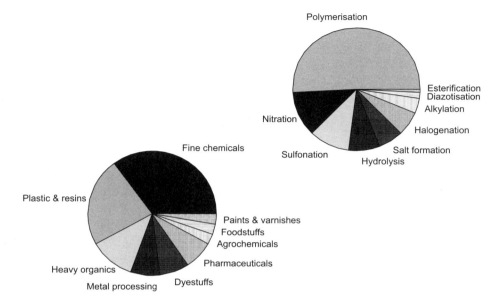

Figure 7.1   *Division of recorded incidents according to the type of reaction and to the sector of industry*

systems in the United Kingdom in the period 1962–1987. Figure 7.1 shows the proportions by types of reactions and per sector. Table 7.1 lists the causes of the runaways.

7.2   Active Relief Methods

The safety measures that are employed to protect reactor vessels are usually classified as either active or passive. *Active relief* techniques can include: designing out (inherent safety), operational procedures, process control (often high integrity), crash cooling, containment, interrupting the chemical reaction (sometimes known as a killer injection or anti-catalyst) and dumping/dilution through use of control valves. The passive methods are what are termed venting, the release of vessel contents into larger low pressure containment by self-actuated devices. Maddison and Rogers [2] discussed the implications of these. Process control alone as a basis of safety would only be used if *all* the hazards and likely maloperations of the system have been identified. However, as this level of knowledge is not normally available, process control is inevitably backed up by one of the other measures. To be able to use crash cooling, quantitative information is required as to the rate and magnitude of any excursion. If containment is to be used in conjunction with process control, the reactor and all associated and interlinked equipment must be able to withstand the peak pressure. Obviously, this must be known accurately. Dumping implies emptying the vessel contents into a large tank normally containing sufficient amounts of a liquid, often water, to provide a heat sink and to act as a diluent. Here, compatibility between process liquids and the dump tank liquid is important and all possible further problems, such as secondary reactions, foaming or inflammable atmospheres, should have been examined.

Table 7.1   Causes of runaways

Cause	Number	Cause	Number
Lack of knowledge of chemistry			
Material out of specification	14	Unintended oxidation	3
No heat of reaction study	8	Reactant concentration too high	2
Mixture decomposed	7	Temperature too low (accumulation)	1
Unstable by-products	6	Catalysed reaction	2
Reaction carried out *en masse*	4	Phase change	1
Incidents caused by maloperation involving poor control of temperature			
Failure of control system	7	Error in taking reading	4
Excessive steam	6	Radiant heating	3
Temperature probe incorrectly sited	6	Heating too rapid	1
Loss of cooling	5	Coated thermocouples	1
Incidents caused by maloperation involving agitation			
Agitator not started or started late	6	Loss of power supply	2
Inadequate agitator specification	4	Agitator stopped, localised heating	2
Mechanical failure	3	—	—
Incidents caused by maloperation involving mischarging of reactants			
Overcharging	12	Undercharging	3
Addition too rapid	8	Improper control	2
Wrong material	5	Addition too slow	1
Wrong addition sequence	4	—	—
Incidents caused by maloperation involving maintenance			
Equipment leaks	7	*In situ* maintenance during reaction	2
Blockages	6	Residues from previous batch	2
Closed reflux valve	3	Unauthorised modifications	1
Water in transfer lines	3	Loss of instrument air	1

Whatever method is adopted, all possible causes of overpressure should be considered and the resultant required relief rate evaluated (approximately). In this evaluation, the type of flow, single or two phase must be identified. It is noted that even when active methods are implemented, they are often backed up by provision of passive methods.

7.3   Passive Relief Methods

The devices to effect *passive* protection are either pressure relief valves or bursting discs. Pressure relief valves are usually of the globe-valve format with the stem held shut by a spring. The tightness of the spring can be adjusted to ensure opening at different pressures as required. It is then usually sealed to ensure that it cannot be tampered with. Many of the valves have the outlet mounted at right angles to the inlet. Typical features of relief valves are illustrated in Figure 7.2a. Bursting discs are thin sheets of metal that have been carefully machined to provide lines of failure, Figure 7.2b. The machining can be of a cruciform shape or around the majority of the circumference. The former allows the disc to divide into four sectors to bend back out of the way of the flow. The latter creates a circular flap, which can be pushed to one side by the flow. The disc can be flat or bowed.

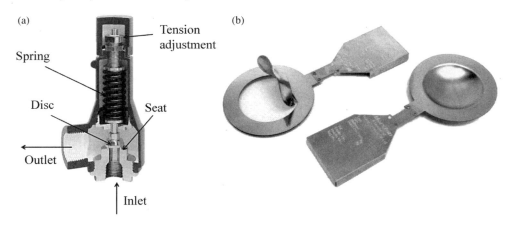

Figure 7.2 *Schematic of: (a) pressure relief valve (Courtesy of Safety Systems UK Ltd.) and (b) bursting disc*

Usually one or the other is deployed, but occasionally both are used in series. This is typically when ingress of material would be detrimental, for example a biochemical reaction where air-borne material could cause contamination. In these instances a bursting disc would be placed on the vessel side of the relief valve.

For two-phase reactors such as bubble columns or sparged stirred vessels, there are some systems where the injected gas is completely consumed in the reaction. Other systems have a steady gas outflow. In both types there is a need to provide venting to protect the vessel. The former type has all the hazards and needs of all liquid reactions, whilst for the latter the normal outlet might not be large enough for venting purposes and so an additional specific vent outlet must be provided.

This and the subsequent two chapters will concentrate on passive methods of protection. This process is often referred to as venting. When this measure is used, consideration must be given to: the compatibility with the design and operation of the process; the certainty with which the 'worse case' situation can be identified; can reaction parameters during runaway be obtained; and design of the relief system including vent and downstream piping and equipment sizing.

An uncontrolled release to the atmosphere usually occurs due to vessel failure, which often results from the fact that the emergency relief system is not correctly sized. Traditionally, emergency relief systems were designed assuming single-phase conditions in the vent line, whereas in reality often two-phase conditions prevail, the reasons for this are discussed in Chapter 8. Compared with relief systems designed to handle single-phase vapour, two-phase flow requires vent sizes that are 2–10 times larger. A further complication has been produced by very correct environmental awareness. Historically, the relief system was the relief device followed by a short pipe to atmosphere. The impact of the piping was minimal and could be incorporated in the vent design. Additional complications arise from the requirement to install special equipment to treat the relieved fluids if they are toxic or flammable. This special equipment may consist of knockout drums, vapour-liquid separations, catch tanks, condensers, scrubbers, and so on. Whatever the system the cost of such equipment is not insignificant. Therefore, it is important to optimise the size of the relief lines and ducting. They should be large enough to ensure that the peak pressure stays within safe limits during relief, but not too large, in order to minimise the amount of fluids emitted from the reactor that will then need to be treated.

In view of this interest, much effort has been expended over the last decade in an attempt to improve the understanding of the basic phenomena associated with emergency relief. For example, the chemical industry has performed many experiments, some of which have been described in the open literature, and a major effort has been the work performed in the DIERS (Design Institute for Emergency Relief Systems) project. The DIERS project consisted of a consortium of 29 companies under the auspices of the American Institute of Chemical Engineers and was formed to: (i) generate experimental data on large-scale vessels undergoing either runaway reactions or external heating (e.g. by external fires or loss of cooling) and (ii) to develop methods for the safe design of emergency relief systems to handle these events. The project ran from 1976 to 1984. Since then the DIERS User Group has continued to meet and share information and developments in the field of process safety and relief both in Europe and the United States. Current membership exceeds 120 companies worldwide. Work was also carried out at the Joint Research Centre (JRC) of the European Community at Ispra in Italy in a programme dealing with 'Industrial Hazards', although this project has now stopped. Significant progress has been made, but the knowledge obtained is by no means complete.

Chemical reactions can be the cause of a rise in pressure in a closed system through increasing the vapour pressure of the system and/or by generating non-condensable gases as reaction or unwanted decomposition products. Even endothermic reactions can cause a pressure increase if the reaction products are gases, or liquids, which are more volatile than the reactants. Exothermic reactions are potentially more dangerous, because in addition they raise the temperature of the reactants and hence accelerate the chemical reaction.

In the literature, distinction is often made between the mechanisms of this pressure rise so that simplifications can be made to the mathematical treatment of the relief system sizing. Following the pioneering work of the DIERS project, reactions are usually classified into:

a. **'Vapour pressure' or 'tempered' systems.** In these, pressure is generated during the reaction due to the increase in vapour pressure of the reactants, products and inert solvent (if present) caused by a temperature rise. Production of vapour requires heat (latent heat of vaporisation). It is possible that the temperature is approximately constant over part of the process. During this, the reaction rate does not increase.
b. **'Gassy' systems.** For these, pressure increase results from permanent gas being produced by the main or side reactions occurring.
c. **'Hybrid' systems.** This is a combination of the previous two cases. In fact hybrid systems are often classified as either being 'hybrid/tempered' or 'hybrid/non-tempered' systems.

The rise in pressure and temperature with time and the pressure–temperature relationship for these systems are illustrated in Figure 7.3.

For such classifications, a number of analytical tools and formulae can be used to calculate the vent size for a particular overpressure. These 'hand calculation' methods

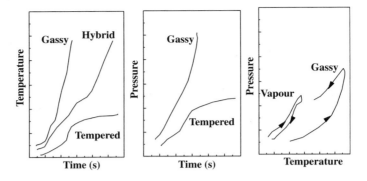

Figure 7.3   *Pressure and temperature behaviour for different systems*

usually treat the vessel as a single node having uniform properties. The obvious difficulty arises when this assumption is not valid and when it is not known *a priori* what type of system is expected. Excellent early review of existing vent sizing methods of this type are available [3].

In the work on relief systems, it is important to define certain terms. A pressure vessel has a design pressure, which may only be exceeded in emergency events and then only by 10% of the gauge pressure. Vessels for low pressure have other requirements, for example API 650 allows no overpressure. A relief device has a set pressure at which it starts to open (for relief valve) or bursts (for bursting disc). This is the pressure at which the relief device is known to be fully open. For safety valves, usually snap or pop action valves, a flow rate will be passed as certified by the manufacturer at a gauge pressure 10% above its set pressure. The term overpressure will be used to mean the difference between the set pressure and the maximum pressure attained during the venting sequence divided by the set pressure. Both pressures are in absolute terms. Typical values of overpressure are 20%. It is noted that some workers specify overpressure in terms of gauge pressures and others consider absolute pressures. These terms are laid down by relevant mechanical engineering codes of practice.

It is often preferable to set the relief device on a reactor to operate at as low a pressure as possible consistent with avoiding spurious operation of the relief device. For a vapour pressure or tempered hybrid system, this allows the vent to open at a low temperature, when the runaway reaction is still at a relatively low rate. A low set pressure also makes it more likely that a hybrid system will temper. Another advantage of a low set pressure, and hence temperature, is protection of the vessel, as an increasing temperature lowers the safe working pressure of the vessel.

For vapour pressure and tempered hybrid systems, a set pressure below the vessel design pressure is advantageous. This is because the pressure may now be allowed to rise during venting. Thus, if two-phase relief occurs, the vent may be sized such that the reactor empties sufficiently for the pressure to pass through a peak and then fall, before the maximum allowable vessel pressure has been reached. In the absence of further information, the design pressure might be used, assuming this is below the maximum vessel pressure allowable.

References

[1] Barton, J.A. and Nolan, P. (1990) Incidents in the chemical industry due to thermal runaway chemical reactions. Paper presented at the Symposium on Chemical Reaction Hazards, December 12–13, 1990, Amsterdam.

[2] (a) Maddison, N. and Rogers, R.L. (1994) Understanding chemical reaction hazards. *Chem. Tech. Europe*, **1**, 13–19; (b) Maddison, N. and Rogers, R.L. (1994) Chemical runaways: incidents and their causes. *Chem. Tech. Europe*, **1**, 28–31.

[3] (a) Duxbury, H.A. and Wilday, A.J. (1989) Efficient design of reactor relief systems. Paper presented at the International Symposium on Runaway Reactions, Boston, Massachusetts, March 1989; (b) Fisher, H.G. (1991) An overview of emergency relief system design practice. *Plant Oper. Progr.*, **10**, 1–12.

8

Behaviour of Vessel Contents and Outflow Calculations

8.1 Introduction

8.1.1 Physics of Venting Processes

This chapter considers the physical processes that occur around the operation of passive pressure protection systems for vessels, particularly those employed as chemical reactors. It presents the evidence of the behavioural trends and the equations required to determine the dimensions of the required device, that is the diameter of the safety relief valve or bursting

---

Hydrodynamics of Gas-Liquid Reactors: Normal Operation and Upset Conditions, First Edition.
B. J. Azzopardi, R. F. Mudde, S. Lo, H. Morvan, Y. Yan and D. Zhao.
© 2011 John Wiley & Sons, Ltd. Published 2011 by John Wiley & Sons, Ltd.

disc. The methodology employed to determine the data necessary for those calculations will be summarised and, finally, how well these calculation methods work will be reported.

When the rate of heat input, whether from the exothermic heat of reaction from an occurring runaway or from an external source such as a fire, exceeds the cooling capability of the equipment, the temperature of the contents will rise. This results in vaporisation, particularly when the saturation temperature or bubble point is reached. Permanent gases can also be formed in some instances when those gases are products of the main reaction or of side reactions that can occur at the higher temperatures during the runaway.

As with carbonated drinks or champagne, during their residence in a liquid, the bubbles will occupy volume and so cause the liquid level to rise or 'swell'. When the set pressure is reached and the reactor or storage vessel relieves, then the removal of fluids permits further evaporation or 'flashing'. This causes the liquid level, or to be more precise the 'two-phase mixture level', to rise further and if this level reaches the vent position, then two-phase venting will occur. The two-phase mixture level separates the region that is predominately liquid (containing vapour bubbles) from the region which is predominately vapour (containing liquid droplets) and is usually defined as the position where there is a discontinuity in the axial void fraction profile. Attention has to be given to the occurrence and extent of level swell. If the reactor is a bubble column or a sparged stirred vessel, there are bubbles already present and evaporation goes into increasing the numbers of bubbles and making the bubbles larger. If the reactor contents are an aerated liquid, more bubbles will quickly form and then the process will follow the same path, that is produced vapour or gas goes into making the bubbles larger. Although most available information starts from a single-phase liquid, there has been some work on venting of bubble columns [1].

The bubbles will rise through the liquid and increase the cumulative gas content up the vessel. The void fraction will increase with vertical position. As the bubbles rise they will drag liquid up with them. Consequently, as with the bubble columns discussed in Chapter 2, there has to be a down flow of liquid. Although some workers have presented this as a global recirculation, the reality is much more complex. It is noted that small bubbles can be carried down with the liquid. These will grow by evaporation, rise and then the cycle is repeated.

At the top of the swollen two-phase pool the bubbles will burst. This is usually known as disengagement. The bursting process can produce a mist of very fine drops. However, at large vapour production rates the bubble density is high and the two-phase mixture can reach the outlet connecting to the vent line. Figure 8.1 shows this phenomenon qualitatively.

As the outflow of material, under critical flow conditions, is inversely proportional to the mixture density at the vent line entrance, the capacity to reduce the system pressure by venting is strongly diminished when the mixture level reaches the vent position. In a runaway situation if the volume production rate due to evaporation and/or gas production is greater than the vented volume flow rate the system pressure will increase. Therefore, the ability to describe the motion of the two-phase mixture level is one of the most important aspects of reactor relief modelling.

8.1.2 Typical Reactions

The behaviour of the vessel contents, in addition to its discharge, has been studied using a number of well characterised reactions. Some, such as hydrolysis of anhydrides (acetic/proprionic), the methanol/acetic anhydride esterification or the polymerisation of styrene,

Normal operation   **Runaway situation**

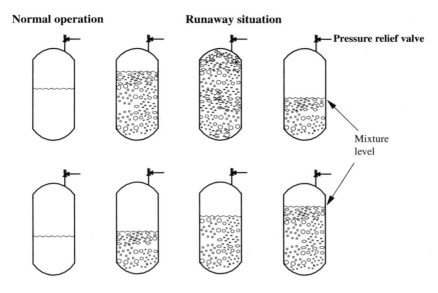

Figure 8.1   *Reactor level swell processes at fast and slow vapour or gas production rates*

are of the tempered type. The decomposition of peroxides represents the gassy type. To ensure pure gassy reaction some work has been carried out using a high boiling point solvent to minimise vaporisation, for example 2,2,4-trimethyl-1,2-pentanediol diisobutyrate. In some instances the mixtures have had extra chemicals added to either increase the viscosity of the liquid or produce foaming properties. Examples of the reactions used and the parameters employed in the tests are given in Table 8.1. These reactions have well known reaction kinetics and as they are carried out with pure reagents, there are no complicating side reactions. It should be noted that this is not comprehensive but merely illustrative. In addition to these, venting experiments have been carried out using the boiling of liquids or by exsolution of gases such as carbon dioxide.

8.1.3   Trends and Observations

When the relief device opens, the pressure will drop rapidly and then come back up to just below the original value before falling again, although more slowly. Gebbeken and Eggers [8] showed this occurring for initially supercritical carbon dioxide. In one series of experiments [9] the changes in void fraction up the vessel were measured using local capacitance probes during the transient boiling following opening of an outlet valve. At short times, the vertical void profile in the lower part of the vessel increases up the vessel due to the accumulation of vapour. Towards the top of the vessel there is a strong gradient of the void fraction with height. This can indicate a strong recirculation with the vapour rising up the middle, carrying liquid up with it whilst part of that liquid returns down the sides. At longer times there is a flattening off with all vapour in the top ~30% of the vessel. Obviously, there is a finite time for the transient; there is only so much excess energy that can be converted into vapour.

Similar behaviour is seen in gas exsolution experiments [10] using water with carbon dioxide dissolved in it, as illustrated in Figure 8.2. This figure contains data from outlet

Table 8.1   *Examples of experiments involving chemical reactions employed in runaway/venting studies*

Reference	Reaction	Vessel volume (l)	Vent restriction diameter (mm)	Notes
Grolmes and Leung [2]	Polymerisation of styrene	32	14.1	—
		2190	76.2	—
Snee *et al.* [3]	Hydrolysis of acetic anhydride	1.5	—	—
Snee *et al.* [4]		340	75	—
		2500	100	—
Friedel and Wehmeier [5]	Methanol-acetic anhydride	1.7	2.5, 4.3	—
Snee *et al.* [4]		105	27	—
		350	15–35	—
Hoff *et al.* [6]	Methanol-propionic anhydride	65	17.5	Added PVP (polyvinylpyrrolidone) to increase liquid viscosity
Vechot *et al.* [7]	Decomposition of cumene hydroperoxide	0.125	0.465	2,2,4-Trimethyl-1,2-pentanediol diisobutyrate used as high boiling point solvent

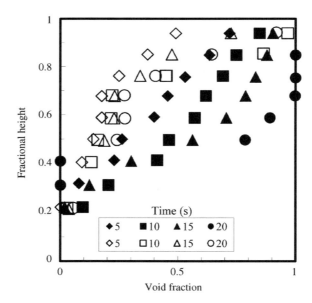

Figure 8.2   *Time evolution of the vertical distribution of a cross-sectionally averaged void fraction during degassing of CO_2 from water in a 35 l vessel. These results show the effect of the vent size on the behaviour. Orifice area/vessel volume: closed symbols, $0.0068\,m^{-1}$; open symbols, $0.0022\,m^{-1}$. Data of Thies and Mewes [10]*

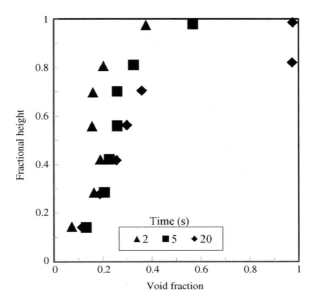

Figure 8.3 *Time evolution of the vertical distribution of a cross-sectionally averaged void fraction during boiling of R12 in a 35 l vessel. Void fractions measured using multiple capacitance probes [9]*

orifices of two different sizes. The larger orifice allows a more rapid pressure decrease in the vessel and hence a higher exsolution rate. The consequently higher void fraction enables more two-phase mixture to reach the vessel outlet and thus to be ejected. The enhanced venting results in the process being essentially over in 20 s, leaving a pool of liquid with just gas above it. It shows that ~35% of the liquid initially present has been ejected. For the small outlet orifice, the pressure drops more slowly and consequently the rate of release of gas is slower and so lower void fractions are observed. After 25 s the process is completed but in this case there is only $< 10\%$ of the liquid ejected. As with the boiling experiments in Figure 8.3, there is a finite time for the venting process as there is a limited amount of gas in solution.

Whether it is gas or a gas-liquid mixture that emerges from the vent has been found to depend on the extent to which the vessel is filled with liquid and the ratio of vent area to vessel volume or cross-sectional area. Note, most vessels used in venting experiments have height to diameters of about two. Wehmeier *et al.* [11] showed that the larger the orifice, the smaller the fractional fill that is possible for there to be only gas emerging.

If instead of boiling or degassing, the heat source is a runaway reaction, which can provide much more energy, the effects are as shown in Figures 8.4 and 8.5. These were carried out with styrene (polymerisation) and ethyl benzene (chosen as having properties similar to styrene but not reacting) for the corresponding boiling tests. These is an obvious higher rate of vapour generation in the reacting case and hence a higher void fraction. Initially there is an increase in pressure when there is a reaction occurring, and it only starts to decrease some 20 s into the transient. This is usually called the turnaround point.

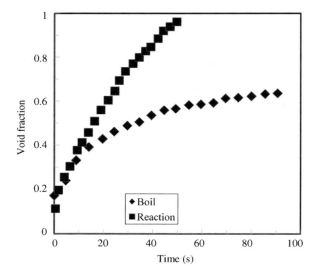

Figure 8.4   *Difference between boiling and reacting systems in the transient behaviour of vessel average void fraction. Data of Grolmes and Leung [2]*

The effect of the restriction in the vent line illustrated in Figure 8.2 is reinforced by data from the esterification of acetic anhydride with methanol shown in Figure 8.6. In these experiments a 350 l vessel 60% full was employed. The set pressure was 2 bar (1 bar = 10^5 Pa) and the vent area/vessel volume ratio was in the range 0.0005–0.0027. As seen, an

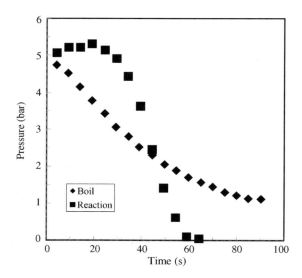

Figure 8.5   *Difference between boiling and reacting systems in the transient behaviour of vessel pressure. Data from Grolmes and Leung [2]*

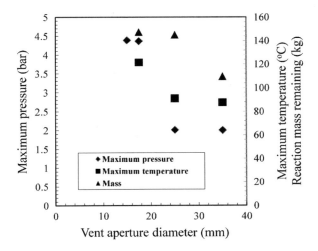

Figure 8.6 *Effect of vent aperture diameter on the maximum temperature and pressure and of reaction mass retained in vessel, reaction of methanol and acetic anhydride. Data from Snee et al. [4]*

increase in the vent aperture causes the pressure at turnaround (the maximum pressure) to drop from ~4.5 to ~2 bar at a diameter of about 20 mm. The maximum pressure follows the pressure trend. The mass of liquid remaining in the vessel, not surprisingly, goes down with increasing orifice size.

Snee *et al.* [4] discussed factors which cause the turnaround. The three major aspects are tempering, emptying (or efflux) and consumption of reactants. Evaporation of either reactants, solvents or products takes heat out of the remaining liquid. However, if it is solvent that is evaporated, then the heat of reaction could heat a smaller mass. Single- or two-phase efflux will remove reactants and thus lower the level of heat created. Similarly, as the reaction goes to completion, the heat generation rate diminishes.

These three aspects will have different magnitudes, the relative importance of which will vary from case to case. The effects are best illustrated via a specific example. Figure 8.7 shows the heat changes during a runaway of the esterification of methanol and acetic anhydride in a 1.7 l vessel [5]. The heat of reaction is given as 66.3 kJ/kg mol. The properties of the reactants and the reaction product for this reaction are given in Table 8.2. The vessel was 60% full and had a 2.5 or 4.3 mm orifice in the vent line.

There is some transfer of heat through the vessel walls. This is initially inwards but latterly outwards. The heat of reaction starts from a low value but increases exponentially as the reaction runs away. The vent opens at 100 s. Thereafter the rate of heat generation from the reaction diminishes. The amount of heat removed by efflux of materials increases rapidly at the opening of the vent, goes through a maximum and then decreases fairly rapidly. These observations can be quantified to be 18 kJ (0.085) in through the wall in the first 80 s and 15.6 kJ (0.074) outwards between 80 and 150 s. For the smaller orifice, the total heat generated by the reaction is 208 kJ (0.99) before opening of the vent and 112 kJ (0.53) after, whilst 210 kJ (1.0) are carried out with the efflux. The figures in parentheses give these

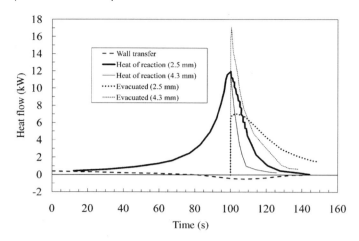

Figure 8.7   *Components of heat flow calculated for the runaway reaction of methanol and acetic anhydride*

values as a percentage of the latent heat required to evaporate the entire original contents of the vessel.

Note that for the smaller sized orifice, only vapour is expected to emerge through it. The larger orifice (4.3 mm diameter) allows not only a higher void fraction in the vessel but also a two-phase discharge through the vent line. Because of the lower boiling points, it is expected that there will be higher proportions of methanol and methyl acetate in the vapour. Therefore more acetic anhydride remains in the vessel when the smaller orifice is used, whereas some of it exits when there is a larger orifice and, not surprisingly, the heat of reaction is noticeably smaller, 55 kJ (0.26), after opening of the vent line. The heat removed through efflux is also lower, 174 kJ (0.83), as there is less evaporation.

The esterification reaction considered above and the related hydrolysis of acetic anhydride to acetic acid by water can be used for further considerations of the relative importance of tempering, emptying and reaction completion. From the data in Table 8.2, it can be seen that to raise the temperature of each of the reactants and products by 1 °C requires as much heat as to vaporise ∼0.3% of the liquid, with the exception of methanol where the figure is 0.13%. This, coupled with the lower volatility of acetic anhydride and acetic acid, means that it is the product, methyl acetate, that is preferentially vaporised. The reaction either goes on to completion or reactants are removed as liquid. For the hydrolysis reaction, increasing

Table 8.2   *Properties of the reactants and reaction product for the esterification of acetic anhydride by methanol or its hydrolysis by water*

Material	Normal boiling point (°C)	Latent heat of evaporation (kJ/kJ)	Sensible heat capacity (kJ/kg °C)
Methanol	64.5	1101	1.47
Acetic anhydride	138.5	407	1.18
Methyl acetate	56.9	400	1.30
Acetic acid	117.8	389	1.34
Water	100.0	2200	4.18

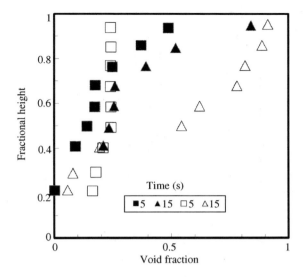

Figure 8.8 *Time evolution of the vertical distribution of cross-sectionally averaged void fraction during degassing of CO_2 from liquids of different viscosities in a 35 l vessel. Information obtained from multiple differential pressure measurements. Closed symbols, viscosity = 0.001 Pa s; open symbols, viscosity = 0.02 Pa s. Data from Thies and Mewes [10]*

the temperature of water by 1 °C requires as much heat as to vaporise ~0.19% of the liquid. Here it is the most volatile (acetic acid) that requires less heat to vaporise and so reactants are removed by tempering. The hydrolysis has similarities to many reactions where larger, less volatile molecules are created.

Increasing the liquid viscosity will mean lower bubble rise velocities. Consequently, higher void fractions are expected. This is clearly seen in Figure 8.8 where nominally equivalent results for water and water with an additive, which increases the viscosity by a factor of 20, are shown. This increased void fraction will result in ejection of more liquid through the vent.

Figure 8.9 shows that the liquid remaining in the vessel after the vent process diminishes systematically as the liquid viscosity increases. The ideas regarding the behaviour of viscous liquids are well illustrated by an unexpected occurrence observed in an experiment. Experiments were being carried out on a 225 mm diameter bubble column in which the gas was injected through a sparger consisting of ten radial pipes, each containing 15 × 2 mm diameter holes. The liquid used was molasses (the unrefined product from the pressing of sugar cane), whose viscosity was ~4 Pa s. In between tests, the gas was not turned off completely. On returning to the column, the researcher found that the molasses had overflowed from the open-topped column. Small bubbles had formed which, because of their negligible rise velocity, did not pass up through the liquid, and thus formed a raft of bubbles, which forcibly pushed the column of liquid upwards. Had this been a sealed vessel, there would have been a significant rise in pressure.

In the above example it is assumed that there could be efficient disengagement of the phases in any freeboard in the vessel. However, a number of systems have been observed

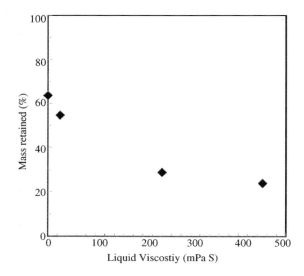

Figure 8.9   *Effect of liquid viscosity of percentage of charge retained in vessel. Orifice area/ vessel volume = 0.0037 m^{-1}. Data from Hoff et al. [6]*

to show foaming behaviour. Foams should be familiar from the head on beer or additives for baths. These are essentially bubbly systems with very high void fractions. Only low concentrations of dissolved materials with surfactant tendencies are required to make water foam, although organic systems have also been observed to foam. The effect of foaming is to fill the reactor vessel and ensure that any discharge is two-phase. For example, Wehmeier *et al.* [11] showed that in experiments with boiling water in a 105 l vessel, water without surfactant did not emerge from the vent until the vessel was at least 60% full. Even with the vessel 95% full, only 35% of the contents were ejected. However, the addition of surfactant caused 75% of the contents to be ejected for a fill of > 25%. The quantification of this effect at a larger vessel size can be seen in Figure 8.10, which shows results from a 2190 l vessel. The water data soon settle out at a value of ∼0.6, whilst the case with surfactant goes on to void fractions in excess of 0.95, that is most of the liquid is ejected. This effect is also seen in organic liquids, see Figure 8.11. Table 8.3 shows how the addition of surfactant significantly increases the total mass of liquid ejected and reduces the time to the turnaround point, but has only a small effect on the duration of the two-phase outflow. In these respects, the effects are similar to increasing the fill to 60%.

8.1.4   Summary of Observations and Measurements of the Level Swell Process

1. The energy provided by the exothermic runaway reaction could make the pressure continue to rise even after the opening of the vent (Figure 8.5). This causes more material to be expelled than without this overpressure.
2. Vaporisation takes energy out of the process (Figure 8.7).
3. Expelling liquid removes reactants and so lowers the rate of heat input (Figure 8.7).

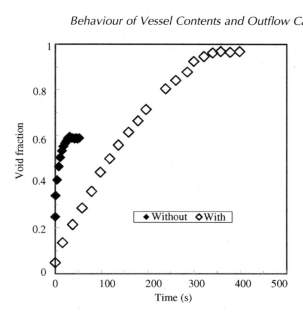

Figure 8.10 *Temporal variation of void fraction for water without and with surfactant; vessel, 2190 l. Data from Grolmes and Leung [2]*

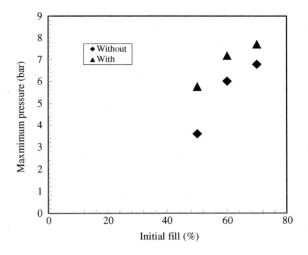

Figure 8.11 *Effect of surfactant on maximum pressure reached during venting, esterification of acetic anhydride with methanol. Data from Snee et al. [4]*

4. The larger the vent size the more quickly the pressure falls and so the more/faster the level will swell (Figure 8.2). Conversely, although smaller vent sizes will keep the pressure higher (Figure 8.6), they could favour gas only venting, which requires a smaller vent diameter than for two-phase venting.

Table 8.3   *Effect of fill level and presence of surfactant on major parameters during the runaway hydrolysis of acetic anhydride in a 2500 l vessel (Snee et al. [4])*

Condition	Mass discharged (%)	Increase in heat rate per kg (%)	Total mass discharged (%)	Duration of two-phase flow (s)	Time from vent opening to P_{max} (s)
50% fill with no surfactant	27.0	129.7	30.2	26.2	42.2
50% fill with surfactant	29.6	161.7	72.1	29.4	19
60% fill with no surfactant	33.2	162.0	49.0	32.6	21.3

5. If the liquid viscosity is higher than that of water, bubble rise velocities will be smaller and so a greater level swell will be seen (Figure 8.8) and more liquid will be expelled (Figure 8.9).

8.2   Modelling of the Level Swell Process

The subject of interfacial momentum transfer in pool systems containing arbitrary fluids, and which are subject to depressurisation, is still not totally understood. There is a paucity of experimental data, particularly for high viscosity fluids, from which realistic models can be developed. The problem reduces to the description of the motion of bubbles within a continuous liquid phase and the motion of droplets within a continuous vapour phase, and how these motions change with void fraction. In a bubbly liquid in a pool where wall effects are negligible, the friction or drag exerted by the liquid on the bubble surface determines how fast the bubbles rise within the liquid. This frictional force decreases markedly in comparison with the buoyancy force as the bubbles increase in size. As the void fraction increases, the rate of slip of the vapour past the liquid increases until the situation arises when the liquid begins to break up into droplets and the vapour becomes the continuous phases. The drag between the droplets and the vapour increases as the droplet size decreases and so the slip decreases. At the extremes of all liquid and all vapour flow the slip goes to the slip velocity of a single bubble (all liquid) or droplet (all vapour). The flux will go to zero. At some intermediate void fraction the slip will be a maximum where the bubbles or vapour 'packets' have their maximum size.

Phase slip is closely related to phasic momentum transfer, a process usually described in terms of a mixture or drift flux model [12] or a two-fluid model [13].

In the drift flux model the phasic momentum equations are replaced by a mixture momentum equation and the actual gas velocity can be determined from the steady-state correlation for void fraction which is given by:

$$\varepsilon_g = \frac{u_{gs}}{C_o(u_{gs} + u_{ls}) + u_{dr}} \tag{8.1}$$

where

ε_g is the void fraction
C_o is the radial distribution parameter
u_{gs} and u_{ls} are the superficial velocities of the gas and liquid
u_{dr} represents the drift velocity $(= \langle \varepsilon_g[u_d - \{u_{ls} + u_{gs}\}]\rangle / \langle \varepsilon_g\rangle)$ where $\langle\ \rangle$ denotes a cross-
  sectional average.

The validity of this is limited to situations where acceleration and wall friction forces can be neglected. It can be shown that u_{dr} can be written as $u_\infty(1 - \varepsilon_g)$. However, some workers have represented the drift velocity as equal to V_T, the terminal rise velocity of a single bubble. The radial distribution parameter is defined as:

$$C_o = \frac{\langle \varepsilon_g(r)u_g(r)\rangle}{\langle \varepsilon_g(r)\rangle\langle u_g(r)\rangle} \tag{8.2}$$

If the local void fractions and gas velocities can be described by a power law relationship, for example

$$\varepsilon_g = \varepsilon_{go}\left[1 - \left(\frac{r}{R}\right)^m\right]; \ u_g = u_{go}\left[1 - \left(\frac{r}{R}\right)^n\right] \tag{8.3}$$

then C_o would range from 1.5 when $m = n = 1$ to 1.0 when $m = n = \infty$. This latter implies flat profiles, that is homogeneous flows. However, it has been reported that when Equation 8.1 is fitted to bubble column data for air-water, C_o took a value of 2 [14]. In the DIERS methodology, C_o is used as an adjustable constant. Values of 1.5 have been used in some of their papers. Kataoka and Ishii [15] suggested $C_o = 1.2 - 0.2\sqrt{(\rho_g/\rho_l)}$, whilst Cumber [16] proposed $C_o = (\rho_l/\rho_g)^{0.05}$, where ρ_g and ρ_l are the gas and liquid densities, respectively. V_T has been related to physical properties through $G\ (= [\sigma g(\rho_l - \rho_g)/\rho_l^2]^{0.25}$, a group which has dimensions of velocity. Here, σ is the surface tension and g is the gravitational acceleration. Early work used relationships of the form:

$$V_T = KG^a \tag{8.4}$$

For example, Peebles and Garber [17] suggested $K = 1.53, a = 1$. These values are proposed in the DIERS method [18]. Subsequently, Cumber proposed that the equations suggested by Kataoka and Ishii [15] should be used. In the form used here these are:

$$K = 0.03\left(\frac{\rho_l}{\rho_g}\right)^{0.157}\left(\frac{\sigma}{\eta_l}\right)^{0.562}; \quad a = 0.438 \quad \text{for} \quad 0.1 = \eta_l G/\sigma > 0.00225 \tag{8.5}$$

$$K = 0.92\left(\frac{\rho_l}{\rho_g}\right)^{0.157}; \quad a = 1 \quad \text{for} \quad \eta_l G/\sigma = 0.00225 \tag{8.6}$$

where η_l is the liquid viscosity and the dimensionless group $\eta_l G/\sigma$ is called the viscosity number. These equations give a value of $K = 2.65$ at their transition for atmospheric air-water. They also seem to imply that the value of K increases as the liquid viscosity increases. This appears counter-intuitive as bubbles would be expected to have a lower velocity in more viscous liquids Indeed, simple drag–buoyancy calculations assuming rigid bubbles confirms this. Examination of the database used by Kataoka and Ishii [15] showed very few points with viscosities greater than water. What is happening in these equations is that the viscosity number is getting larger because of decreases in surface tension. In the DIERS work, values of $K = 1.18$, $a = 1$ and $C_o = 1$ have been suggested [18]. More elaborate equations are given in Chapter 2. These have been developed for steady flow and are functions of the bubble size. However, the last parameter is not usually known in venting systems.

Equation 8.1 with the assumption of zero net liquid flow can be rearranged to give:

$$u_{gs} = \frac{\varepsilon_g V_T}{(1 - C_o \varepsilon_g)} \tag{8.7}$$

For viscous liquids that are more usually in bubbly flow, an alternative equation is proposed:

$$u_{gs} = \frac{V_T \varepsilon_g (1 - \varepsilon_g)}{(1 - \varepsilon_g^3)} \tag{8.8}$$

From an energy balance on a thin horizontal slice of the vertical cylindrical vessel, the differential equation is:

$$\rho_g \frac{d\varepsilon_g}{dt} = \frac{\dot{q} \rho_l (1 - \varepsilon_g)}{\Delta h_v} - \rho_g \frac{du_g}{dz} \tag{8.9}$$

where

t is time
q is the heat input per unit mass of liquid
z is the vertical direction in the vessel.

Sheppard [19] has identified that there are two useful simplifications to this equation: homogeneous flow and pseudo-steady-state flow. In the first, there is no variation of void fraction vertically or radially and Equation 8.9 can be simplified and integrated to give the cumulative volume of vapour produced as:

$$\dot{V}_g = \frac{\dot{q} M t}{\Delta h_v \rho_g} \tag{8.10}$$

where

M is the initial mass in the vessel.

The void fraction can then be written as:

$$\varepsilon_g = \frac{\dot{q} t / \Delta h_v \rho_g}{(\dot{q} t / \Delta h_v \rho_g + 1 / \rho_l)} \tag{8.11}$$

For the pseudo-steady-state case the time term disappears, then rearrangement and using Equation 8.7 gives:

$$\frac{d\varepsilon_g}{dZ} = \psi (1 - C_o \varepsilon_g)^2 (1 - \varepsilon_g) \tag{8.12}$$

In this, Z is the distance up the vessel made dimensionless with the original liquid height and ψ is a dimensionless group, where

$$\psi = \frac{\dot{q}\rho_l H_o}{\Delta h_v \rho_g V_T}$$

which is related to the ratio of gas velocity to the velocity of an isolated bubble. In the DIERS work, $\psi < 0.2$ is taken to be the bubbly flow region, $0.2 < \psi < 20$ is the churn-turbulent region and $\psi > 20$ is termed homogeneous.

For the case of $C_o = 1$, the variation of void fraction up the vessel is calculated from:

$$\varepsilon_g = 1 - \frac{1}{(2\psi Z + 1)^{0.5}} \tag{8.13}$$

The height of the two-phase interface is given by:

$$Z_{\max} = \frac{\psi + 2}{2} \tag{8.14}$$

and the mean void fraction is taken from

$$\bar{\varepsilon}_g = \frac{\psi}{\psi + 2} \tag{8.15}$$

If the radial distribution parameter, C_o, is not equal to 1 then Sheppard and Morris [20] have shown that the equation

$$\bar{\varepsilon}_g = \frac{\psi}{C_o \psi + 2} \tag{8.16}$$

is a good approximation. Sheppard and Morris [20] have also carried out rigorous calculations of the equivalent analysis for horizontal cylinders and spheres. It was suggested that the equation

$$\bar{\varepsilon}_g = \frac{\psi^n}{C_1 + C_o \psi^n} \tag{8.17}$$

with values for C_1 of 2.2 for vertical cylinders, 2.36 for horizontal cylinders and 2.56 for spheres. The exponent n has a value of 0.91 for the three cases.

For vertical cylinders, the initial height can be linked to the initial free volume fraction, ε_{go}, through $H_o = (1 - \varepsilon_g)H_v$, where H_v is the vessel height. We can now derive rules for the effect of level swell:

$$\bar{\varepsilon}_g > \varepsilon_{go} \quad \text{two-phase at outlet} \tag{8.18}$$

$$\bar{\varepsilon}_g < \varepsilon_{go} \quad \text{single-phase at outlet} \tag{8.19}$$

Figure 8.12 shows the results of the models developed above. The time varying analysis, Equation 8.11, predicts a constant void fraction up the vessel and a two-phase level, both of which increase as in Figure 8.12a. The time to fill the entire vessel with two-phase mixture is

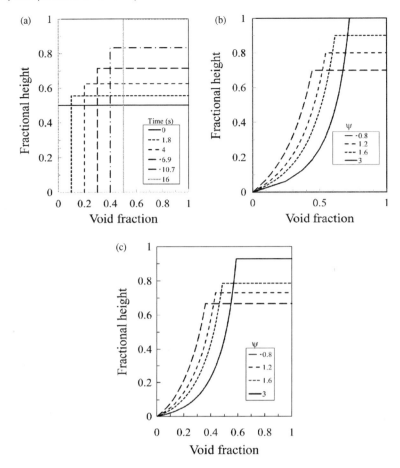

Figure 8.12   *Variation of void fraction up the vessel. (a) Time varying case; (b) pseudo-steady state, $C_o = 1.0$; and (c) pseudo-steady state, $C_o = 1.5$*

given by $\rho_G \Delta h_v (1 - f)/(\rho_L q f)$, where f denotes the initial filling fraction of the vessel. Figure 8.12b and c shows values calculated by the pseudo-steady-state approach, Equations 8.15 and 8.16.

8.3   Vent Sizing and Vent Performance Calculations

In the past, a relief system would have only been a short pipe downstream of the relief device. This meant that a simple value of critical flow rate could be used. Nowadays, with increasing environmental pressure to contain or treat the relieved material, the relief system often includes a considerable length of complex piping with knock out pots and so on. The behaviour of the vessel contents, the relief device (safety relief valve or bursting disc) and vent pipework are very complex and require a computer code to determine the interactions.

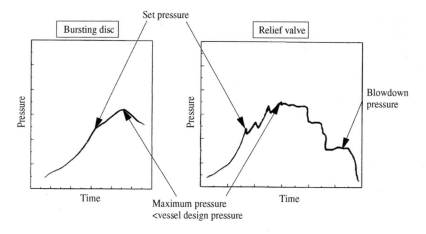

Figure 8.13 *Typical pressure histories during transients*

However, with certain sensible assumptions, hand methods for the determination of vent size and so on can be derived, which show the relative performance of different assumptions. In this section we will first consider these hand methods and then examine the performance of more complex methods incorporated into computer codes.

Before considering derivations, it is useful to examine typical time histories of pressure for systems undergoing a venting process. Figure 8.13 shows curves of the pertinent variables for a tempered system, that is one where vaporisation takes up the heat provided by the runaway reaction or external source (fire) and mitigates the temperature rise that otherwise would have taken place. As seen in the figure, there will be a rise then a fall in pressure (and temperature). However, the maxima might not occur at the same time after the start of the transient for a typical arrangement of a bursting disc and a relief valve. In the case of a spring loaded relief valve, there might be some additional oscillations, as shown. These would be due to the valve opening and closing during the process. This is often called valve chatter.

For those cases in which disengagement takes place in the vessel, the required vent size is determined from single flow analysis and

$$S = \frac{\dot{Q}}{\Delta h_v \dot{m}} \qquad (8.20)$$

where

S the cross-sectional area of the vent.

The mass flux is obtained from the methods suggested in Chapter 9.

For two-phase discharge, the simplest case to consider is that where there is zero overpressure. This might not be a real option in practice as some overpressure is employed to enhance the rate of discharge and, consequently, decrease the size of the vent aperture of the vent. However, this approach serves as a reference case. We can derive a simple formula by assuming that the process is tempered, the vapour density is much lower than that of the liquid, there is no further vaporisation in the vent line and that the vessel contents and vented material can be described by homogeneous relationships. Thus, we can write that the rate of vapour generation (volumetric) is equal to the two-phase discharge rate or:

$$\frac{\dot{Q}}{\Delta h_v \rho_g} = \frac{x \dot{m} S}{\rho_g} + \frac{(1-x)\dot{m} S}{\rho_l} \approx \frac{\dot{m} S}{\rho_l (1-\varepsilon_g)} \tag{8.21}$$

where

x is the vapour mass fraction
S is the cross-sectional area of the vent.

For the void fraction the value at the initial conditions can be used. With the assumption of homogeneous flow this can be equated to $M/(\rho_l V)$, M being the vessel contents (mass) and V the vessel volume. Thence,

$$S_0 = \frac{\dot{Q} v_{lg} M}{\Delta h_v \dot{m} V} \tag{8.22}$$

in this

$$v_{lg} = \frac{1}{\rho_g} - \frac{1}{\rho_l} \approx \frac{1}{\rho_g} \tag{8.23}$$

The next, more complicated description is the emptying time approach. Here it is assumed that the vessel is to be emptied by the time the system reaches its specified overpressure. Assuming that the discharge rate is constant, we can specify a ΔT for the overpressure Δp and the adiabatic rate of temperature rise can be approximated by this, at the set conditions. The emptying time is then:

$$\Delta t_a = \frac{c_p \Delta T}{c_p \frac{dT}{dt}} \tag{8.24}$$

and the vent area is then:

$$S = \frac{M}{\dot{m} \, \Delta t_a} = \frac{\dot{Q}}{\dot{m} \, \Delta T c_p} \tag{8.25}$$

or if considered relative to the area at zero overpressure

$$\frac{S}{S_0} = S' = \frac{V \Delta h_v}{M v_{lg} \, \Delta T \, c_p} \tag{8.26}$$

The right-hand side of the above equation is dimensionless. As it will appear again let us call it the inverse of the overpressure number, $1/O^*$.

DIERS have developed a more complex approach for external heating cases. It is assumed that there is no free volume at the start of the transient and that discharge makes room for vapour in the vessel. From this the void fraction after time Δt is:

$$\varepsilon_g = \frac{\dot{m} S \Delta t}{\rho_l V} \tag{8.27}$$

As it is also assumed that the time to heat up the vessel contents by ΔT corresponds to the adiabatic rise time

$$\Delta t = \frac{c_p \Delta T \rho_l V}{\dot{Q}} \tag{8.28}$$

combining with Equations 8.21 and 8.27 yields:

$$S = \frac{\dot{Q}\rho_l}{\dot{m}\left(\frac{\Delta h_v}{v_{lg}} + \rho_l c_p \Delta T\right)} \tag{8.29}$$

and the value relative to S_0 is:

$$\frac{S}{S_0} = S' = \frac{1}{1 + \frac{c_p \Delta T M v_{lg}}{V \Delta h_v}} = \frac{1}{1 + O^*} \tag{8.30}$$

Banerjee [21] provided a more rigorous approach, which yields

$$\frac{1}{S'}\left[1 - \ln\left(\frac{1}{S'}\right)\right] = 1 - O^* \tag{8.31}$$

For the case with an internal heat source [22] it is assumed that the mass flux from the vessel does not vary significantly over the venting transient. A mass balance over the vessel gives:

$$\frac{d(\rho V)}{dt} = -\dot{m}S \tag{8.32}$$

Similarly, the energy balance is:

$$\frac{d(\rho V h)}{dt} = \bar{q}\,\rho V - \dot{m}Sh \tag{8.33}$$

Following manipulations, which involve time to turnround and so on, the analysis yields:

$$S = \frac{\bar{q}M}{\dot{m}\left[\left(\frac{\Delta h_v V}{v_{lg} M}\right)^{\frac{1}{2}} + (c_p \Delta T)^{\frac{1}{2}}\right]^2} \tag{8.33}$$

Relative to S_0

$$\frac{S}{S_0} = S' = \frac{1}{\left(1 + \sqrt{O^*}\right)^2} \tag{8.34}$$

The relative performance of Equations and 8.31 is given in Figure 8.14. The more rigorous analyses obviously show that smaller vents than those indicated by the more simple approaches are feasible. However, in these important calculations, great care must be taken.

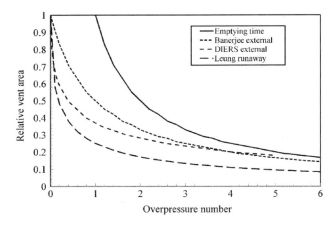

Figure 8.14   *Effect of overpressure number on relative vent area for different models*

For a hybrid system, it is more difficult to specify the term $\Delta h_v/v_{lg}$. The assumption is made that the gas generation rate and the vapour generation rates are both proportional to the reaction rate. This, after some analysis gives, at turnaround,

$$\frac{\Delta h_v}{v_{lg}} = \frac{p_v T}{p_v + p_g}\frac{dp}{dT} \tag{8.35}$$

The expression for the area then becomes:

$$S = \frac{\bar{q}M}{\dot{m}\left[\left(\frac{VT}{M}\frac{p_v}{(p_v+p_g)}\frac{dp}{dT}\right)^{0.5} + (c_p\Delta T)^{0.5}\right]^2} \tag{8.36}$$

For gassy systems, Duxbury and Wilday [23] suggested

$$S = \frac{Q_g M}{\dot{m}_c V} \tag{8.37}$$

where

Q_g is the *maximum* rate of gas production (m³/s)
M is the mass of charge in the vessel of volume V
$\dot{m}_c$ is the critical mass flux calculated using a homogeneous frozen model, see Chapter 9.

8.4   Computer Codes for Level Swell and Venting Calculations

The requirements to combine the hydrodynamics together with the chemical kinetics and (exothermic) heat production rates means that the simple calculations described above are only suitable for scoping calculations. For more detailed assessments, it is more helpful to

Table 8.4 *Characteristics of codes used in level swell/venting calculations*

Code	Nodes in Vessel	Nodes in Vent Line	Number of Chemical Reactions	Number of Components	Two-Phase Models
SAFIRE	1	8 or 50	10	10	Drift flux two versions
RELIEF	n	1	10	10	Drift flux
DEERS	n	n	4	6	Drift flux

have methods that combine the behaviour of the vessel and the piping downstream of it and which carry out a full transient analysis. Here, three commercially available codes are examined and their links to the models described above are highlighted. Details of their characteristics are given in Table 8.4. Although there are other products available, these particular ones have been chosen because there are journal papers describing their basis and discussing their predictions [2, 5, 24].

SAFIRE was developed under the DIERS project [2]. It employs the drift flux approach in the vessel and uses Equation 8.4 with $K = 1.18$ or 1.53 and either Equation 8.7 or 8.8. In the vent piping beyond the vessel, methodologies as discussed in Chapter 9 are employed.

The RELIEF code was developed at the Joint Research Centre of the European Commission at Ispra, Italy [24]. It also incorporates a drift flux approach and describes the relative velocity of the phases as a function of void fraction. The expression chosen is:

$$u_g - u_l = u_{pool} \frac{\varepsilon_g^m (1-\varepsilon_g)^n}{\varepsilon_{g_{max}}^m (1-\varepsilon_{g_{max}})^n} \tag{8.38}$$

where $\varepsilon_{g_{max}}$ is the void fraction that gives the term $\varepsilon_g^m (1-\varepsilon_g)^n$ its maximum value. The denominator is a scaling factor, which ensures that the maximum value of the slip is given by u_{pool}, irrespective of the value of the void fraction at which it occurs. The coefficients m and n describe bubbly flow and droplet flow, respectively; they have been fitted to experimental data; u_{pool} is closely related to Equation 8.4.

The DEERS (Design and Evaluation of Emergency Relief Systems) code [25] models both the vessel and the outlet pipe simultaneously. Because of the expectation that during blowdown the gas-liquid flows would remain near equilibrium, with respect to phase velocities and temperatures and with pressure near the local saturation conditions, mixture variables were used and equations were provided that would describe the deviations from equilibrium. Time varying mixture mass, momentum and energy equations were obtained from formal averaging. In addition, deviation equations were employed for quality, velocity and temperature. Different friction factors were utilised for low and high viscosity liquids. The equations were then simplified by eliminating terms that, by argument, could be considered negligible. The venting rate was found to be sensitive to the relative velocity between the phases, as that term in the energy equation governs the rate at which energy emerges from the vessel. Different forms of the interfacial drag relationship were used, according to whether the liquid or gas was the main continuous phase and for foamy or non-foamy liquids.

The codes also employ models for the venting piping, which are considered in Chapter 9.

8.5   Obtaining Necessary Data

The methodology described in the last section requires a number of pieces of information, such as: critical mass flux, vapour pressure, the gradient of pressure with temperature, the pressure between set and maximum pressures and the rate of heat provision (either from external sources such as fires or the heat generated by a runaway reaction). This section discusses the sources of this information.

Critical mass flux is determined using the methods in Chapter 9. In particular, the homogeneous frozen model, the ω model and the equilibrium rate model are recommended by workers such as Fauske and coworkers [27].

For sub-cooled and gassy systems, data are required on the vapour pressure and its variation with temperature. This might have to be determined experimentally. Information can be tabulated or shown graphically. In some cases, the p_v/T relationship is described through an Antione type of equation.

$$\ln(p_v) = A - \frac{B}{T} \tag{8.39}$$

where

p_v is in Pa
T is in K
B is the ratio of latent heat to gas constant, $\Delta h_v/R$.

Although, theoretically, it is possible to predict the temporal variations of temperature and pressure from the kinetics of the system, in many cases it is not practicable, as the required information is not available. An alternative approach is to use scaled down tests to obtain the data. Such equipment is now available commercially from a number of sources, Tharmalingam [26] gives details, for example. Here we will consider them in general terms and examine the way in which the required information is abstracted from the data produced [27].

The need to have special tests is illustrated in Figure 8.15, created by extending the data presented by Waldram [28]. This shows that the heat loss rate in laboratory glassware is much larger than that which occurs in large metal vessels. Thus tests are required where heat loss rates are more in line with the large (plant) vessels. In practice, this implies that the test needs to be carried out adiabatically, no heat loss or gain. Another important aspect in these bench-top tests is the relative heat capacity of the container and its contents. Although, in some systems the thickness, and hence mass, of the container is minimised, for all systems we need a correction factor. This is usually called the thermal or adiabaticity factor and is defined as:

$$\phi = 1 + \frac{(Mc)_{holder}}{(Mc)_{contents}} \tag{8.40}$$

where

Mc is the product of the mass and the specific heat capacity of either the holder or its contents.

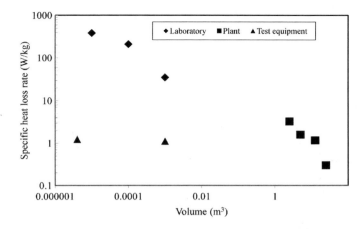

Figure 8.15   *Effect of volume on heat loss rate*

Ideally ϕ should tend towards 1. This value can be approached with a larger sample in a thin walled, low heat capacity container. However, this can lead to difficulty in containing a runaway reaction. It is noted that the equipment has to be very sensitive as some of the chemical systems can give rise to runaway conditions with very low heat production rates. Extrapolation could lead to significant errors. Steps must be taken to ensure that the cell contents are homogeneous and the materials of construction must be compatible with the reactants.

To obtain the necessary data a number of different of techniques have been utilised. These include: differential scanning calorimetry (DSC), differential thermal analysis (DTA), accelerating rate calorimetry (ARC) [29], reaction calorimetry (RC), Dewar calorimetry (DC), microcalorimetry (MC) and vent sizing packages (VSP) [27, 30]. These vary in the amount of test material required: ARC only requires 5–7 ml, RC and DC need 0.5–2 l and VSPs work with the order of 0.1 l. For the ARC, the adiabaticity factor can be high, typically 1.7–5. In contrast the VSP often has values < 1.05. For the DTA type of equipment, the thermal activity is measured as the difference in temperature between the sample and an inert reference of similar heat capacity. The temperature–time data are converted into a power versus temperature plot. In an ARC, the mixture is placed in a calorimeter in an enclosure to eliminate heat losses and the pressure and temperature are monitored continuously. The equipment is operated on a heat–wait search operation: the calorimeter and its content are heated so that its temperature rises by one or two degrees and the temperature is held for a short period. Eventually a temperature is reached when self-heating becomes important. From the measured pressure and temperature data, rates of temperature rise and pressure/temperature relationships can be determined.

The equipment developed by DIERS [27] consists of a test cell of volume 120 ml, wall thickness 3 mm. It is enclosed in a larger vessel rated to a pressure of 138 bar. The weight of the cell including the PTFE [poly(tetrafluoroethylene)] encapsulated magnetic stirrer is 21 g, giving a thermal factor of 1.05. The exterior vessel can be pressurised to minimise the forces on the cell walls. The temperature and pressure in the cell are monitored continuously. The inner heater can simulate the effects of an external heat source, such as fire. It can also be

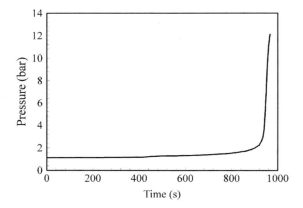

Figure 8.16 *Time history of pressure for hydrolysis of acetic anhydride. Data from Snee et al. [4]*

used in the search and wait approach to increase the temperature of the cell contents in finite steps until the conditions are reached for the onset of the runaway.

The equipment can be used in three modes. The first of these is a closed system from which thermal data can be obtained. A second mode involves an open system and is used to obtain scaling data for vent sizing. The third mode, another open test, has the vent line protruding into the cell. It is used to obtain data on viscosity effects. Examples of the results obtained with similar equipment are shown in Figures 8.16 and 8.17. These are for a methanol-acetic anhydride mixture. This was an adiabatic test and the results are given from the start of the runaway. The time gradient of the temperature curve is given in Figure 8.18. Tharmalingam [26] suggested that temperature rises of 0.01, 0.005 and 0.0018 °C/min should be taken as indicators of runaways for 0.2, 1.0 and 2.0 m^3 vessels when the contents are liquids. For solids or pastes values of 0.003, 0.0005 and 0.0001 °C/min were

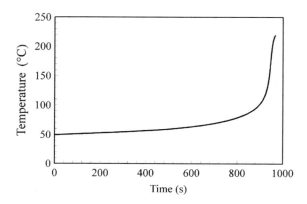

Figure 8.17 *Time history of temperature for hydrolysis of acetic anhydride, data from Snee et al. [4]*

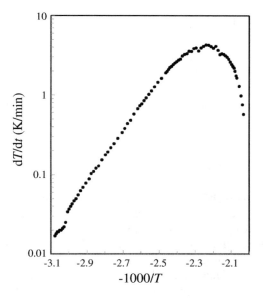

Figure 8.18 *Variation of self-heat rate with temperature for hydrolysis of acetic anhydride. Data from Snee et al. [4]*

recommended. For a system with external heating, deviations from a steady gradient would be the indicator of the start of exothermicity.

Figure 8.19 illustrates how the necessary information can be obtained from data such as that shown in Figures 8.16–8.18. The times at which the set and maximum pressures are reached are identified. The temperatures at these times are then deduced. From the gradients at these times (or from the equivalent of Figure 8.18) we obtain values of dT/dt.

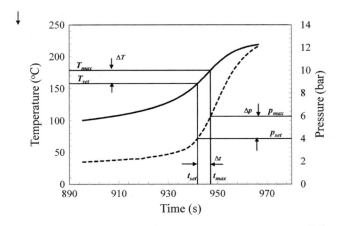

Figure 8.19 *Abstraction of information from temperature and pressure time histories*

If no other data are available the pressure difference, Δp, and the temperature difference, ΔT, can be used to approximate $\mathrm{d}p/\mathrm{d}T$. The rate of heat provision is determined from:

$$\bar{q} = \frac{1}{2} c_p \left[\left(\frac{\mathrm{d}T}{\mathrm{d}t}\right)_{set} + \left(\frac{\mathrm{d}T}{\mathrm{d}t}\right)_{max} \right] \phi \tag{8.37}$$

Other types of averages have been suggested [22], but this arithmetic approach is the one most commonly used.

8.6 Performance of Models and Codes

This section examines how the predictions of the hand methods/computer codes perform against experimental data. It is not intended to be a comprehensive validation, merely an indication of the information that can be obtained. The void fraction variations up the vessel obtained during gas exsolution have been measured by Thies and Mewes [10] and compared by them with the DIERS method. Figure 8.20 shows that there is reasonable prediction of the void fraction up the vessel. However, the model gives a very clear top surface to the bubbly pool with a sudden change from two-phase to single-phase gas. The reality is much more gradual.

For prediction of the boundary between single- and two-phase venting, the effect of the initial fill level and the ratio of the vent orifice area to vessel volume have been studied [16]. In this instance the data of Wehmeier *et al.* [11] were employed. Figure 8.21 shows that there is reasonable prediction when the drift flux parameters of Zuber and Findlay [31] are used,

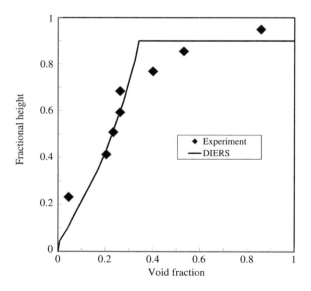

Figure 8.20 *Comparison between measured vertical variation of void fraction and predicition of the DIERS method. Data from Thies and Mewes [10]*

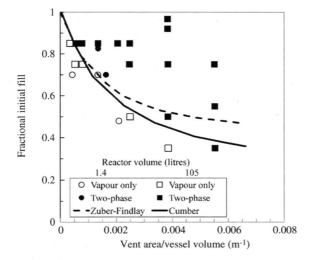

Figure 8.21 *Capability of models to predict the single-/two-phase venting boundary. Data from Wehmeier et al. [11], predictions of Cumber [16]*

but the predictions of Cumber [16], employing the drift velocity equations of Kataoka and Ishii [15] and his own version of C_o, gives better predictions.

Grolmes and Leung [2] presented the predictions of the SAFIRE code against small- and large-scale data for the polymerisation of styrene. These are illustrated in Figure 8.22. They noted that the good predictions shown in the figure were achieved using the bubbly flow description with $C_o = 1.3$ for the small-scale data but that the churn-turbulent approach with

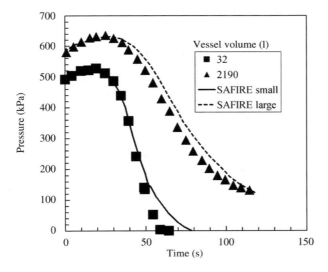

Figure 8.22 *Comparison between measurements of time variation of vessel pressure and predictions of SAFIRE*

$C_o = 1.0$ was used for the larger-scale test. For both cases, the homogeneous model gave a delayed pressure turnaround with higher maximum pressure than observed experimentally and predicted by the optimal models. In their conclusions of the exercise using SAFIRE to model the runaway esterification of acetic anhydride with methanol, Friedel and Wehmeier [11] noted that the transient reactor pressure can be qualitatively predicted in addition to the effects of the initial liquid level and the vent cross-sectional area. More quantitatively, the predicted values are greater than those measured.

Appendix 8.A

Rigorous derivation of Equation 8.15 using assumption of $C_o = 1$:

If it can be assumed that the heat from the external (fire) or internal (runaway reaction) source, $d\dot{Q}$, goes to convert liquid into vapour, the differential volume of vapour produced in a thin horizontal slice of the vessel, $d\dot{V}$, is given by $d\dot{V} = d\dot{Q}/\Delta h_v$. Here, Δh_v is the latent heat of evaporation. In terms of the internal heat generation rate per unit mass, $\dot{q}$, the differential heat input can be expressed as:

$$d\dot{Q} = Adz\rho_l(1-\varepsilon_g)\dot{q} \tag{8.A.1}$$

where

A is the cross-sectional area of the vessel
dz is the differential height of the element
$1 - \varepsilon_g$ is required as the heat goes into the liquid part of the slice.

(see Figure 8.A.1). The differential volume of vapour produced can be linked to the increase in superficial velocity of the vapour by:

$$d\dot{V} = Adu_r \tag{8.A.2}$$

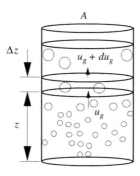

Figure 8.A.1 *Geometry for level swell calculation*

Note that the differential velocity can be obtained from Equation 8.7 and is

$$du_r = \frac{V_T d\varepsilon_g}{(1-\varepsilon_g)^2} \qquad (8.A.3)$$

The energy balance is then:

$$A dz \rho_l (1-\varepsilon_g)\dot{q} = A du_r \rho_g \Delta h_v \qquad (8.A.4)$$

Using $Z = z/H_o$ and substituting for du_r from Equation 8.A.3 results in:

$$\frac{\dot{q}\rho_l H_o}{\Delta h_v \rho_g V_T} dZ = \frac{d\varepsilon_g}{(1-\varepsilon_g)^3} \qquad (8.A.5)$$

where

H_o is the liquid height before bubbling started
V_T is given by Equation 8.4.

Now the group

$$\frac{\dot{q}\rho_l H_o}{\Delta h_v \rho_g V_T} = \psi \qquad (8.A.6)$$

is dimensionless, it is the ratio of produced gas velocity to bubble velocity. Equation 8.A.5 can be rewritten as

$$\psi dZ = \frac{d\varepsilon_g}{(1-\varepsilon_g)^3} \qquad (8.A.7)$$

and can be integrated to give the variation of void fraction with height:

$$\int_0^{Z_{max}} \psi dZ = \psi Z = \int_0^{\varepsilon_g} \frac{d\varepsilon_g}{(1-\varepsilon_g)^3} = \left|\frac{1}{2(1-\varepsilon_g)^2}\right|_0^{\varepsilon_g} = \frac{1}{2(1-\varepsilon_g)^2} - \frac{1}{2} \qquad (8.A.8)$$

resulting in:

$$\varepsilon_g = 1 - \frac{1}{(2\psi Z + 1)^{0.5}} \qquad (8.A.9)$$

Equation 8.A.8 can be then be integrated to give the vessel average void fraction:

$$\begin{aligned}
\bar{\varepsilon}_g &= \frac{1}{Z_{max}} \int_0^{Z_{max}} \varepsilon_g dZ = \frac{1}{Z_{max}} \left[\int_0^{Z_{max}} dZ - \int_0^{Z_{max}} \frac{dZ}{(2\psi Z + 1)^{0.5}} \right] \\
&= \frac{1}{Z_{max}} \left[Z_{max} - \left(\frac{(1 + 2\psi Z_{max})^{0.5}}{\psi} - \frac{1}{\psi} \right) \right]
\end{aligned} \qquad (8.A.10)$$

This contains the maximum dimensionless height, Z_{max}, which is not known *a priori*.

However it can be eliminated from the definition of the void fraction: (volume of gas and liquid − volume of liquid)/volume of gas and liquid

$$\bar{\varepsilon}_g = \frac{Z_{max}H_oA - H_oA}{Z_{max}H_oA} = 1 - \frac{1}{Z_{max}} \tag{8.A.11}$$

Equating (8.A.10) and (8.A.11) combined gives

$$\frac{1}{Z_{max}}\left[Z_{max} - \left(\frac{\langle 1 + 2\psi Z_{max}\rangle^{0.5}}{\psi} - \frac{1}{\psi}\right)\right] = 1 - \frac{1}{Z_{max}} \tag{8.A.12}$$

which gives:

$$Z_{max} = \frac{\psi + 2}{2} \tag{8.A.13}$$

Substituting into Equation 8.A.11 yields:

$$\bar{\varepsilon}_g = \frac{\psi}{\psi + 2} \tag{8.A.14}$$

Appendix 8.B

Detailed derivation of Equation 8.33 originally published by Leung [22]:
  A mass balance over the vessel gives:

$$\frac{d(\rho V)}{dt} = -\dot{m}S \tag{8.B.1}$$

Similarly, the energy balance is:

$$\frac{d(\rho Vh)}{dt} = \bar{q}\rho V - \dot{m}Sh \tag{8.B.2}$$

Expanding the left-hand side of Equation 8.B.2 and using Equation 8.B.1:

$$h\frac{d(\rho V)}{dt} + \rho V\frac{dh}{dt} = \bar{q}\rho V - \dot{m}Sh$$

$$-\dot{m}Sh + \rho V\frac{dh}{dt} = \bar{q}\rho V - \dot{m}Sh \tag{8.B.3}$$

$$\frac{dh}{dt} = \bar{q}$$

which on integration gives

$$h - h_o = \bar{q}t \tag{8.B.4}$$

The increase in enthalpy goes into increasing vessel quality and temperature. Therefore, the relationship can be written in terms of the vessel quality:

$$(x_g - x_{go})\Delta h_v + c_p(T - T_o) = \bar{q}t \tag{8.B.5}$$

Differentiating with respect to time yields:

$$\Delta h_v \frac{dx_g}{dt} + c_p \frac{dT}{dt} = \bar{q}$$

Now, if the temperature turnaround, at which $dT/dt = 0$, occurs at time t_m, we obtain

$$\frac{dx_g}{dt} = \frac{\bar{q}}{\Delta h_v} \tag{8.B.6}$$

From the mass balance, Equation 8.B.1,

$$\frac{d(\rho V)}{dt} = \rho \frac{dV}{dt} + V \frac{d\rho}{dt} = -\dot{m}S$$

but $dV/dt = 0$, so that

$$\frac{d\rho}{dt} = -\frac{\dot{m}S}{V} \tag{8.B.7}$$

If homogeneous conditions are assumed:

$$\frac{1}{\rho} = \frac{x_g}{\rho_g} + \frac{(1 - x_g)}{\rho_l} = x\left(\frac{1}{\rho_g} - \frac{1}{\rho_l}\right) + \frac{1}{\rho_l}$$

$$\frac{1}{\rho} = x_g v_{lg} + v_l \tag{8.B.8}$$

Then

$$\frac{d\rho}{dt} = \frac{-v_{lg}}{(x_g v_{lg} + vl)^2} \frac{dx_g}{dt} = -\dot{m}S$$

substituting from Equation 8.B.6

$$\frac{\dot{m}S(x_{gm}v_{lg} + v)^2}{v_{lg}V} = \frac{\bar{q}}{\Delta h_v} \tag{8.B.9}$$

which yields

$$x_m = \frac{1}{v_{lg}}\sqrt{\frac{\bar{q}Vv_{lg}}{\Delta h_v \dot{m}S}} - \frac{v_l}{v_{lg}}$$

using the definition of initial quality $\rho_o = M/V$

$$\rho_0 = \frac{1}{x_{g0}v_{lg} + v_l} = \frac{M}{V} \quad \Rightarrow$$

$$x_{g0} = \frac{1}{v_{lg}}\left[\frac{V}{M} - v_l\right] = \frac{V}{v_{lg}M} - \frac{v_l}{v_{lg}}$$

gives

$$\Delta x_m = x_{gm} - x_{go} = \frac{1}{v_{lg}}\left[\left(\frac{v_{lg}\,\bar{q}V}{\Delta h_v \dot{m}S}\right)^{\frac{1}{2}} - \frac{V}{M}\right] \tag{8.B.10}$$

Integrating Equation 8.B.7

$$\rho - \rho_0 = -\frac{\dot{m}St}{V}$$

substituting from (8.B.8) for ρ and using $\rho_0 = M/V$

$$\frac{1}{x_g v_{lg} + v_l} - \frac{M}{V} = -\frac{\dot{m}St}{V}$$

$$x_g v_{lg} + v_l = \frac{V}{M - \dot{m}St}$$

$$\frac{1}{v_{lg}}\left[\frac{V}{M - \dot{m}St}\right] - \frac{v_l}{v_{lg}}$$

Using the equation for x_0

$$x_{gm} - x_{go} = \frac{1}{v_{lg}}\left[\frac{V}{M - \dot{m}St_m} - \frac{V}{M}\right] \tag{8.B.11}$$

This can be rearranged as:

$$t_m = \frac{1}{\dot{m}S}\left[M - \frac{V}{(x_m - x_o)\,v_{lg} + \frac{V}{M}}\right]$$

which can be combined with Equation 8.B.10 to yield:

$$t_m = \frac{1}{\dot{m}S}\left[M - \frac{V}{\frac{1}{v_{lg}}\left(\sqrt{\frac{\bar{q}Vv_{LG}}{\Delta h_v \dot{m}S}} - \frac{V}{M}\right)v_{lg} + \frac{V}{M}}\right] \tag{8.B.12}$$

$$t_m = \frac{1}{\dot{m}S}\left[M - V\left(\frac{\Delta h_v \dot{m}S}{v_{lg}\bar{q}V}\right)^{0.5}\right]$$

$$\frac{\Delta h_v}{v_{lg}}\left[\left(\frac{\bar{q}Vv_{lg}}{\Delta h_v \dot{m}S}\right)^{0.5} - \frac{V}{M}\right] + c_p\Delta T = \frac{\bar{q}}{\dot{m}S}\left[M - V\left(\frac{\Delta h_v \dot{m}S}{\bar{q}Vv_{lg}}\right)^{0.5}\right]$$

These results, Equations 8.B.10 and 8.B.12, can be substituted into Equation 8.B.5, which can then be solved for the vent area.

Questions

8.1 What is maximum fractional fill that can be used to ensure that there is single-phase discharge through the vent for a $1.5\,m^3$ vessel? The physical properties: gas and liquid densities are 2 and $700\,kg/m^3$, respectively; the latent heat of evaporation is $1000\,kJ/kg$; and the surface tension is $0.02\,N/m$. The internal heat release rate is $1\,kW/kg$.

Maximum fractional fill $= 0.525$.

8.2 A reaction is carried out in batch mode in a $2.5\,m^3$ vessel (cylindrical, aspect ratio $= 2$). The charge is 1.375 tonnes (1 tonne $= 10^3\,kg$). A value of Ψ of 2 has been calculated. Would you expect single- or two-phase flow through the relief device? The liquid density is $1000\,kg/m^3$. Examine the effect of using different values of C_o. What else should you consider?

If $C_o = 1.0$ – two-phase; if $C_o = 1.5$ – single phase; foaming – if present, assume two-phase venting.

8.3 The phenol-formaldehyde polymerisation reaction is one which has been the cause of a number of incidents. Correct sizing of the passive protection system is therefore very important. What size should the emergency relief vent be if a set pressure of 2 bara (absolute pressure) is to be used and a gauge overpressure of 20% is acceptable? What is the effect of varying the set pressure and overpressure on vent diameter?

See Table 8.Q.1 for typical charge.

Reactor volume $= 7.5\,m^3$ Height $=$ twice diameter

Table 8.Q.1   *Contents of reactor after all chemicals added*

Reactant	Mass (kg)	Weight (%)	Mole (%)
Phenol	2200	35.67	11.58
Formaldehyde	1332	21.59	21.97
Caustic soda	110.4	1.79	1.36
Methanol	360	5.84	5.57
Water	2165.6	35.11	59.52
Total	6168	100.0	100.0

Physical properties:

Liquid density $= 1024 - 0.67T$ kg/m^3 Vapour density $= mp/RT$ kg/m^3
Specific heat $= 2950$ J/kgK   Gas constant $= 8310$ Pa m/kg mol K
Molecular weight $= 18$ kg/kg mol   Surface tension $= 0.0566$ N/m

For the liquid density T is in °C:

$$\ln\left(\frac{\mathrm{d}T}{\mathrm{d}t}\right) = 14.88 - \frac{4398}{T}$$

From tests the following relationships between pressure and temperature and the gradient of temperature with time against temperature were obtained:

$$\ln(p) = 28.29 - \frac{6311}{T}$$

Take the critical mass flux to be 2992 kg/m^2 s.

Vent diameter 0.56 m.

8.4 A 2.1 m^3 reactor contains 1500 kg of reactants. Relief is to be via a bursting disc with a maximum specified bursting pressure of 3.2 bara. The vessel nominal pressure is 2.78 barg (gauge pressure) and the maximum allowable pressure during relief is the design pressure $+ 10\%$ in absolute terms. Use the information given in Table 8.Q.2 to determine the vent size.

Assume a critical mass flux of 2384 kg/m^2 s.

Vent diameter 0.16 m.

8.5 The hydrolysis of acetic anhydride undergoes a runaway. The vessel volume is 2.5 m^3 and the charge is 1520 kg. The relief device has a set pressure of 2 bara and a diameter of 0.1 m. What is the maximum pressure expected in the vessel during the runaway according to the method of Leung?

Tests have generated the following useful relationships:

$$\ln(p) = 16.08 - \frac{1451}{T} \quad \frac{\mathrm{d}p}{\mathrm{d}T} = \frac{1451p}{T^2}$$

Table 8.Q.2 *Data for Question 4*

Parameter	Set	Max
Pressure (bara)	3.2	4.16
Bubble point temperature (°C)	110	120.5
Heat release rate (W/kg)	1150	1660
Liquid density (kg/m^3)	847	835
Vapour density (kg/m^3)	3.75	4.62
Latent heat (kJ/kg)	674.9	663.0
Liquid specific heat (kJ/kg K)	1.96	1.96
$\mathrm{d}p/\mathrm{d}T$ (N/m^2 K)	8300	9500

For $p > 2.5$ bar:

$$\ln(p) = 21.49 - \frac{3696}{T} \qquad \frac{dp}{dT} = \frac{3696p}{T^2}$$

For $p < 2.5$ bar:

$$\frac{dT}{dt} = -1.0575 10^{-7}T^4 + 1.61 10^{-4}T^3 - 9.076 10^{-2}T^2 + 22.636T - 2104.6$$

where p is in Pa and T in K. The specific heat capacity of the liquid can be taken as 1500 J/kg K. Assume that the critical mass flux can be specified by $416.4p + 982$ kg/ m^2 s, where in this equation, pressure is in bar.

Maximum pressure calculated 7 bar.

References

[1] Schmitz, D. and Mewes, D. (2000) Tomographic imaging of transient multiphase flow in bubble columns. *Chem. Eng. J.*, **77**, 99–104.

[2] Grolmes, M.A. and Leung, J.C. (1985) Code method for evaluating integrated relief phenomena. *Chem. Eng. Prog.*, **81**, 47–52.

[3] Snee, T.J., Butler, C., Hare, J.A. *et al.* (1999) Venting studies of the hydrolysis of acetic anhydride with and without surfactant (Vapour System 3), report No. PS/99/13, HSL.

[4] Snee, T.J., Bosch Pagans, J., Cusco, L. *et al.* (2006) Large-scale evaluation of vent-sizing methodology for vapour pressure reactions. Paper presented at IChemE Hazards XIX Symposium, Process Safety and Environmental Protection, What do we know? Where are we going?, Manchester, March 28–30, 2006.

[5] Friedel, L. and Wehmeier, G. (1991) Modelling the vented methanol/acetic anhydride runaway reaction using SAFIRE. *J. Loss Prev. Process Ind.*, **4**, 110–119.

[6] Hoff, A., Neumann, J., Deerberg, G. *et al.* (2001) Pilot plant tests of pressure relief with high-viscosity, reactive and non-reactive systems. *Chem. Eng. Technol.*, **24**, 996–1000.

[7] Vechot, L., Bigot, J.-P., Testa, D. *et al.* (2008) Runaway reaction of non-tempered chemical systems: development of a similarity vent-sizing tool at laboratory scale. *J. Loss Prev. Process Ind.*, **21**, 359–366.

[8] Gebbeken, B. and Eggers, R. (1996) Blowdown of carbon dioxide from initially supercritical conditions. *J. Loss Prev. Process Ind.*, **9**, 285–293.

[9] Friedel and, L. and Purps, S. (1984) Phase distribution in vessels during depressurisation. *Int. J. Heat Fluid Flow*, **5**, 229–234.

[10] Thies, A. and Mewes, D. (1993) The phase distribution in a reactor vessel during depressurisation. Paper presented at European Two-Phase Flow Group Meeting, Hannover, June 7–10, 1993.

[11] Wehmeier, G., Westphal, F. and Friedel, L. (1994) Pressure relief system design for vapour or two-phase flow? *Proc. Safety Environ. Prot.*, **72**, 142–148.

[12] Wallis, G.B. (1969) *One-dimensional Two-phase Flow*, McGraw-Hill, New York.

[13] Ishii, M. (1975) *Thermal-Fluid Dynamical Theory of Two-Phase Flow*, Eyrolles, Paris.

[14] Mashelkar, R.A. (1970) Bubble columns. *Br. Chem. Eng.*, **15**, 1297–1304.

[15] Kataoka, I. and Ishii, M. (1987) Drift flux model for large diameter pipe and a new correlation for pool void fraction. *Int. J. Heat Mass Transf.*, **30**, 1927–1939.

[16] Cumber, P. (2002) Modelling top venting vessels undergoing level swell. *J. Hazard. Mater.*, **A89**, 109–125.

[17] Peebles, F.N. and Garber, H.J. (1953) Studies of the motion of gas bubbles in liquids. *Chem. Eng. Prog.*, **49**, 88–97.

[18] Fisher, H.G. (1991) An overview of emergency relief system design practice. *Plant/Operations Prog.*, **10**, 1–12.

[19] Sheppard, C.M. (1992) Disengagement predictions via drift flux correlation vertical, horizontal and spherical vessels. *Plant/Oper. Prog.*, **11**, 229–237.

[20] (a) Sheppard, C.M. and Morris, S.D. (1995) Drift flux correlation disengagement models: Part I – Theory: analytical and numerical integration details. *J. Haz. Mater.*, **44**, 111–125; (b) Sheppard, C.M. (1995) Drift flux correlation disengagement models: Part II – shape-based correlations for disengagement prediction via churn turbulent drift-flux correlation. *J. Hazard. Mater.*, **44**, 127–139.

[21] Banerjee, S. (1989) *Lecture Series Loss Prevention/Plant Safety*, Swiss Institute of Architects and Engineers, Basel.

[22] Leung, J.C. (1986) Simplified vent sizing equations for emergency relief requirements in reactor and storage vessels. *AIChE J.*, **32**, 1622–1634.

[23] Duxbury, H.A. and Wilday, A.J. (1989) Efficient design of reactor relief systems. Paper presented at International Symposium on Runaway Reactions, Boston, Massachusetts, March 1989.

[24] Skouloudis, A.N. (1992) Benchmark excercises on the emergency venting of vessels. *J. Loss Prev. Process Ind.*, **5**, 89–103.

[25] Klein, H.H. (1986) Analysis of Diers venting tests: Validation of a tool for sizing emergency relief systems for runaway chemical reactions. *Plant/Oper. Prog.*, **5**, 1–10.

[26] Tharmalingam, S. (1989) Assessing runaway reactions and sizing vents. *Chem. Eng.*, **August**, 33–41.

[27] (a) Fauske, H.K. and Leung, J.C. (1985) New experimental technique for characterizing runaway chemical reactions. *Chem. Eng. Prog.*, **81**, 39–46; (b) Leung, J.C., Fauske, H.K. and Fisher, H.G. (1987) Thermal runaway reactions in a low thermal inertia apparatus. *Thermochim. Acta*, **104**, 13–29.

[28] Waldram, S. (1998) Batch runaway rethink. *Chem. Eng.*, (657), 15–16.

[29] Whitmore, M.W. and Wilberforce, J.K. (1993) Use of the accelerating rate calorimeter and the thermal activity monitor to estimate stability temperatures. *J. Loss Prev. Process Ind.*, **6**, 95–101.

[30] Waldram, S. (1994) Toll manufacturing: rapid assessment of reactor relief systems for exothermic batch reactions. *Proc. Safety Environ. Prot.*, **72**, 149–156.

[31] Zuber, N. and Findlay, J.A. (1965) Average volumetric concentration in two-phase flow systems. *Trans. ASME J. Heat Transf.*, **87**, 453–468.

9

Choked Flow

9.1 Introduction
9.2 Single-Phase Flow
9.3 Two-Phase Flow
9.4 Effect of Vent Pipework
Questions
References

9.1 Introduction

The flow rate of a compressible fluid in a nozzle or a pipe cannot, for a given set of upstream conditions, be increased above some maximum 'critical' value. For gas flows, this phenomenon has been studied extensively and may be regarded as being well known. It occurs when a large expansion takes place somewhere in the flow path at: safety (or relief) valves; restrictions; nozzles and orifices. The topic is much studied in chemical engineering because of the need to predict the flow rate of discharge from a pressure vessel. There are two reasons why this might be required. Firstly, there is a need to know the critical flow rate/ outlet pipe or orifice size relationship, in order to design the outlet piping to remove material so that the pressure in the vessel is always kept below the 'safe' operating pressure and to prevent the vessel rupturing. Secondly, the maximum flow rate is required as the source term for any dispersion calculations. It is interesting to reflect that in the first instance, a conservative value is one, which is below the real value. In contrast, for dispersion calculations the conservative (or safe) value is one that is above the real value.

The basic idea of choked flow can be illustrated in Figure 9.1. As the exit pressure, p_e, is reduced below the reservoir pressure, p_o, the fluid will start to flow from the reservoir. As the exit pressure is further reduced, the mass flux (mass flow rate/cross-sectional area of the

Hydrodynamics of Gas-Liquid Reactors: Normal Operation and Upset Conditions, First Edition.
B. J. Azzopardi, R. F. Mudde, S. Lo, H. Morvan, Y. Yan and D. Zhao.
© 2011 John Wiley & Sons, Ltd. Published 2011 by John Wiley & Sons, Ltd.

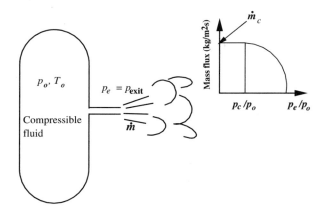

Figure 9.1   *Critical flow – basic concept*

outlet pipe), m, increases to a critical value, m_e, beyond which further reductions in the exit pressure will have no effect. Under such conditions the flow is 'choked.'

A further complication can arise that has significance in reactor venting applications. Figure 9.2 illustrates an arrangement in which the vent lines from several reactors could be connected via a manifold to a knock out drum or dump tank. There are a number of points where choking might take place: the safety valve, the outlet from the downstream pipe into the larger diameter manifold pipe or the outlet into the dump tank. In fact, choking could occur at any or all of these places. Experiments have shown that multiple chokes are possible both for single- and for two-phase gas/liquid flows.

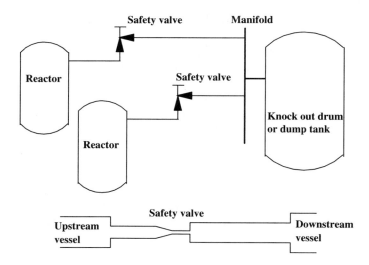

Figure 9.2   *Typical manifold arrangement and generalised schematic of geometry where multiple chokes might occur*

In instances where the two-phase flow is compressible, for example a gas-liquid flow, choking may also occur. This chapter considers the background information to this subject. Much has been worked out in terms of equations to calculate a critical (or limiting) flow rate. However, with the increasing use of complex piping systems in vent applications, brought on by environmental pressures, calculation methods that deal with the whole system are required. These are examined towards the end of the chapter.

The phenomena of critical flow and the choking condition are related to the propagation of information in the flow. For sub-critical conditions, information can propagate upstream and downstream. At the critical condition, information can no longer propagate upstream and thus changing the downstream condition can have no influence on the upstream flow.

For single-phase gas flow, it can be shown that the velocity of information, critical flow and choking form a coherent theory and result in several equivalent definitions of a critical flow. In contrast, in two-phase flow there is not the coherence of theories. This chapter first considers single-phase flow and then two-phase (gas-liquid) flow from the point of view of a single choke point. Finally, the methods to calculate multiple chokes are explored.

9.2 Single-Phase Flow

For a fluid, the velocity of propagation of the disturbance can be derived by considering a plane, infinitesimal pressure wave moving with velocity, δv, along a duct of constant cross-sectional area. The changes in pressure and density across the wave are δp and $\delta \rho$, respectively. The flow will be steady when referred to a coordinate system moving with velocity that brings the pressure pulse to a rest. This is the flow relative to an observer moving with the pressure pulse. The fluid flows towards the observer with velocity c, pressure p and density ρ and then slows down to a velocity $(c - \delta v)$, with corresponding pressures $p + \delta p$ and density $\rho + \delta \rho$. Applying the conservation of mass in the limit as δv and $\delta \rho$ go to zero, gives:

$$\frac{d\rho}{\rho} = \frac{dv}{c} \tag{9.1}$$

Applying the conservation of momentum through the wave front in limit gives:

$$dp = \rho c \, dv \tag{9.2}$$

Eliminating dv from Equations 9.1 and 9.2:

$$c^2 = \left(\frac{dp}{d\rho}\right) \tag{9.3}$$

This gradient is determined by the type of process that causes the change in the fluid properties as the fluid goes to the wave front. This is assumed to be adiabatic, therefore if we apply the energy equation to the flow through the wave front we obtain:

$$h + \frac{c^2}{2} = (h + dh) + \frac{(c + dv)^2}{2} \tag{9.4}$$

where

h is the enthalpy.

In the limit this gives:

$$dh = c\,dv \tag{9.5}$$

If we compare Equations 9.2 and 9.5 we see that they imply

$$dh - \frac{dp}{\rho} = 0 \tag{9.6}$$

which, when compared with the familiar thermodynamic relationship

$$T\,ds = dh - \frac{dp}{\rho} \tag{9.7}$$

where

T is the temperature
s is the entropy.

this implies that $Tds = 0$. The flow is therefore isentropic, that is

$$c^2 = \left(\frac{\delta p}{\delta \rho}\right)_s \tag{9.8}$$

where the subscript s denotes constant entropy. If the fluid in the duct is flowing with finite velocity, then the above analysis gives c as the speed of disturbance relative to the flow. Sound waves are infinitesimal pressure disturbances and therefore Equation 9.8 gives the velocity of sound in the fluid in the tube.

The general behaviour of a flow depends on whether the fluid velocity is greater or less than the local velocity of sound. If the velocity is less than the local velocity, that is the flow is subsonic, then disturbances can be communicated throughout the flow. If the flow is sonic, then the disturbances form a plane front, containing the source of the disturbances, which separates the undisturbed fluid from the disturbed fluid behind it. If the flow is supersonic, then any infinitesimal disturbances are swept downstream.

For an ideal gas undergoing an isentropic process, the term p/ρ^γ is constant with γ being the isentropic coefficient. Using this in Equation 9.8

$$c = \sqrt{\frac{dp}{d\rho}} = \sqrt{\frac{\gamma p}{\rho}} \tag{9.9}$$

When the value of atmospheric pressure (10^5 Pa) and air density ($1.2\,\text{kg/m}^3$) are substituted into this equation, a velocity of sound of $\sim340\,\text{m/s}$ is obtained, which is a recognisable value.

If flow in a converging nozzle is considered and the mechanical energy balance is applied between the entry, where $p = p_o$, $\rho = \rho_o$ and v is assumed to be very small and can be set to zero, and an arbitrary plane

$$\frac{v^2}{2} = \int_{p_o}^{p} \frac{dp}{\rho} \tag{9.10}$$

using the isentropic relationship and integrating:

$$\frac{v^2}{2} = \frac{\gamma p_o}{(\gamma-1)\rho_o}\left[1-\left(\frac{p}{p_o}\right)^{\frac{\gamma-1}{\gamma}}\right] \tag{9.11}$$

Defining the mass balance as $\dot{M} = \rho S v$ and using $\rho = \rho_o(p/p_o)^{1/\gamma}$ gives:

$$\frac{\dot{M}}{S} = \sqrt{\frac{2p_o\rho_o\gamma}{\gamma-1}\left(\frac{p}{p_o}\right)^{\frac{2}{\gamma}}\left[1-\left(\frac{p}{p_o}\right)^{\frac{\gamma-1}{\gamma}}\right]} \tag{9.12}$$

This equation has a maximum, which occurs at a dimensionless pressure of

$$\frac{p_c}{p_o} = \left(\frac{2}{\gamma+1}\right)^{\frac{\gamma}{\gamma-1}} \tag{9.13}$$

This pressure has the subscript c as it is the critical pressure ratio, which can be seen to depend only on γ. For typical values of γ, that is 1.3–1.4, this ratio has values of 0.528–0.546.

The maximum value of mass flux is the choked flow rate. It is obtained by substituting Equation 9.13 into Equation 9.12:

$$\dot{m} = \frac{\dot{M}}{S} = \sqrt{p_o\rho_o\gamma\left(\frac{2}{\gamma+1}\right)^{\frac{\gamma+1}{\gamma-1}}} \tag{9.14}$$

It is important to note that the maximum flow rate, which determines the flow capacity of the nozzle, depends on the entrance conditions and γ, but not on the downstream conditions.

9.3   Two-Phase Flow

Two-phase gas-liquid flow is the combination of a compressible fluid (gas) and a slightly compressible one (liquid). These can travel at different velocities and, as the pressure drops, there can be transfer of mass from the liquid to the gas phase. The methodologies developed for determining whether choking takes place and what is the limiting flow rate at choking needs to take into account the impact of these two features.

The *sonic velocity* for a two-phase mixture can be obtained using methods similar to those for single-phase flow but allowing for the phases occupying different fractions of the channel. The mass and momentum balances across the wave front for the phases can be written as:

$$\text{Mass} : cd[\varepsilon_g\rho_g + (1-\varepsilon_g)\rho_l] = \varepsilon_g\rho_g du_g + (1-\varepsilon_g)\rho_l du_l \tag{9.15}$$

$$\text{Mom} : c[\varepsilon_g\rho_g du_g + (1-\varepsilon_g)\rho_l du_l] = dp \tag{9.16}$$

where

ε_g is the void fraction

ρ_g and ρ_l and u_g and u_l are the densities of the liquid and gaseous phases, respectively.

Rearranging Equations 9.15 and 9.16, results in an expression for the two-phase sonic velocity:

$$c^2 = \frac{dp}{d[\varepsilon_g \rho_g + (1-\varepsilon_g)\rho_l]} \tag{9.17}$$

The void fraction, ε_g, is related to the quality, x_g, and the slip ratio, $U_R = u_g/u_l$, by

$$\varepsilon_g = \frac{1}{1 + U_R \frac{(1-x_g)\rho_g}{x_g \; \rho_l}} \tag{9.18}$$

Combining Equations 9.17 and 9.18, after some manipulation yields:

$$c^2 = \frac{[(1-x_g)\rho_g + x_g\rho_l]}{x_g\rho_l^2 \frac{d\rho_g}{dp} + (1-x_g)\rho_g^2 \frac{d\rho_l}{dp} - (\rho_l-\rho_g)\frac{dx_g}{dp} + x_g(1-x_g)(\rho_l+\rho_g)\rho_l\rho_g \frac{dU_R}{dp}} \tag{9.19}$$

The differential terms in Equation 9.19 will be determined by various interfacial processes, which will depend on the specific interfacial configurations, flow patterns and the time available for reaching the equilibrium of these processes. Henry [1], who also provided a more rigorous derivation of Equation 9.17, discussed these processes in detail. The derivation given here is somewhat simplified and a number of subtle assumptions have been implicitly assumed.

It is clear from Equations 9.17 and 9.19 that the velocity of sound in a two-phase mixture is not a simple function. It includes derivatives $d\rho_g/dp$, $d\rho_l/dp$, dx_g/dp and dU_R/dp in Equation 9.19, which will be determined by the rates of inter-phase, heat, mass and momentum transfer, respectively. The magnitude of these transfer rates will in general depend on the flow regime, the wave frequency and the type of disturbance. A full discussion can be found in Hsu [2] and Henry [1]. To illustrate the complexity of the phenomena, one might consider data from experimental measurements of the velocity of sound in a gas-liquid mixture for the void fraction range from 1 to 67%. It was found that the velocity was strongly dependent on frequency, void fraction and pressure and that there was a well-defined minimum velocity associated with a void fraction of 50%.

The effect of flow pattern in sonic velocity is illustrated in Figure 9.3. These values were measured for different configurations of the phases about the cross-section of a pipe. Stratified flow in a horizontal pipe implies that there is gas flowing above a layer of liquid, that is there is continuous gas path. Consequently, the velocity sound is that of the gas. In bubbly flow, more common in vertical pipes, the gas is dispersed as bubbles in the liquid.

As with single-phase gas flow, the analysis of two-phase *critical flow rate* starts with the steady-state mass, momentum and energy balances.

$$\frac{d(\rho_m SW)}{dz} = 0 \tag{9.20}$$

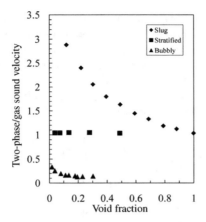

Figure 9.3 *Measured and calculated pressure wave propagation velocity in air-water mixtures at 1.7 bar*

$$\frac{\mathrm{d}(\rho_m SW^2)}{\mathrm{d}z} + S\frac{\mathrm{d}p}{\mathrm{d}z} = -P_w \tau_{TP} + S\rho_m g \tag{9.21}$$

$$\frac{\mathrm{d}}{\mathrm{d}z}\left(S\rho_m W\left[h_m + \frac{W^2}{2}\right]\right) = P_H \dot{q} + S\rho_m Wg \tag{9.22}$$

where

ρ_m is the mixture density
S is the channel cross-sectional area
P_w is the perimeter of the wall
τ_{TP} is the wall shear stress
g is the gravitational acceleration
$\dot{q}$ is the heat flux
W is a characteristic velocity.

Problems arise in the solution of these equations, particularly in the specification of the two-phase density. Complications arise from the terms noted in the last section, the changes with pressure of phase densities, quality and slip ratio. The density of the gas or vapour phase will vary with pressure but also with any temperature change due to the rapid expansion. The quality will also vary as flashing occurs. However, the processes can be very fast with the fluids passing through the constriction, which causes the large pressure drop in a very short time. Given this short time, is there an opportunity for the fluids to reach a new equilibrium? Obviously, assumptions are necessary to simplify the problem into a form that is more easily solved. A start will be made with the easiest case, and then the assumptions will be examined in turn to establish their validity and to consider alternatives.

From the testing of pressure drop equations for gas-liquid pipe flow [3], it was seen that the homogeneous model, $U_R = 1$, became most accurate at higher mass fluxes, the type that might be expected for choked flows. Therefore, the first assumption that can be made is that the flow is homogeneous. A further assumption is to expect the heat and mass transfer processes to occur rapidly enough for the phase equilibrium to always be valid for local conditions. This combination is commonly known as the homogeneous equilibrium model (HEM). With these assumptions, Equations 9.20–9.22 can be developed to the following matrix of equations.

$$\begin{pmatrix} x_g/\rho_g + \dfrac{(1-x_g)}{\rho_l} & \dfrac{1}{\rho_m W} & \dfrac{1}{\rho_g} - \dfrac{1}{\rho_l} \\ 1 & \dfrac{1}{\rho_m W} & 0 \\ x_g h_g + (1-x_g)h_l & \dfrac{1}{W} & h_g - h_l \end{pmatrix} \begin{pmatrix} \dfrac{dp}{dz} \\ \dfrac{dW}{dz} \\ \dfrac{dx_g}{dz} \end{pmatrix} = \begin{pmatrix} \dfrac{1}{\rho_m S}\dfrac{dS}{dz} \\ -\dfrac{P_W}{S}\tau_{TP} + \rho_m g \\ g \end{pmatrix} \qquad (9.23)$$

For orifices, where wall shear and gravity can be ignored, this reduces to:

$$\dot{m}_{crit} = \sqrt{\left[\dfrac{d}{dp}\left(\dfrac{(1-x_g)}{\rho_l} + \dfrac{x_g}{\rho_g} \right) \right]^{-1}} \qquad (9.24)$$

Here, although the derivatives of density with respect to pressure are obtainable for water and, with more difficulty for other fluids, there are problems determining the changes of quality with pressure. If isentropy is assumed, the quality x_g, and the derivative of quality with respect to pressure dx_g/dp are evaluated via the definition of two-phase entropy.

However, the full matrix enables the behaviour to be calculated along a pipework system. This is discussed further below.

Entropy data might be available for steam-water and other common fluids. However, they might not be available for many of the materials to be dealt with. An alternative method is the ω approach used by Leung [4]. This uses a correlating group defined by:

$$\omega = x_o \dfrac{v_v(p_o)}{v_{hom}(p_o)} + c_{pL}T(p_o)\dfrac{p_o}{v_{hom}(p_o)}\left[\dfrac{v_v(p_o) - v_l(p_o)}{h_v(p_o) - h_l(p_o)} \right]^2 \qquad (9.25)$$

where

v is the specific volume (m^3/kg)
h is the specific enthalpy (J/kg).

The subscripts v, l and *hom* refer to vapour, liquid and a two-phase mixture when calculated assuming that flow is homogeneous. The critical pressure ratio is then given by:

$$\eta = \dfrac{p_c}{p_o} = 0.55 + 0.217\ln(\omega) - 0.046\left[\ln(\omega)\right]^2 + 0.004\left[\ln(\omega)\right]^3 \qquad (9.26)$$

whilst the critical mass flux is given by:

$$\dot{m}_c = \sqrt{\dfrac{p_o}{v_o}}\dfrac{0.6055 + 0.1356\ln(\omega) - 0.0131\left[\ln(\omega)\right]^2}{\sqrt{\omega}} \quad \text{for } \omega \geq 4 \qquad (9.27)$$

and

$$\dot{m}_c = \sqrt{\frac{p_o}{v_o}} \frac{0.66}{\omega^{0.39}} \quad \text{for } 4 \geq \omega \geq 1 \tag{9.28}$$

The differences between this method and the HEM have been tested for vapour-liquid mixtures of water, refrigerant 12 (with properties similar to many light hydrocarbons), chlorine and air-water [5]. If the method is being used for pressures above 0.5 of the thermodynamic critical pressure, the liquid specific heat, c_{pl}, is not always known accurately. Although differences of up to 50% have been reported, particularly at higher system pressures, correct use of the liquid enthalpy/temperature gradient on the liquid enthalpy curve at constant pressure can give improved results [6].

It is important to note that for gas-liquid flows the critical pressure ratio is usually higher than that for a single-phase gas flow, which can be calculated from Equation 9.13. Calculations using the ω method have been shown to give critical pressure ratios of 0.85 as the quality tends to zero, and of $\sim$0.6 as the quality tends to 1.0 for fluids with properties similar to many light hydrocarbons [7].

The assumption of the homogeneous equilibrium model that the slip ratio $U_R = 1$ is clearly a deficiency for those interfacial configurations, or flow patterns, which allow relative motion between the phases and will be a source of error. Fauske [8] and Moody [9] have proposed two particularly simple relationships for the slip, U_R, which can be used to modify the homogeneous equilibrium model. Fauske [8] proposed that the slip ratio was equal to the square root of the density ratio. This expression was derived from arguments and analysis, which maximised the two-phase momentum flux at the critical location. In contrast, Moody [9], using arguments that maximise the mass flux with respect to pressure and slip, deduced that the slip ratio was equal to the cube root of the density ratio, which is equivalent to maximising the energy flux at the critical location.

Comparisons of the Fauske [8] and Moody [9] models with experimental data show that they tend to over-predict the critical mass flux. Data from experiments [10], some of which are shown in Figure 9.4 illustrate the variation of slip ratio, U_R, with quality, x_g. For the conditions of these experiments, the Fauske model predicts that $U_R = 20.4$ and the Moody model predicts that $U_R = 9.3$. Thus, both the Fauske model and the Moody model grossly over-predict the slip ratio. It can also be seen from Figure 9.4 that at high quality the approximation that the slip ratio, $U_R = 1$ is reasonable and this in part explains the success of the homogeneous equilibrium model at high qualities.

The simplest non-equilibrium model to adopt the limiting assumption that the residence time of the fluid in the nozzle is so short, in comparison with the timescales for interfacial heat and mass transfer, that the quality remains constant throughout the expansion, is termed the homogeneous frozen model (HFM) [11]. This starts from the momentum Equation 9.20. If the wall effects and gravity terms in that equation are ignored and concentrating on the throat where there is no change in area with axial distance and where the mass flux is a maximum (i.e. $d\dot{m}/dp = 0$), the momentum equation can be rewritten as:

$$\dot{m}_c^2 = \left[U_R \left(A \frac{dv_g}{dp} + B \frac{dx_g}{dp} + C \frac{dv_l}{dp} + D \frac{dU_R}{dp} \right) \right]^{-1} \tag{9.29}$$

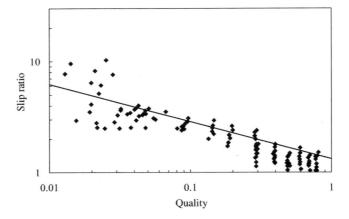

Figure 9.4 *Measured slip ratio [10]*

where

$$A = [1 + x_g(U_R - 1)]x_g$$
$$B = v_g[1 + 2x_g(U_R - 1) + 2U_R v_l(x_g - 1) + U_R(1 - 2x_g)]$$
$$C = U_R\left[1 + x_g(U_R - 2) - x_g^2(U_R - 1)\right], \quad D = x_g(1 - x_g)\left[U_R v_l - \frac{v_g}{U_R}\right]$$

Using the assumption of homogeneous flow, that is $U_R = u_g/u_l = 1$, and the flow is frozen, x_g is constant, then this equation reduced to

$$\dot{m}_c^2 = \left[x_g \mathrm{d}v_g/\mathrm{d}p + (1 - x_g)\,\mathrm{d}v_l/\mathrm{d}p\right]^{-1}$$

If the liquid is assumed to be incompressible, that is $\mathrm{d}v_l/\mathrm{d}p = 0$, and that the flow of the gas is isentropic, that is $pv^\gamma = \mathrm{constant}$, this equation further reduces to:

$$\dot{m}_c = \sqrt{\frac{n\gamma p}{x_g v_g}} \tag{9.30}$$

where

n is a parameter to allow for heat transfer, developed by Tangren *et al.* [12].

It takes a value of one if heat transfer can be ignored. Values of critical mass flux calculated using this general frozen approach were compared with experimental data [11], and reasonable agreement was obtained at higher qualities, but there appeared to be deviation from the data at lower qualities.

Henry and Fauske [11], whilst using the assumptions of homogeneous (slip ratio $U_R = 1$) frozen quality, that is $x_g = \mathrm{constant}$ and isentropic flux, made an additional assumption concerning the gas-phase expansion. They assumed that the kinetic energy of the two-phase mixture is due solely to the vapour expansion and that the critical flow-rate is defined by gas-dynamic principles. Using these assumptions, Henry and Fauske derived the following expression for the homogeneous frozen critical two-phase mass flux.

$$\dot{m}_c = \frac{1}{v}\left\{2x_{go}v_{go}p_o\left(\frac{\gamma}{\gamma-1}\right)\left[1-\left(\frac{p}{p_o}\right)^{\frac{\gamma-1}{\gamma}}\right]\right\}^{1/2}$$

(9.31)

$$v = [1-x_o]\,v_{lo} + x_{go}v_{go}\left(\frac{p_c}{p_o}\right)^{1/\gamma}$$

(9.32)

Comparisons of the homogeneous frozen model with experimental data shows that the model tends to underestimate the critical two-phase mass flux at very low inlet quality and that the accuracy prediction can also be poor at higher qualities.

Leung and Fauske [13], following Tangren *et al.* [12], gave an alternative approach. This calculates the critical pressure ratio from

$$\eta = \left[2.016 + \left(\frac{1-\varepsilon_{go}}{2\varepsilon_{go}}\right)^{0.7}\right]^{-0.714}$$

(9.33)

and the critical mass flux from

$$\dot{m}_c = \dot{m}^*\sqrt{\frac{p}{v}}$$

(9.34)

where

$$\dot{m}^* = \frac{\left\{\frac{2}{\varepsilon_{go}}\left[\left(\frac{1-\varepsilon_{go}}{\varepsilon_{go}}\right)(1-\eta)-\ln\eta\right]\right\}^{\frac{1}{2}}}{\frac{1}{\eta}+\left(\frac{1-\varepsilon_{go}}{\varepsilon_{go}}\right)}$$

(9.35)

In reality, the flow will be somewhere between frozen and equilibrium. A number of models have been developed to try and produce better approximations to reality.

Henry *et al.* [14] have developed a non-equilibrium slip model for the low quality regions. They argued that the derivative of the liquid specific volume with respect to pressure will be very small and wrote an expression for the critical mass flux:

$$\dot{m}_c = \sqrt{-U_R\left\{[1+x_g(U_R-1)]x_g\frac{dv_g}{dp} + v_g[1+2x_g(U_R-1)]\frac{dx_g}{dp} - x_g(x_g-1)\frac{v_g}{U_R}\frac{dU_R}{dp}\right\}^{-1}}$$

(9.36)

Non-equilibrium was introduced through an empirical parameter *N*:

$$N = \frac{x_g}{U_R}\frac{1}{x_{geq}}$$

(9.37)

In Equation 9.37 x_{geq} is the equilibrium quality. Differentiating Equation 9.37 with respect to pressure gives the following relationship for the change of quality:

$$\frac{dx_g}{dp} = U_R N\frac{dx_{geq}}{dp} + U_R x_{geq}\frac{dN}{dp} + Nx_{geq}\frac{dU_R}{dp}$$

(9.38)

Henry, Fauske and coworkers restricted their analysis to the low-quality region and assumed an isothermal expansion for the gas (vapour phase). Using Equations 9.37 and 9.38. They simplify (9.36) to give:

$$\dot{m}_c^2 = \left(\frac{N}{(\dot{m}_c^2)_{HEM}} - v_g x_{geq} \frac{dN}{dp} \right)^{-1} \tag{9.39}$$

where

$$(\dot{m}_c^2)_{HEM} = - \left(x_{geq} \frac{dv_g}{dp} + v_l \frac{dx_{geq}}{dp} \right)^{-1} \tag{9.40}$$

Equation 9.40 was obtained from (9.36) by substituting $U_R = 1$. The quality, slip and void fraction are, of course, related and for low quality

$$x_g \approx U_R \frac{\varepsilon_g}{(1-\varepsilon_g)} \frac{\rho_g}{\rho_l} \tag{9.41}$$

which, when combined with Equation 9.37, gives

$$N = \frac{\rho_g \varepsilon_g}{x_{geq} (1-\varepsilon_g)} \rho_l \tag{9.42}$$

From Equation 9.42, the empirical value of N can be obtained from the void fraction data.

Henry, Fauske and coworkers further argued that from experimental data the term dN/dp is very small and therefore the expression for the critical mass flux will reduce to

$$\dot{m}_c^2 = \frac{(\dot{m}_c^2)_{HEM}}{N} \tag{9.43}$$

It should be noted that (9.43) is only valid if

$$\frac{x_{geq}}{\rho_g} \frac{dN}{dp} \gg \frac{N}{(\dot{m}_c^2)_{HEM}} \tag{9.44}$$

Equation 9.44 implies that Equation 9.43 should be restricted to low void fraction flow, that is flows with low quality or under high pressures. They proposed that for long tubes

$$\begin{aligned} N &= 20\,x_{geq} & 0 < x_{geq} < 0.05 \\ N &= 1 & x_{geq} > 0.05 \end{aligned} \tag{9.45}$$

that is in higher void fraction flows, thermodynamic equilibrium is achieved and as they observed a strong length effect, they restricted the analysis to the case where $L/D < 12$.

Another approach has been termed the delayed equilibrium model (DEM) [15]. It assumed that the fluids could be divided into two parts:

- $(1 - y)$ of metastable liquid per kg of mixture
- y kg of saturated fluids with:
  - x kg of saturated vapour (steam quality)
  - $(y - x)$ kg of saturated liquid

They suggested that the variation of y can be described by:

$$\frac{dy}{dz} = 0.01 \frac{P_W}{S} (1-y)^2 \left(\frac{w_{lo}}{w}\right)^{0.1} \left[\frac{p_{sat}(T_{LM})-p}{p_{crit}-p_{sat}(T_{LM})}\right]^{0.25} \qquad (9.46)$$

where

P_W is the wetted perimeter
S is the channel cross-section
w_{lo} is the velocity of the liquid phase flowing alone
w is the mean velocity of the mixture
T_{LM} is the temperature of the metastable liquid
p_{crit} is the critical pressure of the fluid.

Another approach suitable for hand calculations was given by Fauske [16], who noted that in the absence of significant friction losses (generally the case in the non-equilibrium regime), flashing flows are generally always choked and can be, to a first order, approximated by:

$$\dot{m}_c \approx \frac{\Delta h_v}{v_{lg}} \left(\frac{1}{NTc_l}\right)^{0.25} \qquad (9.47)$$

where

$$v_{lg} = \frac{1}{\rho_g} - \frac{1}{\rho_l} \qquad (9.48)$$

and N is a non-equilibrium parameter given by:

$$N \approx \frac{\Delta h_v^2}{2\Delta p \rho_l K^2 v_{lg}^2 Tc_l} + 10L \qquad (9.49)$$

where

K is the discharge coefficient
Δp is the total pressure drop available
L is the duct length.

When $L/D = 0$, the residence time is zero and there can be no flashing, so that Equation 9.47 reduces to that for flow through an orifice

$$\dot{m} = 0.61 \sqrt{2\Delta p \rho_l} \qquad (9.50)$$

For $L > 0.1$ m, the fluid residence time is probably sufficiently large so that equilibrium flashing conditions are reached and the equation reduces to:

$$\dot{m}_c \approx \frac{\Delta h_v}{v_{lg}} \sqrt{\frac{1}{Tc_l}} \qquad (9.51)$$

approximated using stagnation properties. This is known as the equilibrium rate model (ERM). If properties are not known, Equation 9.51 can be restated as:

$$\dot{m}_c \approx \frac{dp}{dT}\sqrt{\frac{T}{c_l}} \tag{9.52}$$

For cases where the discharge is sub-cooled, account is taken of the increased pressure due to the sub-cooling $[p_o - p(T_o)]$. If the equilibrium rate model were appropriate for the flow under saturated conditions ($L > 0.1$ m) then

$$\dot{m}_c \approx \sqrt{2[p_o - p(T_o)] + \dot{m}_{cERM}^2} \tag{9.53}$$

For a gas-vapour-liquid equilibrium system, we can write

$$\dot{m}_c \approx \sqrt{2p_g\rho_l + \dot{m}_{cERMg}^2} \tag{9.54}$$

where

p_g is the partial pressure of the gas

and

$$\dot{m}_{cERMg} \approx \frac{p_v}{p}\frac{dp}{dT}\sqrt{\frac{T}{c_l}} \tag{9.55}$$

where

p_v is the partial pressure of the vapour
$p = p_g + p_v$.

9.4 Effect of Vent Pipework

With the introduction of more complex pipework into the vent system, the approach of considering *only* what occurs at a single choke point, such as a relief valve, becomes less realistic. The data presented in Figure 9.5 show how the pressure gradient increases sharply at the sudden expansion, at 1.4 m from the start and at the outlet. The large pressure gradients characterise the occurrence of choking. The pipe diameters are 17 and 28 mm, respectively, and each pipe is 1.4 m long. The inlet pressure is 5 bar (1 bar $= 10^5$ Pa) and the temperature is 151 °C.

Here the conditions for multiple chokes will be discussed and methodology to handle the complex pipework system indicated [15]. The simplest system where more than one choke could be encountered is illustrated in Figure 9.6. Examination of Figure 9.2 shows where this could fit into a practical system. The criteria necessary for chokes to occur at planes 1 and *b* are developed in terms of gas-only flow to illustrate the salient feature. There are equivalent and more complex versions of the equations for gas-liquid flows. Consider a proposed pipe of length, L, and diameter, D, which is connected upstream to a smaller diameter pipe and downstream to a larger diameter pipe or vessel.

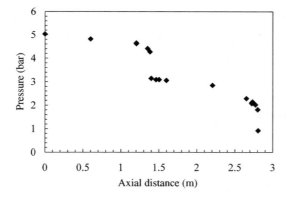

Figure 9.5 *Pressure profile along pipework consisting of two pipes of different diameter joined at a sudden expansion and illustrating the steep pressure gradient at the two positions of choking [15]*

The planes designated 1 and 2 are just outside this pipe, whilst a and b are just inside. Because there are sudden expansions at the upstream and downstream connections and there are singular pressure drops connected with these, the conditions at a are taken as the state and average velocity extrapolated from the fully developed region in the pipe. The area ratio between plane a and plane 1 is σ. The velocity reaching the speed of sound defines critical flow. It could occur at 1. The sudden expansions cause the velocity to fall. However, the pressure drop due to friction in the pipe will cause the velocity to rise again and there could once again be critical flow at plane b. This could be repeated at subsequent expansions. To define the necessary conditions for a double (or multiple) choke, mass and energy balances are written:

$$\rho_1 u_1 = \sigma \rho_b u_b \tag{9.56}$$

$$h_1 + \frac{u_1^2}{2} = h_b + \frac{u_b^2}{2} \tag{9.57}$$

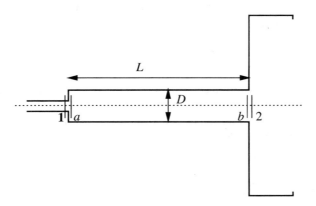

Figure 9.6 *Geometry for development of double-choke criterion*

If the flow is choked at both 1 and b, from the definition of the critical velocity,

$$u_1 = c = \sqrt{\gamma RT_1} \quad u_b = c = \sqrt{\gamma RT_b} \tag{9.58}$$

where

γ is the adiabatic expansion coefficient.

With the assumption of constant specific heat, Equation 9.57 can be written as:

$$c_p T_1 + \frac{\gamma RT_1}{2} = c_p T_b + \frac{\gamma RT_b}{2} \tag{9.59}$$

From Equation 9.58 it can be seen that:

$$T_1 = T_b \tag{9.60}$$

and

$$u_1 = u_b \tag{9.61}$$

Substiuting into Equation 9.56 yields:

$$\rho_1 = \sigma \rho_b \tag{9.62}$$

and from the ideal gas assumption

$$p_1 = \sigma p_b \tag{9.63}$$

This gives the *first* necessary condition for a double choke:

$$p_2 < \frac{p_1}{\sigma} \tag{9.64}$$

The increase in entropy can also be calculated to be:

$$s_b - s_1 = c_p \ln\left(\frac{T_b}{T_1}\right) - R \ln\left(\frac{p_b}{p_1}\right) = R \ln(\sigma) \tag{9.65}$$

where

R is the gas constant.

The increase in entropy arises from contributions of the irreversible processes in the sudden expansion and the wall friction in the downstream pipe, the latter being known as Fanno flow. The effect of the expansion can be calculated from the mass, momentum and energy balance of Equations 9.56–9.58 using the relationships in Equations 9.59–9.63. With the assumption of choking at plane 1, that is $Ma_1 = 1$, this results in

$$\left[(\gamma - 1) - \frac{\gamma^2(\gamma + 1)}{(\gamma + \sigma)^2}\right] Ma_a^4 + 2\left[1 - \frac{\gamma(\gamma + 1)}{(\gamma + \sigma)^2}\right] Ma_a^2 - \frac{\gamma + 1}{(\gamma - \sigma)^2} = 0 \tag{9.66}$$

The maximum pipe length that can be used with the second choke just occurring is L_c. This can be determined from the standard analysis of Fanno flow, which relates the change in Mach number to incremental increase in pipe length via

$$dz = \frac{1-Ma^2}{\gamma Ma^2 \left(1 + \frac{\gamma-1}{1} Ma^2\right)} \left(\frac{D}{4f}\right) \frac{dMa^2}{Ma^2} \tag{9.67}$$

Assuming a constant friction factor, f, and integrating between sections a and b results in:

$$\left(\frac{4fL}{D}\right) = \frac{2+(\gamma-1)Ma_a^2}{2\gamma\, M_a^2} + \frac{2+(\gamma-1)\, Ma_b^2}{2\gamma\, Ma_b^2} + \frac{\gamma+1}{2\gamma}\ln\left[\frac{(2+(\gamma-1)\, Ma_b^2)\, Ma_a^2}{(2+(\gamma-1)\, Ma_a^2)\, Ma_b^2}\right] \tag{9.68}$$

For the case where $Ma_b=1$, this reduces to:

$$\left(\frac{4fL}{D}\right)_c = \frac{1-Ma_a^2}{\gamma\, Ma_a^2} + \frac{\gamma+1}{2\gamma}\ln\left[\frac{(\gamma+1)\, Ma_a^2}{2+(\gamma-1)\, Ma_a^2}\right] \tag{9.69}$$

Ma_a, the Mach number at section a, can be eliminated from a combination of Equations 9.56 and 9.69 to yield a relationship between the dimensionless maximum length and the area ratio, σ. This gives the second necessary condition for multiple choking, the length must be less than the possible maximum length. The possible cases that could arise are illustrated by the flow chart, Figure 9.7. If there is no choke at section 1 and the proposed length of pipe $L<L_c$, then the flow is always subsonic (case A). If, in contrast, $L>L_c$, the flow is impossible and a lower flow rate is required to meet the limit $L=L_c$. This results in $Ma_b=1$, that is choking at section b (case B).

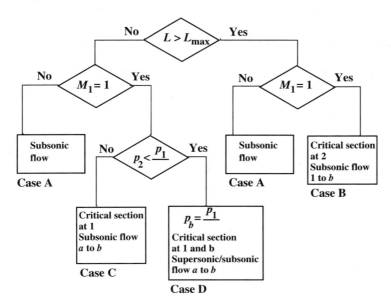

Figure 9.7 *Flow diagram for calculation of two chokes in series*

For choked flow at section 1, the proposed length must again be compared with that which is the maximum before choke occurs at its end. Lengths greater than this will have choking. For shorter lengths, the flow is subsonic in the pipe. A check with the first criterion yields: for $p_2 > p_1/\sigma$, the flow is subsonic at section b (case C); for $p_2 < p_1/\sigma$, the flow is sonic at section b (case D).

By the very nature of the problem, the solution method for a pipe system is iterative. Equations derived from the mass, momentum and energy balances together with appropriate ancillary equations are reorganised into a suitable form. For example, for the homogeneous equilibrium model, they take the form of Equation 9.23. For the delayed equilibrium model, there would be an extra variable, y, and terms relating to dy/dx. In solving the equations, the following are selected: geometry of upstream pipe or nozzle of safety valve; upstream fluid conditions (temperature, pressure and whether sub-cooled or saturated); downstream pressure. A mass flow rate is chosen and the balance equations (e.g. Equation 9.23) are integrated along the pipe with track being kept of the determinant of the set of equations. If the determinant reaches zero during the integration but before the end of the pipe, then the flow is critical (an impossible flow) and the calculation has to be restarted with a lower flow rate. If the determinant does not vanish along the pipe and the downstream pressure is less than p_2, the flow is non-critical and the mass flow rate must be increased and the integration repeated. When the determinant becomes zero at the end of the pipe section, the correct and critical flow rate is obtained.

To complete the calculation for a double choke case, the above is carried out for the first pipe section. Then with knowledge of the downstream pipe geometry, upstream conditions at critical section 1 (pressure, temperature, quality and flow rate) and the downstream pressure, the following procedure is applied. The pressure just downstream of the sudden enlargement, p_e, is chosen. From mass and energy balances, the variables at this section are determined. The flow is then calculated in a stepwise manner until the flow becomes critical or the end of the pipe is reached. The pressure, p_e, is adjusted until the length of pipe required for choked flow is equal to the actual length. Both the HEM and the DEM have been tested using this approach. It was seen that both models give the same trends, but the DEM is more accurate than the HEM. The ability of the DEM to calculate the pressure profiles is shown in Figure 9.8. The characteristic very steep gradient characterising choking can be seen to occur twice and be well predicted by the model.

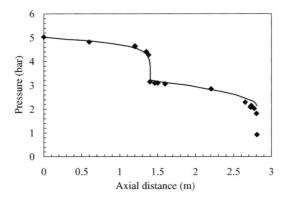

Figure 9.8   *Performance of DEM based calculations for a double choke [15]*

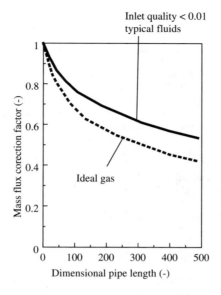

Figure 9.9   *Effect of length to diameter ratio of vent line on flow reduction factor*

A short-cut approach to allow for the effect of a tail pipe uses a simple correction factor to the critical mass flux. This is assumed to depend on the downstream pipe length (in number of pipe diameters). The relationship is shown graphically in Figure 9.9. However, it is noted that this implicitly assumes the occurrence of a single choke.

Questions

9.1 For the example given in Question 8.3 of Chapter 8, calculate the critical mass flux if a set pressure of 2 bara is to be used. The specific heat capacity is 2950 J/kg °C. The variation of vapour pressure with temperature can be represented by:

$$\ln(p) = 28.29 - \frac{6311}{T}$$

where p is in Pa and T in K.

2992 kg/m² s.

9.2 It is planned to carry out a reaction that is non-tempered and which produces gas in a 3 m³ cylindrical vessel. Planned charge of reactants is 2700 kg and it is intended to operate at 1.7 barg. A 10% overpressure is to be allowed for, based on gauge pressure. The void fraction has been determined to be 0.04. Calculate the critical mass flux and compare with the value corresponding to liquid only flow.

16715 kg/m² s; 19340 kg/m² s.

References

[1] Henry, R.E. (1979) Calculational techniques for two-phase critical flow. In: (eds A.E. Bergles and S. Ishigai) Two-Phase Fluid Dynamics, Japan-US Seminar, pp. 415–436.

[2] Hsu, Y.-Y. (1972) Review of critical flow rate, propagation of pressure pulse, and sonic velocity in two-phase media, report TN D-6814, NASA.

[3] Azzopardi, B.J. and Hills, J.H. (2003) One dimensional models for pressure drop, empirical equations for void fraction and frictional pressure drop and pressure drop and other effects in fittings, in *Modelling and Experimentation in Two-Phase Flow* (ed. V. Bertola), Springer-Verlag Wien, New York, pp. 157–220, isbn 3-211-20757-0.

[4] Leung, J.C. (1986) Generalized correlation for one-component homogeneous equilibrium flashing choked flow. *AIChE J.*, **32**, 1743–1746.

[5] Lenzing, T., Friedel, L. and Alhusein, M. (1998) Critical mass flow rate in accordance with the omega-method of DIERS and the Homogeneous Equilibrium Model. *J. Loss Prev. Proc. Inds.*, **11**, 391–395.

[6] Friedel, L., Kranz, N.J., Lenzing, T. and Westphal, F. (1997) Impact of the reproduction accuracy of the fluid properties on the formulations of the homogeneous equilibrium critical mass flow rate model. *J. Loss Prev. Proc. Inds.*, **10**, 43–53.

[7] Hardekopf, F. and Mewes, D. (1988) Critical pressure ratio in two-phase flows. *J. Loss Prev. Proc. Inds.*, **1**, 134–140.

[8] Fauske, H.K. (1962) Contribution to the theory of two-phase, one-component critical flow, report ANL-6633, Argonne National Laboratory.

[9] Moody, F.J. (1965) Maximum flow rate of a single component two-phase mixture. *Trans. ASME, J. Heat Transf.*, **87**, 134–142.

[10] Klingebiel, W.J. and Moulton, R.W. (1971) Analysis of choking two-phase one-component mixtures. *AIChE J.*, **17**, 383–390.

[11] Henry, R.E. and Fauske, H.K. (1971) The two-phase critical flow of one-component mixtures in nozzles, orifices, and short tubes. *Trans. ASME, J. Heat Transf.*, **93**, 179–187.

[12] Tangren, R., Dodge, C.H. and Seifert, H.S. (1949) Compressibility effects in two-phase flow. *J. Appl. Phys.*, **20**, 637–645.

[13] Leung, J.C. and Fauske, H.K. (1987) Runaway systems characterisation and vent sizing based on DIERS methodology. *Plant/Oper. Progr.*, **6**, 77–83.

[14] Henry, R.E., Fauske, H.K. and McComas, S.T. (1970) Two-phase critical flow at low qualities, Parrt II: analysis. *Nucl. Sci Eng.*, **41**, 335–342.

[15] (a) Attou, A. and Seynhaeve, J.M. (1999) Steady-state critical two-phase flashing flow with possible multiple choking phenomenon. Part 1: Physical modelling and numerical procedure. *J. Loss Prev. Proc. Inds.*, **12**, 335–345; (b) Attou, A. and Seynhaeve, J.M. (1999) Steady-state critical two-phase flashing flow with possible multiple choking phenomenon Part 2: comparison with experimental results and physical interpretations. *J. Loss Prev. Proc. Inds.*, **12**, 347–359; (c) Seynhaeve, J.M. and Giot, M. (1996) Choked flashing flow at multiple simultaneous locations. paper presented at the European Two-Phase Flow Group Meeting, Grenoble, June 3–5, 1996, Paper G3.

[16] Fauske, H.K. (1985) Flashing flows or: some practical guidelines for emergency releases. *Plant/Oper. Progr.*, **4**, 132–134.

Part Three

10

Measurement Techniques

Hydrodynamics of Gas-Liquid Reactors: Normal Operation and Upset Conditions, First Edition.
B. J. Azzopardi, R. F. Mudde, S. Lo, H. Morvan, Y. Yan and D. Zhao.
© 2011 John Wiley & Sons, Ltd. Published 2011 by John Wiley & Sons, Ltd.

10.3 Falling Film Reactors
        10.3.1 Film Thickness
        10.3.2 Heat and Mass Transfer
Questions
References

10.1   Bubble Columns

10.1.1   Gas Hold-Up

For bubble columns, the gas hold-up is one of the most important quantities. It determines the flow regime and the mass transfer from and to the gas phase. The gas hold-up can be interpreted as a global, volume averaged quantity. Measurement of the global or over-all gas hold-up can be carried out via 'liquid swell': the difference between the gassed and ungassed height of the liquid represents the total gas volume present in the column. From this the gas hold-up (or void fraction) follows directly as:

$$\varepsilon_g = \frac{V_g}{V_{tot}} = \frac{H - H_0}{H} \tag{10.1}$$

where H and H_0 are the gassed and ungassed height of the liquid in the bubble column V_g and V_{tot} denote the total volume occupied by the gas phase and the total volume of the bubbly mixture, respectively.

The advantage of using visual observation is obviously its simplicity. However, especially for foaming systems, the determination of the hold-up in this way is inaccurate.

The gas hold-up can also be found from pressure measurements. A pressure sensor mounted flush with the wall (see Figure 10.1) will pick up various contributions: flow induced and gravity induced. The latter dominates and allows measuring the hold-up from pressure sensing. It is important to understand that a pressure sensor measures the weight of the bubbly mixture above its location. In a bubble column, this does not mean that the

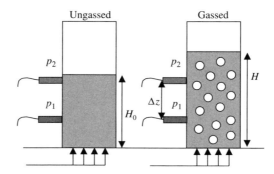

Figure 10.1   *Pressure sensors in ungassed and gas bubble column*

hold-up in the volume above the sensor is obtained from the measurement. On the contrary: a pressure sensor in a bubble column is sensitive to the hold-up below the sensor! This is a consequence of the fixed amount of liquid in the bubble column. Any additional bubble above the pressure sensor does not change the amount of liquid above the sensor. However, if more gas is put in below the sensor, this automatically implies that liquid is moved upwards and the total amount of liquid above the sensor increases. More local gas fractions can be found by measuring the pressure difference over a distance Δz of the column:

$$p_i(\varepsilon_g) = p_{atm} + \rho(\varepsilon_g)g(H - z_i) \rightarrow \Delta p = p_1 - p_2 = \rho(\varepsilon_g)g(z_2 - z_1) \tag{10.2}$$

where

z_i is the height of pressure sensor i
g is the acceleration of gravity.

The ungassed pressure difference is $\Delta p(\varepsilon_g = 0) = \rho_l g(z_2 - z_1)$. Combining these two equations gives the gas fraction between the two sensors:

$$\varepsilon_g = \frac{\left[1 - \frac{\Delta p}{\Delta p(\varepsilon g = 0)}\right]}{1 - \frac{\rho_g}{\rho_l}} \approx \left[1 - \frac{\Delta p}{\Delta p(\varepsilon g = 0)}\right] \tag{10.3}$$

where the second equality is valid if $\rho_g \ll \rho_l$, for example at atmospheric conditions. Intrusive probes, based on conductance or refraction index differences are used for local information on the gas phase. These will be discussed in the next section.

10.1.2　Local Probes: Conductance or Refraction Index

10.1.2.1　Gas Fraction
The differences in conductance or refraction index can be exploited for local measurements. In essence, the probes determine which of the phases is present at the location of the probe. The optical probe [1] consists of a glass fibre. One end is connected to a light source (e.g. an LED), the other end is submerged in the bubbly flow. If the probe tip (100–200 μm diameter) is surrounded by the liquid phase, most of the light will propagate into the liquid. If the tip has pierced a bubble, the light will be internally reflected at the glass-gas interface. The light will travel back and, via a Y-splitter, be detected. A typical signal from a bubble is given in Figure 10.2. The signal is low in the liquid phase, rises when the bubble is pierced to the gas level and drops quickly when the probe leaves the bubble.

From a time trace, the time-averaged void fraction can be obtained. This goes in two steps: (i) the signal is made binary and (ii) the mean value of the binary signal is calculated. The latter is the average time the tip is inside the gas phase and is thus the time-averaged void fraction at the position of the tip:

$$\varepsilon_g = \frac{1}{N}\sum_{i=1}^{N} b(i) \text{ with } b(i) = 1 \text{ if } s(i) > s_l + \delta_t[s_g - s_l]; b(i) = 0 \text{ else} \tag{10.4}$$

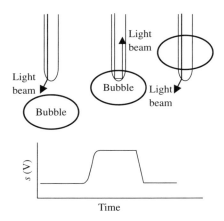

Figure 10.2 *Path of light beam (top); bubble signal from probe (bottom)*

where

$s(i)$ is the probe signal at time t_i
s_l and s_g are the signal values in the liquid and gas phase, respectively
δ_{th} is the threshold value (between 0 and 1)
N is the total number of samples.

As the rise of the signal when a bubble is pierced is not immediate, the choice of the threshold value influences the outcome. In general, a low threshold should be taken. Another form of bias comes from the interaction of the probe with a bubble. For instance, owing to the surface tension the bubble-liquid interface will resist deformation, which is required to pierce the bubble. If a bubble is hit by the probe, close to the bubble's edge, the surface tension force has a relatively large horizontal component, which may deflect the bubble. Consequently, the probe signal may be too low and too short. This leads to underestimation of the void fraction. For bubbles of several millimetres an underestimate of 10% is usually found. This can be understood by considering the probability of hitting a bubble (assumed spherical, with radius R) at a distance r from the central axis (see Figure 10.3). This probability is $p(r) = 2r/R^2$. If a bubble bounces off if hit at $r/R > \beta$, the void fraction obtained with the probe is: $\varepsilon_{g\ meas}/\varepsilon_{g\ true} = 1 - (1 - \beta^2)^{3/2}$.

Conduction probes can measure the conductance from a single probe tip to a ground plate or electrode. By making the electrode relatively large (e.g. the sparger plate), the sensitivity

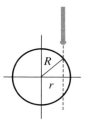

Figure 10.3 *Piercing a bubble at the side*

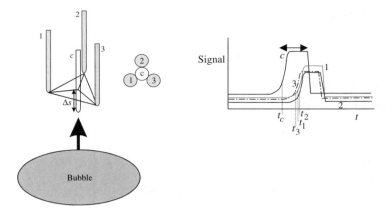

Figure 10.4   *Four point probe and signals*

is localised at the probe tip in much the same way as the optical fibre probe. Similar algorithms can be used.

10.1.2.2   Bubble Size and Velocity

The conductive and optical probes can be used to measure bubble size and velocity. Multiple point probes are used for this purpose. For instance, four-point optical probes have been used [2–4]. Such a probe is sketched in Figure 10.4 and illustrated photographically in Figure 10.5 and illustrated photographically in Figure 10.5. Three tips form an equilateral triangle; a fourth probe goes through the centre of this triangle and is a distance Δs longer ($\Delta s \sim 1.5$ mm). A bubble will now generate four signals, shown on the right in Figure 10.4. Each of the tips hits the bubble at a specific time: (t_c, t_1, t_2, t_3). Provided the time differences $\tau_1 = t_1 - t_c$, $\tau_2 = t_2 - t_c$, $\tau_3 = t_3 - t_c$ do not deviate too much, it can be assumed that the bubble is pierced symmetrically and rises in line with the probe. Then, the bubble velocity is calculated from:

$$v_b = \frac{\Delta s}{\tau} \text{ with } \tau = \frac{1}{3}(\tau_1 + \tau_2 + \tau_3) \tag{10.5}$$

Figure 10.5   *Photograph of a bubble pierced by a four-point probe*

The vertical bubble length follows as:

$$D_v = v_b T \tag{10.6}$$

where

T is the time the central tip is in the bubble.

Corrections can be made for the curvature of the bubble interface [4].

10.1.3 Wire Mesh Sensors

The intrusive probes discussed in the previous section provide, in essence, data measured in a point. In many cases it is more convenient to have information in a plane or a volume. For 2D information on the gas fraction distribution, wire mesh sensors (Figure 10.6) have been developed by, for example Prasser *et al.* [5]. This has recently been extended to use capacitance measurements so that the instrument can be employed with non-conducting liquids, such as organic chemicals or oils [6]. The sensor is made of two parallel planes of conducting wires, with a plane–plane distance in the order of 1.5 mm. The wires themselves have a diameter of 100 μm. In each plane the wires are parallel to each other; for the two planes they are at 90°. The number of wires per plane is 8, 12, 16, 24, 32 and 64, creating 64, 144, 256, 576, 1024 and 4048 measuring points (if the geometry is a square). By quickly measuring the conductance between wire i in plane 1 and wire j in plane 2, the conductivity of the multiphase flow is mapped out. As the wire-sets are at 90°, the measured conduction is dictated by the volume around the crossing of the wires. In a first step, a cubical volume can be assumed through which the current runs. Fast electronics allow for scanning of the entire plane at 5000 frames/s. If the bubbles are large compared with the wire–wire spacing, the measurements range from the levels obtained in a water only case, to virtually no current if a bubble surrounds a crossing completely. In this case, the bubble shape can be reconstructed from the images, by stacking a sequence of images. If the bubbles are small (or if for example small, non-conducting oil droplets are present) a local, average volume fraction is obtained from the measured data.

Figure 10.6 *Wire mesh sensor [5] (Reprinted from Prasser, H.-M., Scholz, D. and Zippe, C., Bubble size measurement using wire-mesh sensors, Flow Meas. & Instr., **12**, 299–312. Copyright (2001) with permission from Elsevier.)*

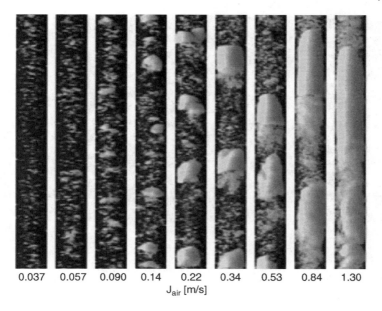

A linear dependence between the void fraction and the conductivity is assumed. For the wire pair $\{i, j\}$, this gives:

$$\varepsilon_g = 1 - \frac{\sigma_{ij} - \sigma_{g,ij}}{\sigma_{l,ij} - \sigma_{g,ij}} \tag{10.7}$$

where

ε_{gij} is the void fraction associated with the volume around the crossing point of the wires i and j

σ_{ij} is the measured conductivity

$\sigma_{g,ij}$ and $\sigma_{l,ij}$ are the conductivity in pure gas and liquid, respectively.

So far, this technique has mainly been used in small-scale pipes (diameter about 2 in; 1 in = 2.54×10^{-2} m). By using a double set of this sensor, the velocity perpendicular to the measuring plane can also be obtained. In addition, this allows turning the time information that is obtained in the flow direction (rather than the actual size) into size information. An example of a sequence of data of a pipe flow is shown in Figure 10.7. It shows bubbly flow for small air flow rates, churn–turbulent flow with small and larger bubbles for higher gas flow rates and slug flow and, finally, almost annular flow for the highest gas flow rates used.

The wires are, of course, intrusive and tend to break up the bubbles. However, right after the downstream wire plane the fragments of big bubbles coalesce again, more or less to the original bubble.

Figure 10.8 *Photograph of bubbles at $\varepsilon_g \sim 50\%$*

For small bubbles or droplets, the signals are more difficult to interpret and it is not possible to uniquely find the bubble size.

10.1.4 Photographic Techniques

Photographs can be taken if the column is transparent. Obviously, only the bubbles on the outside can be seen, especially at the higher void fractions (see Figure 10.8). Nevertheless, they can provide a direct measure of the size of the bubbles, giving both the horizontal and vertical dimensions. Simiano *et al.* [7] used photography to study the size and shape of bubbles on a large scale, 3D, oscillating bubble plume. They used back lightening from LED screens. The camera was set at a focal depth of 2–3 cm. The resolution was 40 μm/pixel, see Figure 10.9. Each bubble was detected and analysed using pattern recognition techniques.

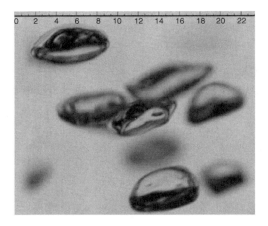

Figure 10.9 *Photograph of bubbles at $\varepsilon_g \sim 0.5\%$ [7] (Reprinted from Simiano, M., et al., Comprehensive experimental investigation of the hydrodynamics of larg-scale, 3D, oscillating bubble plumes, Int. J. Multiphase Flow, **32**, 1160–1181. Copyright (2006) with permission from Elsevier.)*

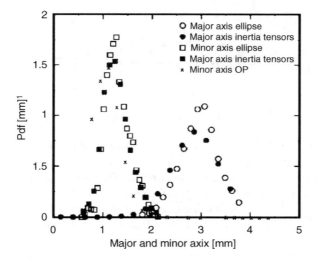

Figure 10.10   *Bubbles distribution at $\varepsilon_g \sim 0.5\%$ [7] (Reprinted from Simiano, M., et al., Comprehensive experimental investigation of the hydrodynamics of larg-scale, 3D, oscillating bubble plumes, Int. J. Multiphase Flow,* **32***, 1160–1181. Copyright (2006) with permission from Elsevier.)*

In this way the distribution of the major and minor axes of the bubbles were found (Figure 10.10).

10.1.5   Laser Doppler Anemometry (LDA)

Laser Doppler anemometry (LDA) has been successfully used in bubbly flows and bubble columns [2, 8, 9]. The working principle is the interference of two light beams coming from the same laser. These two beams intersect inside the bubbly flow, creating a measuring volume. Small, neutrally buoyant, seeding particles (e.g. hollow spheres diameter $\sim 10 \, \mu m$) scatter light that is picked up by a photomultiplier. An easy way to interpret LDA is via the fringe model. The two intersecting light beams form straight fringes, with fringe spacing d_f (depending on the optics). The frequency, f_s, of the scattered light and the velocity of the scattering particle are related as $v = f_s \times d_f$. With bubbly flows, the path of each of the light beams can be obstructed by a bubble. In such instances, a measuring volume is not formed and a gap in the data is created. Therefore, LDA is restricted to low void fractions or the regions close to the wall. Groen *et al.* [8] have shown that the probability that a light beam reaches a certain distance, l, into a bubbly flow is given by $\exp[-\alpha l/d_v]$, with d_v as the vertical dimension of the bubbles. For LDA, the two light beams need to reach the measuring position simultaneously. Therefore, the data rate drops exponentially when measuring deeper into the bubble flow:

$$\frac{f}{f_0} = \exp\left(-C\frac{\alpha l}{d_v}\right) \tag{10.8}$$

where

f and f_0 denote the actual data rate and the data rate close to the wall, respectively
C is about 2.2.

Owing to the presence of the bubbles, the back-scatter mode (Figure 10.11) is usually more convenient.

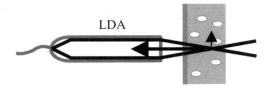

Figure 10.11 *LDA set-up in back-scatter mode*

Figure 10.12 shows a time trace of the vertical velocity of the liquid phase measured in a 15 cm bubble column. As can be seen, the flow is fluctuating as a consequence of the vertical structures in the heterogeneous regime. In principle, the bubbles can also generate velocity data. However, Groen *et al.* [8] showed that the probability for such data is very low. So, for practical purposes, only the liquid velocity is measured. Figure 10.13 gives the axial liquid velocity for two different void fractions. Furthermore, it shows that from the LDA data the transient character from the flow field is easily obtained by calculating the time-windowed probability density function of the velocity.

Inherent to LDA data sets is that the data are not equidistant. This holds even stronger for bubbly flows, due to the interference of the bubbles. This makes time-series analysis more difficult. Standard Fourier transforms to obtain the power spectra are less suited for the LDA data. Care should be taken, as the spectra have a tendency to show a $-5/3$ power behaviour for the higher frequency range. Slotting techniques [10] perform much better.

10.1.6 Particle Image Velocimetry (PIV)

Particle image velocimetry (PIV) is an optical, non-intrusive technique to measure the instantaneous flow field in a two-dimensional plane [11]. The liquid is seeded with small particles, which act as flow tracers. By illuminating a plane with a laser sheet (or a fast sweeping laser beam) and making two consecutive photographs, the displacement of the recorded seeding particles can be found. In combination with the known time interval

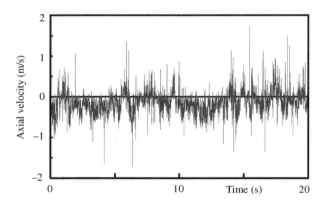

Figure 10.12 *Axial velocity measured by LDA [8] (Reprinted from Groen, J.S., Mudde, R.F., and Van Den Akker, H.E.A., On the application of LDA to bubbly flow in the wobbling regime, Exp. in Fluids,* **27**, *435–449. Copyright (1999) with permission from Springer.)*

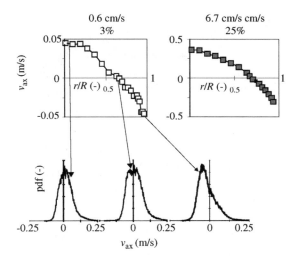

Figure 10.13 *Axial velocity at low and high gas fraction measured by LDA [8] (Reprinted from Groen, J.S., Mudde, R.F., and Van Den Akker, H.E.A., On the application of LDA to bubbly flow in the wobbling regime, Exp. in Fluids, 27, 435–449. Copyright (1999) with permission from Springer.)*

between the two photographs, the velocity of these particles is established. By averaging in so-called interrogation areas, a smooth liquid velocity field may be obtained. Nowadays, laser light in combination with a CCD camera is the standard. Obviously, optical access is needed, limiting PIV to low void fraction bubbly flows.

Hassan *et al.* [12] used PIV to investigate bubbly pipe flow. The flow in and around a bubble plume was investigated by Deen *et al.* [13], who reported that experiments at void fractions of 4% or more are very difficult to perform. Lindken and Merzkirch [14] combined laser light and fluorescent particles with LED back-illumination. The light fluorescent particles could be separated from the light scattered by the bubbles. The back-light generated shadow images from the bubbles. In this way the velocity of the bubble and liquid phase could be found simultaneously. Figure 10.14 shows the 2D liquid velocity field with a few bubbles.

Sommerfeld and coworkers [15] used combined PIV–PTV to study bubbly flows (PTV stands for particle tracking velocimetry, in which individual bubbles are tracked). They studied bubbles flowing and colliding in a turbulent flow. Their experiments not only revealed the average velocity, but for both phases they could measure the fluctuating velocities. Figure 10.15 shows one of their velocity fields. Measurements were taken in a 14 cm bubble column up to a void fraction of 5%. The tumbling and zig-zag motion of the bubbles could be measured. It was found that the turbulence in the liquid phase was anisotropic and that the fluctuations of the bubble velocity were higher than those of the liquid.

10.1.7 Electrical Tomography Methods (ECT and ERT)

Electrical capacitance tomography (ECT) and electrical resistance tomography (ERT) are non-intrusive tomographic measurement techniques that measure, in essence, the electrical

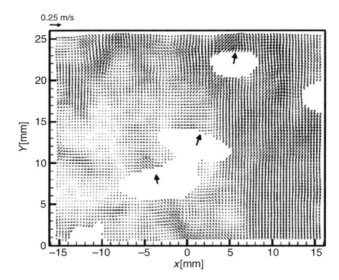

Figure 10.14 *PIV results of a bubbly flow [13] (Reprinted from Deen, N.G., Westerweel, J., and Delnoij, E., Two-phase PIV in bubbly flows: status and trends, Chem. Eng. Technol., 25, 97–101. Copyright (2002) with permission from John Wiley & Sons.)*

permittivity or conductivity distribution in a slab of finite thickness. To this end, a ring of sensors is mounted at the wall of the bubble column. Above and below, guard electrodes are positioned to confine the electric field to the slab created by the electrodes (Figure 10.16). In most cases 8 or 12 electrodes are used. Fast electronics measure the capacitance between each pair of electrodes, while the others are guarded. The capacitance of each pair depends on the distribution of the phases inside the slab. For a 12 electrode system, 66 independent

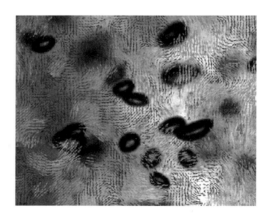

Figure 10.15 *PIV/PTV results of a bubbly flow at $\varepsilon_g \sim$ 1% [14] (Reprinted from Lindken, R., and Merzkirch, W., A novel PIV technique for measurements in multiphase flows and its application to two-phase bubbly flows, Exp. in Fluids, 33, 814–825. Copyright (2002) with permission from Springer.)*

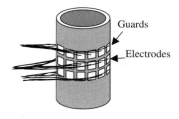

Guards

Electrodes

Figure 10.16   *ECT electrodes and guards*

pairs can be measured. The electronics circuit is certainly fast enough to complete such a cycle in 1–10 ms. For a given distribution of the phases, and thus a given permittivity distribution $\varepsilon(x,y)$, the electric potential, ϕ, is given by:

$$\nabla \cdot [\varepsilon(x, y)\nabla\phi] = 0 \qquad (10.9)$$

The measured capacitance between two sensors $\{i,j\}$ is calculated from:

$$C_{ij} = \frac{1}{V_{ij}} \int_{A_i} \varepsilon\nabla\phi \cdot \vec{n} dA \qquad (10.10)$$

where

V_{ij} is the voltage difference over the sensors
A_i is the surface area of sensor i with normal $\vec{n}$.

Reconstruction algorithms are available to generate the distribution of the phases from the measured data. The linear back-projection (LBP) method is often used. With these reconstruction techniques perturbations of a known, uniform situation are calculated beforehand. To this end, the cross-section to be reconstructed is divided into pixels. The result is the so-called sensitivity matrix $S_{k,p}$, which represents the change in capacitor k (the pairs $\{i,j\}$ are renumbered to a single index k) when pixel p is perturbed while all other pixels are kept constant. In formula:

$$S_{k,p} = \frac{1}{\varepsilon_{max} - \varepsilon_{min}} \frac{C_k(\varepsilon_p) - C_k(\varepsilon_{min})}{C_k(\varepsilon_{max}) - C_k(\varepsilon_{min})} \qquad (10.11)$$

where

ε_{min} and ε_{max} correspond to the permittivity of an empty and liquid-only bubble column.

The void fraction is now calculated from the relative change, λ_k, of the measured capacitances, C_k, with respect to the empty and full column:

$$\lambda_k = \frac{C_k - C_k(\varepsilon_{min})}{C_k(\varepsilon_{max}) - C_k(\varepsilon_{min})} \qquad (10.12)$$

and

$$\alpha_p = \frac{\sum_{i=1}^{N} \lambda_k S_{k,p}}{\sum_{i=1}^{N} S_{k,p}} \tag{10.13}$$

where

N is the total number of independent capacitances.

The entire system is relatively cheap (especially compared with γ and X-ray based tomographic systems, discussed below) and fast. However, the tomographic images suffer from poor resolution. There are two reasons for this: (i) the number of independent data is limited; and (ii) the electric fields used are so-called soft fields, that is the field between a sending and receiving sensor does not move in straight lines. In other words, each measured capacitance is sensitive to changes in the distribution of the phases anywhere in the measuring volume. This makes reconstruction a very difficult problem.

Bennet *et al.* [16] reported experiments in a 56 mm diameter bubble column using a 12 electrode ECT system with purified water. The data rate was such that 100 frames/s were reconstructed. Various flow regimes were investigated. These workers reported good agreement for the obtained void fraction and good identification of the flow regimes and transition superficial gas velocities.

An alternative approach was employed by Warsito and Fan [17] who used neural networks to reconstruct images of the void fraction distribution. Their column has a diameter of 10 cm and is filled with paraffin. The capacitance sensors form two planes, placed above each other, with six electrodes per plane. Reconstruction is made for the volume spanned by the two sets of electrodes (two of 8 cm length) digitised in $20 \times 20 \times 20$ voxels (volume elements, the 3D equivalent of pixels). These workers reported that their system, including the neural network approach, is fast and sensitive enough to investigate the spiralling motion of a bubble plume that was injected, both in a two-phase and a three-phase system. The method was extended by applying cross-correlation and auto-correlation techniques to the data [18]. This allowed estimation of velocities of the bubbles. For dilute systems, individual bubble trajectories were obtained, and for more dense systems, the 3D velocity of bubble swarms.

Electrical resistance tomography (ERT) employs the resistance of the bubbly mixture instead of the capacity. For liquids such as water in particular, the resistivity is usually too high to allow capacitance measurements. Jin *et al.* [19] presented ERT experiments on an air-water bubble column of 0.16 m inner diameter. They used two sets of 16 electrodes to measure the resistivity distribution in two planes. The measuring frequency is relatively low: two frames per second were reconstructed. Time averaged data are presented on the radial void fraction profiles in Figure 10.17. The superficial gas velocity ranged up to 0.25 m/s. From the ERT data, the local conductivity, σ_m, can be reconstructed. Once these values are known, the local void fraction follows from Maxwell's equation for the conductivity of a mixture:

$$\varepsilon_g = \frac{2\sigma_1 - 2\sigma_n}{2\sigma_1 + \sigma_m} \tag{10.14}$$

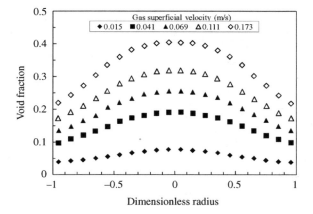

Figure 10.17   *Radial void fraction profiles as measured with ERT [19] (Reprinted from Jin, H., Yang, S., Wang, Mi, and Williams, R.A., Measurement of gas holdup profiles in a gas liquid cocurrent bubble column using electrical resistance tomography, Flow Meas. & Instr., **18**, 191– 196. Copyright (2007) with permission from Elsevier.)*

where

σ_1 is the conductivity of the liquid phase.

In the above equation, the conductivity of the gas bubbles is assumed to be zero.

10.1.8 γ and X-Ray Tomography

In contrast to the ERT/ECT fields discussed in the previous paragraph, γ and X-rays form so-called hard fields. This makes their interpretation easier. However, they are more difficult to use and usually the measurements are slower.

γ and X-rays are electromagnetic radiation, that is photons, as is visible light. The main difference is in their energy, which is much higher. This makes these photons penetrate into and through any form of material (Figure 10.18).

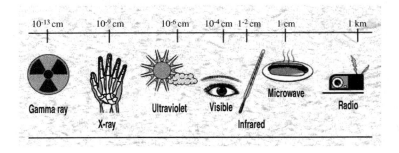

Figure 10.18   *Electromagnetic radiation: high energy, short wavelength γ and X-rays*

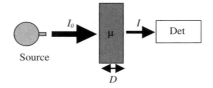

Figure 10.19 *Attenuation of γ or X-rays by a slab*

The working principle is given by the Lambert–Beer law. This describes the attenuation of a monochromatic (i.e. with a single photon energy E) of the γ and X-rays when passing through a homogeneous slab of material with thickness D (Figure 10.19):

$$I(D) = I_0 \exp(-\mu D) \tag{10.15}$$

where

I_0 is the intensity of the incoming beam
μ is the attenuation coefficient.

The latter is a function of the photon energy and the material of the slab. Kumar *et al.* [20] used a single 100 mCi ^{137}Cs source. It was collimated to a horizontal fan beam with a thickness of 6.5 mm. An array of 39 NaI scintillation detectors measured the radiation (Figure 10.20). The source–detector array could be rotated and 90 different views were taken. From all these data the time-averaged void fraction can be reconstructed. To this end a few steps have to be taken. Firstly, the Lambert–Beer law is written for the set-up, including the influence of the column wall:

$$I = I_0 \exp[-\mu_w D_w - (1 - \varepsilon_g)\mu_l D - \varepsilon_g \mu_g D] \tag{10.16}$$

where

μ_w, μ_l, μ_g are the attenuation coefficients of the column wall, liquid and gas, respectively
D_w and D are the thicknesses of the wall as seen by a particular beam and the path length
    through the bubbly mixture by the same beam.

A two-point calibration is sufficient: one with an empty column, that is $\varepsilon_g = 1$, and one with liquid only, that is $\varepsilon_g = 0$. From each beam the line-averaged void fraction is obtained. For reconstruction of the spatial gas distribution, various algorithms are available. Dudukovic

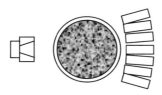

Figure 10.20 *Set-up used by Kumar et al. [20] (Reprinted from Kumar, S.B., Moslemian, D., and Duduković, M.P., Gas-holdup measurements in bubble columns using computed tomography, AIChE J., **43**, 1414–1425. Copyright (1997) with permission from American Institute of Chemical Engineers.)*

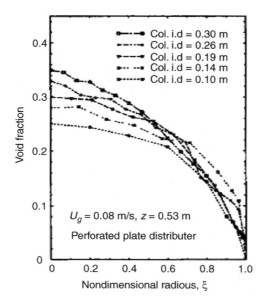

Col. i.d = 0.30 m
Col. i.d = 0.26 m
Col. i.d = 0.19 m
Col. i.d = 0.14 m
Col. i.d = 0.10 m

$U_g = 0.08$ m/s, $z = 0.53$ m

Perforated plate distributer

Figure 10.21   *Radial gas fraction profiles using g tomography [20] (Reprinted from Kumar, S.B., Moslemian, D., and Duduković, M.P., Gas-holdup measurements in bubble columns using computed tomography, AIChE J., **43**, 1414–1425. Copyright (1997) with permission from American Institute of Chemical Engineers.)*

and coworkers [20] used the estimation-maximisation (E-M) algorithm, which they found to be better than the algebraic reconstruction technique (ART).

An illustration of the results obtained for columns of different sizes is given in Figure 10.21. A comparison between the reconstructed void fraction and those obtained from pressure drop measurements and from the bed expansion shows the accuracy of the tomographic method, see Table 10.1.

A recent development is fast X-ray tomography. With X-rays, it should be realized that usually a wide energy spectrum of photons is produced. Hence, the radiation is not monochromatic. As a consequence, the scattering characteristics of photons with different energies are different, as the attenuation coefficient, μ, is a function of the photon energy. Therefore, low energy photons are scattered easier than high energy ones. This effect is called beam hardening. It requires a more complicated calibration procedure.

Table 10.1   *Comparison of gas fraction*

u_{gs} (cm/s)	Gas fraction		
	Tomographic	Δp	Bed expansion
2	0.087	0.088	0.085
5	0.123	0.125	0.129
8	0.147	0.150	0.144
12	0.179	0.182	0.188

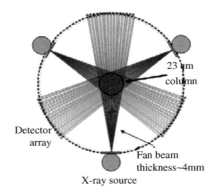

Figure 10.22 *Three-source X-ray set-up*

Two different set-ups are used. Mudde *et al.* developed a three-source system [21]. Each source radiates a fan beam that is detected by 30 detectors (Figure 10.22). Thus, 90 independent data points for each tomogram are obtained. The system can generate data at 400 frames/s [22]. This becomes comparable with the ECT/ERT methods, taking away one of the most important drawbacks of nuclear techniques.

An X-ray based CT scanner is described where the photons are generated from the impact of a fast scanning electron beam, hitting a tungsten target [23]. The arrangement is shown in Figure 10.23. The beam scans the target at a fast pace, creating many independent views per unit time. These researchers showed that high quality images with frame rates of up to 10 000 per second are possible. Various phantoms have been investigated to assess the spatial resolution of the reconstructed images. A specific form of the algebraic reconstruction techniques, binary-ART, gave better results than the standard ART. A resolution of 1 mm can be achieved, even in the presence of 2% Gaussian noise in the data.

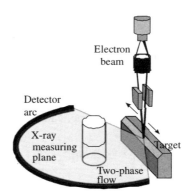

Figure 10.23 *The electronbeam X-ray scanner [23] (Reprinted from Bieberle, M., and Hampel, U., Evaluation of a limited angle scanned electron beam X-ray CT approach for two-phase pipe flows, Meas. Sci. Technol., **17**, 2057–2065. Copyright (2006) with permission from Institute of Physics.)*

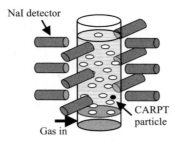

Figure 10.24 *CARPT set-up*

10.1.9 CARPT and PEPT

The tomographic techniques discussed above, in essence, all measure the phase distribution. For measuring the liquid velocity, particle tracking techniques have been developed.

CARPT (computer automated radioactive particle tracking) was developed by the group working with Dudukovic [24]. A single, radioactive, neutrally buoyant particle is put into the liquid phase. The particle is made radioactive and emits γ photons in all directions. The isotope ^{46}Sc is used: it emits high-energy photons of 890 keV and 1.12 MeV, with a half-life of 84 days. A high energy is preferred as then the attenuation coefficient only weekly depends on the scattering materials. The tracer particle is first irradiated in a research reactor to make it radioactive. The particle is observed by several (16 is frequently used) NaI scintillation detectors, located around the column (Figure 10.24). The intensity of the radiation scales is inversely proportional to the square of the distance from the particle to a detector. Thus, from a single detector it follows that the particle is located on a spherical shell centred around the detector. By intersecting all shells, the position of the particle can be found. A sampling frequency of about 100 Hz is usually used. From the time series of the particle positions, the particle velocity is obtained by straightforward time differentiation.

CARPT generates Lagrangian information. In order to turn this into the liquid velocity field, the bubble column is divided in voxels. Each time the particle is in voxel α, the velocity at that time is assigned to that voxel. The averaged velocity is found from ensemble averaging of all data belonging to the voxel:

$$\langle \vec{v}_{liq,\alpha} \rangle = \frac{1}{N_i} \sum_{k=1}^{N_i} \vec{v}_{liq,\alpha}(k) \tag{10.17}$$

From the velocity data, the Reynolds stresses can also be obtained. Firstly, the ensemble averaged velocity in each voxel is computed. Secondly, the fluctuating component is found as $\vec{v}'_{liq,\alpha}(k) = \vec{v}_{liq,\alpha}(k) - \langle \vec{v}_{liq,\alpha} \rangle$. Finally, the stresses are computed as:

$$\tau_{liq,\alpha,ij} = -\rho_{liq} \langle v'_{liq,\alpha,i} v'_{liq,\alpha,j} \rangle \tag{10.18}$$

Positron emission particle tracking (PEPT), is also a particle tracking technique. In this case, the particle contains an element that is radioactive and decays via positron emission. The positron, a positively charged electron, does not travel very far. It is quickly attracted

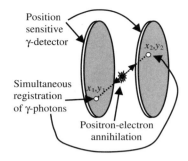

Figure 10.25   *Principle of PEPT*

towards one of the many electrons in the particle. Upon collision, the positron and electron annihilate and two photons are emitted. These photons travel (almost) back to back, that is at 180°. Moreover, their energy is equal to the energy of the original positron–electron pair. The latter is dominated by the mass of the particles, the kinetic energy of the electron and the positron are negligible (Figure 10.25). Thus, each photon has an energy of $E = m_e c^2 = 511\,\text{keV}$. Two γ detectors are used, one at each side of the bubble column. These detectors are position sensitive, that is they provide the $\{x,y\}$ position at which the photon is detected. The two detectors operate in coincidence mode: owing to the high speed at which the photons travel, they reach the detectors simultaneously. This helps remove 'false readings' caused by, for example, background radiation.

Once a pair of photons is detected, it is known that the particle must be positioned on the line connecting the two photon positions of the sensors. By continuously measuring these photon pairs, a set of lines is generated. In theory, the particle is at the intersection of the lines. Of course, in practice these lines never intersect. This is due to measurement noise, motion of the particle and different positions at which the positron is annihilated. Thus, in practice the particle location is the point that has the smallest distance to all lines used to reconstruct a single particle position. By time-stepping through the photon lines measured, the position of the particle in time can be followed; see for example Seville *et al.* [25] for more information and further references.

There are limits to the possible time resolution with which the particle position can be found. This has to do with the strength of the positron emission: if few positrons are emitted per unit time, the number of lines found is low and the time resolution becomes poor. However, if the number of positrons emitted is very high, the number of photons registered per detector is high. The detector needs a certain time to measure a single photon, that is there is a certain dead-time. If the number of photons becomes too high, false pairs are formed and the position detection becomes inaccurate.

PEPT has certain advantages over CARPT. As a specific photon is generated, the detection can be tuned to this photon making the technique sensitive. Moreover, the particle can be very small (or re-phrased, the amount of positron emitting material inside the particle can be very small). Thus the particle can be made smaller than with CARPT, where the intensity of the radiation is followed. Or, alternatively, in three-phase fluidisation, a true particle can be made radioactive. Finally, no calibration of particle positions is needed, making it easy. On the other hand, there are disadvantages. The electronics are much more complicated, as coincidence measurements must be performed. Detection of the photons requires a complete absorption of the photon, as it needs to be validated that the photon was indeed a 511 keV photon. This significantly lowers the amount of photons detected that can be used, making high time resolution difficult.

10.1.10 Acoustic Methods

Sound is a pressure fluctuation. It is usually divided into infra-sound (frequencies below the range of the human ear), audible and ultra-sound (frequencies above the range of the human ear). Acoustic methods can be divided into the active and the passive. The former involves shooting a pulse, usually of ultra-sound into a flow and picking up and analysing the response. Although not used in bubble columns or aerated stirred vessels, it has been employed to measure film thicknesses. More details are given in Section 10.3.1. The basic premise of the measurement of bubble sizes by detection of passive acoustic emissions is that processes, such as reactions, bubble formation, multiphase flow through a pipe, amongst others, emit distinguishing sounds that can be registered and characterised. Indeed the seminal paper of Minnaert [26] discussed bubbles singing. He presented relationships between liquid properties, bubble diameter and resonance frequency. There were not many publications for the next 60 years until another important paper appeared, by Pandit *et al.* [27]. Subsequently, there have been many more reports showing applications to a number of processes: plunging jet systems [28]; bubble columns [29]; and aerated agitated tanks [30]. They are all based on the recording of a time series of local pressure fluctuation and extracting the relative proportions of each frequency present.

Plesset and Prosperetti [31] gave a basic equation for the frequency of volume oscillations of an originally spherical bubble as:

$$f = \sqrt{\frac{3\gamma P_\infty}{\pi \rho_l d_B^2} - \frac{4\sigma}{\pi \rho_l d_B^3}} \tag{10.19}$$

It is not easy to make this equation explicit in bubble diameter. However, inspection of the terms shows that the first term under the square root is 3–4 orders of magnitude larger than the second for typical fluids and bubbles sizes (e.g. 1–3 mm diameter). Consequently, Equation 10.19 reduces to:

$$f = \frac{1}{\pi d_B} \sqrt{\frac{3\gamma P_\infty}{\rho_l}} \tag{10.20}$$

which was the form given by Minnaret [26] in his early paper. Note that for air-water systems, at atmospheric pressure, this reduces to $f = 6.58/d_B$.

It is also considered that the distance from specific bubbles and the presence of other bubbles can attenuate the sound intensity and therefore its amplitude, especially at high frequencies [32]. From Equation 10.20 it is seen that frequency is inversely proportional to the bubble size, that is the greater the bubble, the smaller the frequency produced. Pandit *et al.* [27] stated that bubbles of different sizes do not contribute in the same proportion to the power spectrum. They developed an equation to calculate the contribution of a bubble of determined size:

$$\frac{p}{P_\infty} = \left(\frac{3\gamma P_\infty}{\rho_l}\right)^{1/2} \frac{1}{\pi f d_B} \left[\frac{4}{3}\left(\frac{d_0}{d}\right)^2 + \left(\frac{d_0}{d}\right)\ln\left(\frac{d_0}{d}\right) - \frac{4}{3}\frac{d}{d_0}\right] \tag{10.21}$$

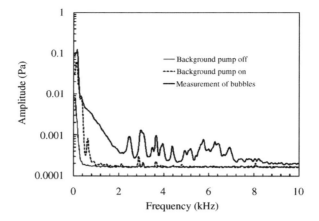

Figure 10.26   *Amplitude of contributions at different frequencies showing effect of switching on a pump*

where

d_0 is the original bubble diameter, which is compressed to a diameter d.

There are two different approaches to analysing the passive acoustic emissions raw data: (i) the windowing technique [33] in which the fundamental frequencies are removed and then studied separately, and (ii) the study of the power spectra as a whole [31].

Figure 10.26, taken from an experiment where the bubbles were created in a pool into which a jet of liquid plunged, illustrates the effect of specific equipment on the background sound. In this case, the pump driving round the liquid has a noticeable effect, but at frequencies that correspond to bubbles larger than are of interest. Note, if this approach is applied to a sparged stirred tank, the impeller will produce background sound [33].

A bubble size distribution is displayed in Figure 10.27.

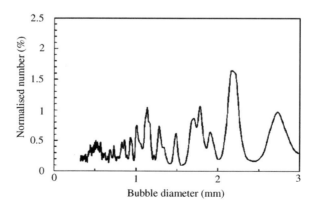

Figure 10.27   *Size distribution of bubbles*

The detector is a microphone if measurements are to be taken in air. In bubble columns it is more likely to be a hydrophone, which is a microphone that can be immersed in water. The detector must have sensitivity for the frequency range of interest. Originally the amplification–filtration and analysis equipment tended to be hard-wired and very specific in its applications. However, advances in computer technology has meant that these functions can be carried out digitally once the raw data from the detector have been digitised and fed into the computer.

Testing of the technique can be carried out by injecting air through a small diameter nozzle into the bottom of a transparent tank, to produce a continuous stream of bubbles. Nozzles of varying sizes can be employed to create streams of bubbles of different sizes. Comparative results can be obtained using photography or by counting the number of bubbles and measuring their volume, by capturing them in an inverted container. Figure 10.28 shows an example obtained from this arrangement together with a photograph of the bubbles formed.

10.1.11 Mass Transfer Coefficient

In many processes in bubble columns, mass transfer is crucial. In most cases, the resistance, against mass transport from gas to liquid or vice versa, is in the liquid phase. Therefore, the mass transfer can be approximated as: $\phi_m = k_l A \Delta c \rightarrow \phi_{m,vol} = k_l a \Delta c$, where $\phi_{m,vol}$ is the mass transfer per unit of volume of the bubbly mixture. Thus, the mass coefficient $k_l a$ needs to be known, where k_l is the actual mass transfer coefficient in the liquid and a is the interface area between gas and liquid per unit volume of the bubble column. In principle a can be found from probe measurements discussed above.

However, in the churn–turbulent regime, this is difficult and not very accurate. In most cases, for the mass transfer only the combination $k_l a$ is relevant. This can be measured directly by measuring the mass transfer itself. Two different approaches are used. One is based on chemical methods. A gaseous species is, via gas-liquid mass transfer, introduced into the liquid phase, where it reacts with a dissolved second species. From the depletion of

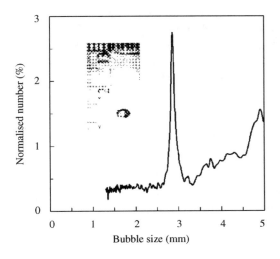

Figure 10.28   *Result of check trails using stream of bubbles from an orifice*

this second species (in combination with the kinetics of the reaction) the mass transfer can be found. A disadvantage of this method is that the reaction may change the properties of the liquid phase. Specifically, the coalescence behaviour of the bubbles is sensitive to dissolved species, for example in the form of ions. This makes the method less useful. Physical methods do not suffer from this drawback. They rely on measuring the concentration of a gas that is being dissolved. Frequently nitrogen-air or nitrogen-oxygen is used. With nitrogen gas, the liquid is freed from any oxygen. Subsequently, air or pure oxygen is bubbled through the reactor and the oxygen concentration in the liquid phase is measured over time. If one can assume that the liquid is perfectly mixed, the oxygen concentration in the liquid phase is given by:

$$\frac{c_{sat} - c(t)}{c_{sat} - c(t = 0)} = \exp(-k_l at) \qquad (10.22)$$

where

c_{sat}, $c(t=0)$ and $c(t)$ are the saturation, initial and actual concentrations, respectively.

Oxygen probes use the following reaction as the basic measurement: $O_2 + 4e^- + 2H_2O \rightarrow 4OH^-$. The dissolved oxygen diffuses through a membrane to the sensor surface, where it is reduced. By measuring the current (picoampere range), the oxygen concentration in the water can be obtained. This type of sensor needs calibration. However, the response of many modern sensors is linear with respect to the dissolved oxygen concentration. Hence, a two-point calibration is sufficient. One point is found by vigorously bubbling air through water, giving the saturation concentration at the temperature of the experiment. The other is obtained by completely removing the oxygen from the water. This can be done by chemical methods (e.g. by preparing a solution of sodium ascorbate and NaOH, both 0.1 M) or by bubbling N_2 through the system (avoiding any contact with air) until all oxygen has been removed.

A probe measures the oxygen concentration in time. If the response time of the probe, τ_p, is much smaller than $1/k_l a$, the analysis is straightforward. If, on the other hand, $\tau_p \gg 1/k_l a$ the method will fail and basically the probe response is measured. If $\tau_p \sim 1/k_l a$, correction is possible. A simple model assumes that the probe has a constant relaxation time, τ, and the probe itself shows an exponential relaxation, that is:

$$\frac{dc_p}{dt} = \frac{[c_{liq}(t) - c_p]}{\tau} \qquad (10.23)$$

where

$c_p(t)$ is the concentration as given by the probe.

If an oxygen depletion experiment is performed, starting from the saturated condition, in an ideal stirred tank, the liquid concentration follows from:

$$\frac{dc_l}{dt} = k_l a[0 - c_l(t)] \qquad (10.24)$$

Solving both equations, with initial conditions, $c_p(0) = c_l(0) = c_{sat}$, gives, for the probe response:

$$\frac{c_p(t)}{c_{sat}} = \left[1 - \frac{1}{1 - k_l a \cdot \tau} \right] e^{-t/\tau} + \frac{1}{1 - k_l a \cdot \tau} e^{-k_l a \cdot t} \tag{10.25}$$

In Figure 10.29 the concentration measured by the probe is shown for various values of $k_l a \tau$. Also the true liquid concentration is given.

The probe response time needs to be measured. This can be done be placing the probe first in oxygen-free liquid and subsequently moving it quickly into a well-mixed liquid of given oxygen concentration. Nowadays, probes can be fast and $k_l a$ values of 1/s can be detected easily. An example of such an experiment (where at about $t = 4$ s the probe is moved from a container with oxygen-saturated water to one free of oxygen) is shown in Figure 10.30. As can be seen, the response time of the probe is about 0.5 s.

10.2 Sparged Stirred Tanks

10.2.1 Power Draw

Impeller power draw, P, is one of the most important parameters in designing and operating stirred tank reactors. Fluid mixing, mass transfer and many scale-up rules heavily depend on

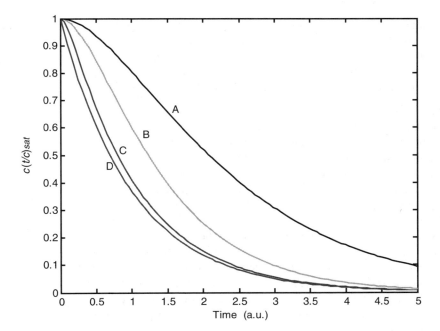

Figure 10.29 *Measured concentration using a probe with a finite response time for various values of $k_l a \tau$: A, 1.5; B, 0.5, and C, 1.5. D is the true liquid concentration*

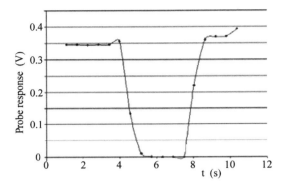

Figure 10.30 *Response time measurement of oxygen probe*

the specific power input. The accurate measurement of power draw is required on many occasions, especially at small scale mixing systems.

Power draw is usually determined through the shaft torque, T_q, measurement by:

$$P = 2\pi N T_q \tag{10.26}$$

where

N is the impeller speed.

There are many methods available to measure the reaction torque. Some allow measurement of the individual torque contribution for each impeller (or any bearings, etc., in the system), while others can only be used to measure the entire system torque. The torque measurement device must be carefully calibrated over the range of measured torque values, be free of errors caused by friction and compensated for any temperature effects.

10.2.1.1 Strain Gauges

One of the most common and reliable devices for torque measurements is strain gauges mounted on the mixer shaft. Generally, strain gauges are used in pairs aligned on the shaft. The alignment of the gauges is extremely important in ensuring reliable readings. With multiple-impeller systems, or where the contribution from a bottom bearing needs to be removed, multiple sets of gauges can be used on the same shaft.

With careful design and choice of shaft material this technique can be used to measure a very wide range of strains. Extremely careful calibration of the strain gauges is required. Known torques must be applied to the shaft and the response measured. Note that no bending loads should be applied during calibration. If a single strain gauge is to be mounted on the shaft, it should be placed towards the top of the shaft where it will not be submerged and above where any impellers are to be mounted. Where multiple strain gauges are to be mounted on the shaft, allowing the measurement of torque contributions from individual impellers and bearings, the gauges should be placed in positions that allow reasonable flexibility for impeller position.

Strain gauges generally produce extremely low voltage signals, which must be amplified before the signal can be logged and recorded. The amplifier must be mounted on the mixer shaft above the water level, while the amplified signal is passed to the recording device through either slip rings or a radio-telemetry system. Gauges of this type are extremely delicate and must be handled and calibrated with extreme care.

Great care must also be taken in the waterproofing of any gauges fitted to the shaft. The waterproofing usually consists of several layers of heat-treated wax and adhesive built-up until the surface of the shaft is smooth. Great care should be taken when using a strain-gauged shaft to ensure that the waterproof seal is not ruptured: for example, when the shaft is being removed from or positioned in the tank, or when the impellers are being moved.

Torque transducer units are available that contain strain-gauged sections of shaft that can be fitted in line with the mixer shaft. They must be suitably isolated from any bending moments and axial loads. They are generally subject to the same calibration and care during use. The same precautions should be used as for sensitive strain gauges.

10.2.1.2   Measurement of Motor Power

In large-scale systems, the power draw can be estimated from the motor power consumption. However, bear in mind that not all the motor power is transferred to the fluid and the losses in the motor, gearbox and any bearings must be considered. In most situations the accuracy of the information on these losses is so poor that this technique is not recommended for power draw measurements on a small scale, particularly where the losses tend to be greater than the power delivered to the fluid.

10.2.1.3   Modified Rheometer Method

Some rheometers can be modified to provide accurate torque readings as they are designed to measure shear stresses at known shear rates. This can be a very good method to use in small-scale vessels.

10.2.2   Velocity Field

Standard techniques that are used today to find the velocity field in equipment such as stirred tanks include: laser Doppler anemometry (LDA) and particle image velocimetry (PIV). Both can provide high-quality data, with good spatial and temporal resolution. However, as both are light-based techniques, they are of limited use for sparged stirred tanks. CARPT and PEPT with a single radioactive tracer particle can be used in opaque systems. As discussed in Section 10.1.8, both techniques use a radioactive particle that follows the liquid flow. Both Lagrangian and Eulerian velocity information is obtained. As the radiation is penetrating through the liquid phase, the presence of gas bubbles does not hamper the measurement of the particle position. The drawback of these techniques is the rather long measuring time to obtain sufficient information for an accurate description of the flow field.

An example of the use of CARPT for sparged stirred tanks can be found in Khopkar *et al.* [34]. A 1 mm diameter neutrally buoyant polypropylene particle with ^{46}Sc as the radioactive compound is submerged in a 20 cm diameter gassed, stirred tank. The tank operates at an impeller Reynolds number of about 15 000. Data are taken during 24 h at a data rate of 200 Hz. The average void fraction was 1.5%. The experiments were part of a study in which CFD simulations were compared with experimental data. The data are in good agreement. In the same experiment, the void fraction was measured using γ radiation. This will be further discussed in the next section. PEPT has been used for similar experiments [35]. In this research gassed and ungassed experiments were carried out. A comparison at the same impeller speed revealed clear differences in the liquid velocity field.

The idea of both techniques is the same: tracking a radioactive particle in time. As discussed before, the major difference is the type of radiation. With CARPT, in essence the signal strength is used, whereas PEPT employs the generation of two back-to-back travelling 511 keV photons to find the particle position. Once the time series of particle positions is obtained, the data analysis for turning positions into velocity and kinetic energies is almost the same for both methods.

10.2.3 Void Fraction

The overall void fraction can be measured from the liquid height in the gassed and ungassed state. However, these heights are difficult to find accurately for stirred systems, as the liquid surface is in constant motion. It has been reported that the accuracy of the void fraction is not better than 15%. As an alternative one could use optical or resistance probes. However, finding the global void fraction from these probes is cumbersome, as they measure the local void fraction basically in a 'point'. Furthermore, in the highly complex flow of a gassed stirred tank, the bubbles will approach the probe from various sides. This will lead to underestimation of the void fraction.

Another approach is the use of tomographic techniques, such as ERT/ECT or γ and X-rays. ERT/ECT (electrical resistance/capacitance tomography) has the advantages of being relatively cheap compared with the nuclear options. Moreover, it can deal with large scales, which is not so easy, due to safety issues, for γ and X-rays. On the downside is the difficulty of interpreting the data. The technique relies on soft-fields making it non-linear and poorly localised. Therefore, a good calibration is required, which is difficult to achieve.

The nuclear techniques are more straightforward to interpret. In practice, they are usually restricted to laboratory scales, although that is not the principle barrier. Ford *et al.* [36] used an X-ray CT (computed tomography) scanner to study the void fraction distribution inside a 21.7 cm diameter gassed, stirred tank. An example of their results is given in Figure 10.31. The scanning time per image is 1 s. A sufficient number of views are needed for high quality data on the void fraction. Therefore, the scanner was rotated over 1° after each measurement creating 360 independent measurements. A typical scan took about 45 min. Thus, only time averaged information could be obtained. Moreover, non-linear averaging becomes an issue: the void fraction as measured over one X-ray line will not be constant in time, due to the passage of the blades. This is especially true in the impeller region. Scans were made for various measuring planes, thereby creating a three-dimensional overview of the void fraction distribution. The maximum error on the local void fraction is reported as 15%, while the uncertainty in the global void fractions is about 5%.

10.2.4 Mixing Time

Various methods exist for assessing the mixing time. One option is changing the colour of the liquid, for example by decolorising the fluid. With this technique, the fluid is initially uniformly coloured by a dye. By injection of a reagent that decolourises the dye, the mixing is observed. When the last streaks of colour disappear, the mixing time is obtained. For example, Kraume and Zehner [37] used an iodine-starch solution and added sodium thiosulfate. In modern approaches, light induced fluorescence techniques are used to make

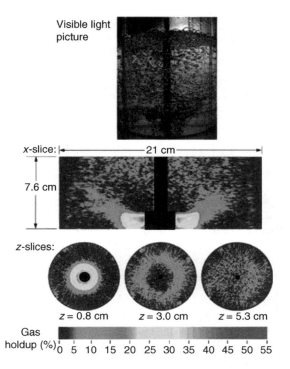

Figure 10.31 *Example of the gas fraction distribution in a gassed, stirred tank (operated in the loading regime) [36] (Reprinted from Ford, J.J., Heindel, Th.J., Jensen, T.C., and Drake, J.B., X-ray computed tomography of a gas-sparged stirred-tank reactor, Chem. Eng. Sci., 63, 2075–2085. Copyright (2008) with permission from Elsevier.)*

quantitative measurements and to study the mixing structures in the flow. For multiphase flows these techniques are not an attractive option, as the flow quickly becomes opaque, making visual-based observation of the disappearance of the last streaks difficult and of other structures virtually impossible.

A second method that can be used is measuring the conductivity, pH or temperature after a pulse-injection of saline liquid, acid or base, or hot fluid into the tank. The measurement will start directly after injection and the measured signal (in a 'point') shows a fluctuating character as mixing progresses with time. Care should be taken when injecting fluids that are of different density than the bulk fluid, as gravity or buoyancy will then influence the mixing.

Complete mixing depends on the definition: should it be mixed at the molecular level, or at the hydrodynamic level, or are the probe volume and response controlling what is meant by mixed? With probes, for example used in conductivity measurements, usually a 95% criterion (or similar) is used: the mixing time is defined as the time lapse after injection of the fluid to be mixed for which the fluctuations of the probe signal stay within a 5% band of the final value, see Figure 10.32. It has been reported that for turbulent flows, the exact position of the injection point and of the measuring point do not influence the final mixing time estimate. Moreover, if the volume added is small compared with the total liquid volume, the added volume does not influence the outcome of the experiment.

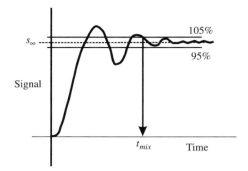

Figure 10.32 *Response of probe in mixing experiment*

Kraume and Zehner [37] found that in their single-phase experiments the decolouring and conductivity measurements yielded the same mixing times.

10.2.5 Mass Transfer Coefficient

The mass transfer in gassed, stirred tanks can be measured using the principles discussed in Section 10.1.11. However, the situation is more complicated due to the gas flow through the reactor. This flow needs to be steady, as the mass transfer depends on the full hydrodynamics inside the stirred tank. Consequently, the gas volume fraction as well as the bubble size distribution needs to be constant during the determination of $k_l a$. A sketch of the situation is given in Figure 10.33.

In this instance we need to set up mass balances for both the gas and the liquid phase. We assume that the initial concentration of the gas that will be transferred from the gas phase to the liquid is zero. This holds both for the liquid and for the bubbles in the reactor. At time $t = 0$, the inlet is switched to a gas flow at the same flow rate, but now with a constant concentration of the gas species, c_{in}. As the gas flow rate is constant during this switch, we can assume that all hydrodynamics in the tank stay the same. The total surface area of all bubbles is A. Furthermore, we assume that the tank is ideally stirred. The mass balances read as:

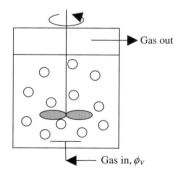

Figure 10.33 *Gas flow through the reactor showing that the gas volume fraction as well as the bubble size distribution needs to be constant during the determination of $k_l a$*

$$\text{Gas} \quad \frac{dV_g c_g}{dt} = \phi_V(c_{in} - c_g) - k_l A \left[\frac{c_g}{m} - c_l\right] \tag{10.27}$$

$$\text{Liquid} \quad \frac{dV_l c_l}{dt} = k_l A \left[\frac{c_g}{m} - c_l\right] \tag{10.28}$$

As the hydrodynamics are kept constant, the volume of gas and liquid are constant and related to the total volume as $\varepsilon_g = V_g/V$, $1 - \varepsilon_g = V_l/V$. We can use this to simplify the above equations:

$$\text{Gas} \quad \frac{dc_g}{dt} = \frac{\phi_V}{\varepsilon_g V}(c_{in} - c_g) - k_l a \frac{1 - \varepsilon_g}{\varepsilon_g} \left[\frac{c_g}{m} c_l\right] \tag{10.29}$$

$$\text{Liquid} \quad \frac{dc_l}{dt} = k_l a \left[\frac{c_g}{m} - c_l\right] \tag{10.30}$$

where $k_l a$ is defined with respect to the volume of the liquid phase. The solution of these equations for the concentration in the liquid is given by:

$$c_l(t) = \frac{c_{in}}{m} \frac{k_l a}{\varepsilon_g} \frac{\phi_V}{V} \left[\frac{1}{r_1 r_2} + \frac{1}{r_1 - r_2}\left(\frac{1}{r_1}e^{r_1 t} - \frac{1}{r_2}e^{r_2 t}\right)\right] \tag{10.31}$$

where

$$r_1 = \frac{-A + \sqrt{A^2 - 4B}}{2}$$

$$r_2 = \frac{-A - \sqrt{A^2 - 4B}}{2}$$

$$A = k_l a + \frac{\phi_V}{\varepsilon_g V} + \frac{k_l a}{m}\left(\frac{1 - \varepsilon_g}{\varepsilon_g}\right)$$

$$B = \frac{\phi_V}{\varepsilon_g V} k_l a$$

Note that the measured liquid concentration is not identical to the actual concentration in the liquid, as the probe has its own response. If we take again a first-order response for the probe, the measured concentration is given by:

$$c_p(t) = \frac{c_{in}}{m} \frac{k_l a}{\varepsilon_g} \frac{\phi_V}{V} \left[\frac{1 - e^{-t/\tau}}{r_1 r_2} + \frac{1}{r_1 - r_2}\left(\frac{e^{r_1 t} - e^{-t/\tau}}{(1 + r_1 \tau)r_1} - \frac{e^{r_2 t} - e^{-t/\tau}}{(1 + r_2 \tau)r_2}\right)\right] \tag{10.32}$$

The differences between the simple model, first-order model of Section 10.1.11 and the more complete, second-order model discussed above can be significant. Bakker [38] presented data on $k_l a$ obtained from experiments in a gassed stirred tank analysed using the first- and the second-order model, see Figure 10.34. It is clear from the figure that large errors can be made if the wrong model is selected.

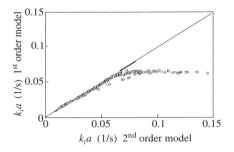

Figure 10.34 k_la *values from measurements analysed by a first- and second-order model [38] (Reprinted from Bakker, A. (1992) Hydrodynamics of stirred gas-liquid dispersions, Ph.D. thesis, Delft University of Technology, The Netherlands.)*

10.3   Falling Film Reactors

10.3.1   Film Thickness

For the measurement of film thickness in falling film type arrangements, selection of which technique to employ depends on what data are sought. Most of the methods developed measure the variation of film thickness with time at one (or more) point(s) in space. Some provide averages around the pipe circumference whilst others are more local. Alternative techniques can give the spatial variation of film thickness at a point in time but they do require geometries that could be different from the traditional pipe. However, as will be pointed out, there are technological developments that are making the gathering of information in time and space more accessible. A thorough review of the published material in this area has been given by Clark [39], which has drawn from falling film research as well as those from other applications.

Here, the methods with most potential are explained, indicating advantages and disadvantages. Those considered are based on electrical (conductance or capacitance), ultrasonic or optical methods.

The version of the *electrical method* to be employed depends on the properties of the liquid studied. For water-based liquids, conductance can be employed otherwise capacitance must be used. The conductance methods considered here essentially utilise a form of Wheatstone bridge. The major differences in the diverse applications have to do with what is actually being measured and in the shape of electrodes employed. The two major variants are aimed at determining: (i) circumferentially averaged film thickness; and (ii) local film thicknesses.

In the case of (i) above, the electrodes tend to consist of rings of metal inserted into a non-conducting wall. Many of the applications of this type are aimed at measuring the void fraction, which this system can do for all gas-liquid flow patterns from bubbly (bubbles in a liquid continuum) through intermittent flows, such as slug flow, to annular flow (film flow on the wall with gas at the pipe centre). Note that there are different relationships between the void fractions and outputs for the two extreme geometries.

For measuring local film thicknesses, the electrodes can be flush on the pipe wall or pin or wire pairs protruding through the film. As will be shown below, the former can be more

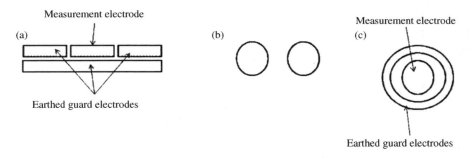

Figure 10.35 *Examples of electrode arrangement for flush mounted type: (a) Coney [40] (Reprinted from Coney, M. W. E., The theory and application of conductance probes for the measurement of liquid film thickness in two-phase flow, J. Phys. E: Sci. Inst.,* **6**, *903–910. Copyright (1973) with permission from Institute of Physics.); (b) Geraci et al. [41] (Reprinted from Geraci, G., Azzopardi, B.J. and van Maanen, H.R.E., Effect of inclination on circumferential film thickness variation in annular gas/liquid flow, Chem. Eng. Sci.,* **62**, *3032–3042. Copyright (2007) with permission from Elsevier.); and (c) Brown [42] (Data from Brown, D.J., Non-equilibrium annular flow, DPhil Thesis, University of Oxford (1978).)*

accurate for thin films, but can saturate and become insensitive for thicker films. The wire pairs can handle thicker films.

A number of different geometries have been employed for flush mounted electrodes. The motivation for the increased complexity in some cases is to minimise bulging of field lines and so make the measurements more truly local. It is known that the greater the area of influence of the electrodes, the greater the averaging of the film thickness that occurs, particularly when a wave passes by. Because of this, measured wave heights can be significantly lower than the true value. Examples of electrode arrangements which have been employed are given in Figure 10.35. The output of the probes/circuit depends on both the film thickness and on the conductivity of the liquid. The latter is a function of the salt concentration and of the system temperature. Although calibration curves are dependent on conductivity, the effect of this can be dealt with by making a measurement at reference conditions, that is with the pipe full of liquid.

Calibration curves are obtained by creating films of known thicknesses. For flush mounted probes, this is achieved by inserting plugs of non-conducting materials of different diameters and introducing liquid between the pipe and the plugs. For the wire pair electrodes, calibration can be achieved by laying the pipe on its side with the wire pairs emerging from the bottom and trapping different amounts of liquid in the pipe. This will give different depths of liquid, that is different film thicknesses. Examples of calibration curves for the two types of probes are illustrated in Figure 10.36. These show how the output from the flush probes is highly non-linear and is beginning to saturate by 3 mm, whilst that from the wire pair arrangement has a linear relationship over a much larger range of thicknesses.

The underlying principle of flushed mounted probes is based on the potential field theory. The electrical potential of the electrodes immersed in two-phase media subjected to AC can be described by the Laplace equation with the proper boundary conditions:

$$\nabla^2 V = 0 \tag{10.33}$$

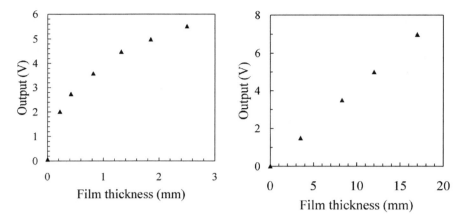

Figure 10.36 *Calibration curves for conductance film thickness measurement: (a) flush mounted electrodes [type (b) in Figure 10.35]; and (b) wire pair*

In 1973 Coney [40] first proposed a theoretical solution to the electrical behaviour of a liquid layer of thickness δ and electrical conductivity λ for strip-shaped flush electrodes at constant potential:

$$G_e = \tilde{G}_e(l\lambda) = K(m)/K(1-m) \tag{10.34}$$

$$m = \frac{\sinh^2(\pi s/2\delta)}{\sinh^2[\pi(s+D_e)/2\delta]} \tag{10.35}$$

where

l, s and D_e are the length, width and spacing of the electrodes, respectively
K represents the complete elliptic integral of the first type [43]
m is a function of the geometrical parameters.

$$K(m) = \int_0^{\frac{\pi}{2}} (1 - m\sin^2\vartheta)^{-1/2} d\vartheta \tag{10.36}$$

It has been confirmed [44] that the Coney's solution can be extended to deal with the cylindrical geometry. For ring electrodes covered by a liquid layer (annular and stratified configurations), the length l should be replaced by the wetted length of the electrodes P and the film height δ by the equivalent film height δ_e in Coney's equation [45]. δ_e is defined as:

$$\delta_e = A(1 - \varepsilon_g)/P \tag{10.37}$$

where

A is the cross-sectional area of the cylindrical duct
$(1 - \varepsilon_g)$ is the volume fraction occupied by the liquid.

Liquid conductivity has a clear effect on G_e. The same workers *suggested* employing a normalised conductance G_e^*, referred to the conductance of the pipe full of liquid in order to exclude the influence of the liquid conductivity on the dependence of the ring probe response to the liquid film thickness. The calibration results have indeed confirmed the success of the approach in accounting for different liquid conductivities.

Tsochatzidis *et al.* [46] provided an analytical solution for the Laplace equation for a ring-probe response to a conducting annulus of film thickness δ:

$$\tilde{G}_e = \frac{\pi^3}{32}\left(\frac{2s}{L}\right)^2\left[\sum_{i=0}^{\infty}\frac{1}{(2i+1)^3}b_i^2f_i\right]^{-1} \tag{10.38}$$

with

$$b_i = \cos\left[\frac{k_i(D_e-2s)}{D_e}\right] - \cos(k_i) \tag{10.39}$$

$$k_i = \frac{\pi D_e(2i+1)}{2L} \tag{10.40}$$

and f_i is a combination of modified Bessel functions I_0, I_1, K_0, K_1:

$$f_i = \frac{I_0(k_iD/D_e)}{I_1(k_iD/D_e)}\left(\frac{1+\dfrac{K_0(k_iD/D_e)I_1\left[2k_i\left(\dfrac{D}{2}-\delta\right)/D_e\right]}{I_0(k_iD/D_e)K_1\left[2k_i\left(\dfrac{D}{2}-\delta\right)/D_e\right]}}{1-\dfrac{K_0(k_iD/D_e)I_1\left[2k_i\left(\dfrac{D}{2}-\delta\right)/D_e\right]}{I_0(k_iD/D_e)K_1\left[2k_i\left(\dfrac{D}{2}-\delta\right)/D_e\right]}}\right) \tag{10.41}$$

This solution is in good agreement with the extension by Andreussi *et al.* [44] of the model developed by Coney [40].

Film thickness measurement using ultrasonics uses the time of flight for an echo of a signal of a pulse transmitted from a probe, mounted on the outside of the pipe wall, to return to a detector, also mounted on the outside of the pipe [47]. Echoes are created by the solid-liquid and liquid-gas interfaces. Knowing the velocity of sound in the liquid allows the film thickness to be extracted. It must be noted that wavy films can reflect the return pulse away from the receiver unless the receiver area is of a large enough.

The simplest optical technique uses direct illumination to cast a shadow of the film onto a screen, which can be photographed [48]. Angling the screen away from the perpendicular to the light beam will give significant amplification of the distance proportional to film thickness. Obviously, this will only work if the film is flowing on the outside of a vertical rod and there is unimpeded optical access. The technique gives a spatial (axial) distribution of thickness at a point in time. However, use of high speed video can provide temporal resolution.

Liquid absorption techniques can give large amounts of information about the spatial distribution of film thickness. However, the approach is limited to a rectangular cross-section channel with liquid flowing on only one wall. Addition of a dye to the liquid and careful positioning of the light and camera enable photographs to be obtained from which

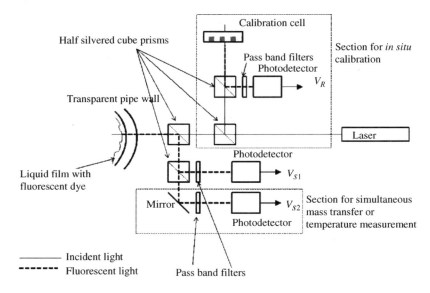

Figure 10.37 *Schematic diagram for fluorescence technique showing arrangement for in situ calibration and possible additional measurements for mass transfer or temperature measurements discussed in the next section*

the light intensity and hence film thickness can be extracted. Modern CCD cameras will do this in one step. Clark [39] used separation of the white light into red, green and blue components to obtain the best results. To minimise the problem of diffraction at the wavy gas-liquid interface a diffuse illumination source is employed. Calibration of the method can be carried out by taking simultaneous photographs of liquid layers of known thicknesses, that is by the use of a wedge cell made up of two glass plates touching at one edge and with a small angle between them. Examples of data obtained using this approach are shown in Plate B.

Another, very powerful optical approach uses a fluorescent dye dissolved in the liquid. When irradiated with blue or UV light, the dye emits light of a higher wavelength. The intensity of this fluorescent light relative to the incoming light is a function of film thickness, though it is a non-linear function. The original version of the technique, which was developed in the 1960s, employed mercury lamps for the source of UV light, spectrometers to separate the incoming and fluorescent light and photomultipliers to determine the intensity of fluorescence [49]. In modern applications these may be replaced by lasers, narrow pass band filters and modern photodetectors [50]. Figure 10.37 shows an arrangement which might be used.

A typical calibration curve is shown in Figure 10.38. The shape of the curve can be explained if the processes occurring during fluorescence are all considered. As the incident light passes into the film, it is absorbed according to the Beer–Lambert law. The light arriving at any position y from the wall will be related to the light level at the wall by e^{-k_1cy}, where k_1 is the absorption coefficient and c is the dye concentration in mole/litre. The quantity of light absorbed in a thin layer of thickness dy at a distance y from the wall is $dI = k_1ce^{-k_1cy}dy$. The quantity of fluorescent light detected from this

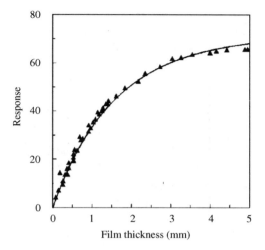

Figure 10.38 *Experimentally obtained calibration data. Line is calculated from Equation 10.43*

layer is then:

$$dI_f = \frac{EaI_ok_1c}{1+k_2c}e^{-k_1cy}e^{-k_3\sqrt{cy}}dy \qquad (10.42)$$

where

E is the efficiency of fluorescence
a is the fraction of fluorescing light (emitted spherically) which enters the detector.

As the detector is far beyond the wall, a can be taken as independent of position within the film. The factor $(1 + k_2c)^{-1}$ accounts for the quenching fluorescence by the molecules of the dye [51]. The second exponential term compensates for absorption of the fluorescent light by the dye solution. It involves a square root in cy as it takes into account fluorescence generated by this absorbed light [52]. For a film of thickness δ, the emerging light is:

$$I_f = \frac{EaI_ok_1c}{1+k_2c}\int_0^{\delta} e^{-k_1cy}e^{-k_3\sqrt{cy}}dy \qquad (10.43)$$

Because of the square-root term, this integral has to be evaluated numerically. Values of the constants have been evaluated for sodium fluorescein excited by light at 436 nm: $k_1 = 1.9 \times 10^4$ litre/mole/m; $k_2 = 310$ litre/mole; $k_3 = 30$ litre$^{0.5}$/mole$^{0.5}$/m$^{0.5}$.

It can be seen in Figure 10.38 that this equation gives a good prediction of the shape of the calibration curve. Figure 10.39 illustrates the trends calculated using Equation 10.43 and that though there can be a distinct gain in sensitivity by increasing the dye concentration, this becomes less useful at larger thicknesses where self-quenching becomes more important, and makes the measured fluorescence insensitive to film thickness at higher thicknesses. Note that oxygen can have a quenching effect on fluorescence and thought should be given to the gas to be used in experiments.

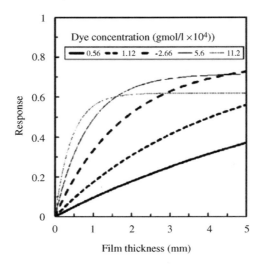

Figure 10.39 *Effect of concentration and film thickness on fluorescence response showing the effect of self quenching at higher concentrations*

This technique can give very accurate time-resolved measurements at one point in space. More recent developments have taken advantage of laser illumination and high-speed video cameras to obtain spatial (and temporal) variations in film thickness. In these versions it is not the intensity of fluorescence but the extent of glowing area that is measured.

10.3.2   Heat and Mass Transfer

In both heat and mass transfer the objective of experiments is usually to determine a transfer coefficient, that is the ratio of a flux and a driving force. This means that several quantities will have to be measured to achieve the required outcome. Another factor to be considered is the scale at which the measurements are to be made. Early work determined transfer coefficient values over the entire falling film unit. Given that finite development lengths are required for waves, see Figure 4.4, and given the difference between wave-free and wavy films, this could bias the data. However, modern measurement methods can provide data resolved in time and space.

For heat transfer, be it heating or cooling, the simplest arrangement is the use of a double tube system with the falling film on the inside of the inner tube and the heating or cooling medium in the annular jacket. Required measurements are then: film and outer flow rates and the inlet and outlet temperature of both streams. Heat balances give:

$$\dot{Q} = \dot{M}_f c_f \left(T_{fi} - T_{fo}\right) = \dot{M}_a c_a \left(T_{ai} - T_{ao}\right) \tag{10.44}$$

where

$\dot{Q}$ is the heat flow (W)
$\dot{M}$ are the flow rates (kg/s)
c are the specific heat capacities (J/kg °C)
T are temperatures (°C)

subscripts f and a refer to the film and the annulus fluid
i and o are inlets and outlets.

The overall heat transfer coefficient can be obtained from:

$$U = \frac{\dot{Q}}{A\Delta T} \tag{10.45}$$

where

A is the heat transfer area (m^2)
ΔT is the driving temperature difference.

If the temperature difference is not uniform along the unit, the log mean temperature difference should be used. Knowing the four temperatures and the flow rates, it is simple to solve for the value of U. The heat transfer coefficient on the falling film side, α_f, can be determined from:

$$\alpha_f = \left(\frac{1}{U} - \frac{1}{\alpha_a} \frac{A}{A_a} - \frac{\Delta r}{\lambda_w} \frac{A}{A_m} \right)^{-1} \tag{10.46}$$

where

Δr is the tube wall thickness
λ_w is the thermal conductivity of the wall material
A_m is the logarithmic wall area.

Obviously, the outside heat transfer coefficient must be known.

Inspection of these equations shows that to get a small variation in outer fluid (and hence wall) temperature requires a large flow rate through the annulus. However, this will make the errors in the temperature difference, $T_{ai} - T_{ao}$, and hence the heat flow, larger. For heating, some of the problems can be avoided by employing condensing steam as the heating medium. This has the advantage of occurring at a constant temperature.

To obtain more local conditions, a number of short annular jackets could be employed. In these cases, the temperature of the film at axially separated positions has to be measured more locally by intrusive thermocouples. Another method of heating is to apply a direct electrical current to the tube wall over which the film flows. The heat produced is given by:

$$\dot{Q} = IV = \frac{V^2}{R} \tag{10.47}$$

where

V is the voltage difference
I is the current
R is the resistance of the tube wall material.

It is expected that the tube wall will be made of metal to provide the strength to contain the pressure within it. This implies a low resistance, which will require low voltages so as not to produce excessive power. By implication this will demand a large current.

Another approach to obtain both the heat flux and the inside wall temperature is to employ a thick walled tube with pairs of thermocouples embedded at accurately known positions

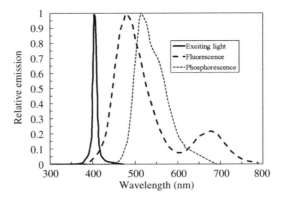

Figure 10.40   *Fluorescence and phosphorescence spectra for biacetyl [55] (Reprinted from Schagen, A., Modigell, M., Dietze, G. And Kneer, R., Simultaneous measurement of local film thickness and temperature distribution in wavy liquid films using a luminescence technique, Int. J. Heat Mass Trans., 49, 5049–5061. Copyright (2006) with permission from Elsevier.)*

along the radii [53]. The temperatures can be used to determine the heat flux using the conduction equation. The information can then be used to obtain the temperature of the solid surface over which the falling film flows. The approach has also been applied to rotating disc systems [54].

In all these methods it is important to minimise heat losses. Adequate insulation must be provided.

Methods related to the fluorescence technique for film thickness, which was presented in Section 10.3.1, can be extended to give data on heat transfer coefficient that are spatially and temporally resolved. This employs the characteristic of some dyes, that they emit both fluorescence and phosphorescence. The former is short lived and can only persist for a very short time after the exciting illumination is switched off. In contrast, phosphorescence persists for a period after the loss of the exciting light. Quantitatively, this can be expressed as the decay period for phosphorescence being of the order 1 ms, $O(10^{-3}$ s) whilst that for fluorescence is $O(10^{-6}$ s) and that for a pulse of exciting light is $O(10^{-8}$ s). Ideally, it would be useful if the fluorescence and phosphorescence were of different colours. Unfortunately, the spectra [55] for a possible dye, 2,3-butanedione (known as biacetyl) show distinct overlap as shown in Figure 10.40. However, the difference in decay times noted above allowed another approach. This was helped by the fact that the characteristic time for the exponential decay of phosphorescence had a linear dependence on the liquid temperature. This could be described through:

$$I = I_o e^{-t/\tau} \tag{10.48}$$

where

I is the instantaneous phosphorescence intensity
I_o is the intensity at the start of phosphorescence
τ is the characteristic time with a temperature relation of $\tau = A\theta + B$.

Schagen *et al.* [55] found that A had a value of $-3.4\cdot10^{-6}$ s/°C and $B = 2.99\cdot10^{-4}$ s. The fluorescence intensity showed negligible effect of temperature. The apparatus employed is similar to that in Figure 10.37 except that there are two of them, which are positioned one slightly below the other. A pulsed laser providing UV light is employed together with very sensitive and fast response detectors. The fluorescence intensity during the pulse is used to provide a value of film thickness. The decaying phosphorescence signal from the time the fluorescence is not present is utilised to provide the temperature profile in the film. Schagen *et al.* noted that though this is a classical inverse operation, because it has problems of ill-posedness and non-linearity, it needs to be carried out using optimisation techniques. Allowance has to be made for the movement of liquid carrying still phosphorescing biacetyl from the measurement zone. Figure 10.41 shows an example of the result that can be obtained using this technique.

It is noted that too high a concentration will result in self-quenching and that the gas must be oxygen-free, as that will also affect the phosphorescence intensity.

Mass transfer can be studied using methods similar to those used above for heat. The prime requirement is to determine concentration in the liquid. For overall tests, the concentration at the inlet and outlet can be measured with oxygen probes if it is oxygen absorption that is being researched. For other parameters, wet chemistry could be employed, for example titration with alkali to measure carbon dioxide take up into water, or with acid if the liquid is already an alkali solution. If measurements are being made over a shorter length of falling film column, careful sampling is required.

For more detailed information, the effect of oxygen in quenching phosphorescence can be used to permit extraction of concentration data in isothermal film [56]. The details are similar to those for the heat transfer application above. Another possible approach utilises the fact that the emitting spectrum of fluorescent dyes varies with pH [57]. At one wavelength the emitted intensity is independent of pH. This is used for film thickness measurement. Other wavelengths where there is emission at one pH can be used to determine the amount of mass transferred.

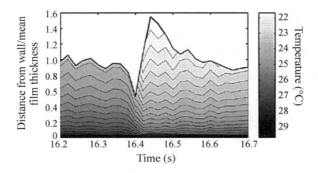

Figure 10.41 *Temperature profiles determined using the luminescence technique [55] (Reprinted from Schagen, A., Modigell, M., Dietze, G. And Kneer, R., Simultaneous measurement of local film thickness and temperature distribution in wavy liquid films using a luminescence technique, Int. J. Heat Mass Trans., **49**, 5049–5061. Copyright (2006) with permission from Elsevier.)*

Table 10.Q.1 *Solutions to Question 10.1*

Parameter	1 bar	10 bar	100 bar
ρ_{mix} (kg/m^3)	849	8.49	849
ε_g (%)	5.7	5.8	7.3

Questions

10.1 In a bubble column, the void fraction is measured via the pressure drop. The two pressure transducers are separated by 1.2 m. The liquid has a density of 900 kg/m^3. The system is operated at 1, 10, 100 bar (1 bar $= 10^5$ Pa). The gas phase has a density of 2.0 kg/m^3 at 1 bar and can be treated as an ideal gas. The pressure difference is 10 kPa. Compute the void fraction in all three cases.

The solutions are given in Table 10.Q.1.

10.2 In Section 10.1.2 the single-point probe is discussed. This probe pierces bubbles and measures the piercing length provided the velocity of the bubbles is known or can be computed from the measurements. In this question we will consider the two-dimensional world: our bubbles here are circular, all of radius R, which travel vertically upwards in a straight line. The probe pierces the bubbles at a distance r from the symmetry line. The pierced length through the bubble is termed λ. This length depends on the radial distance, r, the probe hits the bubble from the symmetry axis through the centre of the bubble (Figure 10.Q.1).

a Derive the probability density function that describes the probability of hitting a bubble at a distance between r, $r + dr$ from the central line.
b What is the relationship between the pierced length, λ, and the position r from the symmetry axis where the bubble is hit?
c Use **a** and **b** to calculate the relationship between the mean pierced length and the radius R of the bubbles.

10.3 An LDA experiment is performed in a bubble column operated at a void fraction of 8%. The bubbles are assumed to be spherical with a diameter of 3 mm. The maximum achievable data rate of the LDA set-up is 250 Hz.

a Estimate the data rate that will be found 10 cm away from the wall of the column.

17 Hz.

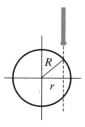

Figure 10.Q.1 *Optical probe piercing a spherical bubble*

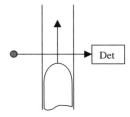

Figure 10.Q.2 *High energy beam radiating through a gas-liquid flow*

b Same question, but now in a stirred tank where the bubbles are only 1.5 mm in diameter.

1.2 Hz.

10.4 In a bubbly flow experiment the researchers consider using PIV. They prefer a large field of view and have chosen a neutrally buoyant particle with a diameter d as flow tracer. Obviously, the response time of the particle to changes in the flow velocity is important. The researchers would like to be able to see fluctuations with time scales of 10 ms. Take water as the working fluid.

a Take only the drag force into account (assume the particle Reynolds number is small, so Stokes drag law can be used) and compute the response time of the particle for $D = 5, 40, 120$ μm. Which of these particles meets their 10 ms criterion?

b The same question, but now incorporate the added mass of the particle in the calculation.

10.5 An ECT system consists of electrodes with a height of 5 cm and a width of 3 cm. The system is mounted on a cylindrical reactor with a diameter of 30 cm. The multiphase flow inside the cylinder has a relative permittivity of 2. Estimate the capacity of two sensors across the diameter.

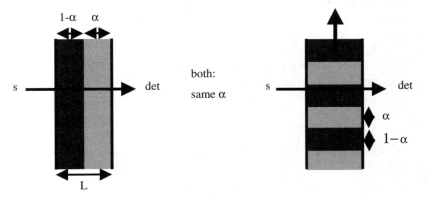

Figure 10.Q.3 *Two different arrangements of the phases at the same volume fraction. Left: the gas-liquid flow consists of two separate, vertical layers; the layer on the left is liquid, the right layer is gas. Right: alternating gas-liquid flow, where the dark areas denote liquid, light areas indicate gas*

10.6 In an experiment to measure the void fraction in a bubble flow, a γ source is used. The calibration uses an empty flow system in addition to one completely filled with liquid. The γ beam measures over a distance d through the bubbly flow. The attenuation coefficient of the liquid phase is μ_{liq}, that of the gas phase is negligible.

    a Show that the void fraction is obtained from

$$\varepsilon_g = \frac{\ln(I_f) - \ln(I)}{\ln(I_f) - \ln(I_e)}$$

    where I is the measured intensity during the experiments, I_f and I_e are are the intensity measured for the pipe completely filled with liquid and completely empty, repectively.

    b γ sources are inherently noisy: the number of photons generated follows a Poison process. This means that the number of photons counted in a sampling time Δt, N, has an inherent uncertainty, σ_N, that is equal to the square root of N: $\sigma_N = \sqrt{N}$. Show that this means that the uncertainty in I is inversely proportional to the square root of Δt.

10.7 Radiation is used to measure the void fraction in a gas-liquid vertical pipe flow. The flow rates are such that the flow is slugging, that is big bubbles that span almost the entire pipe diameter flow through the liquid. Obviously, the detected radiation intensity will vary with time: little attenuation when the bubble passes, much attenuation when the liquid is in between the source and detector. Model this system as: (i) vertical separated flow and (ii) horizontal alternating strips of liquid and gas, both with the same void fraction, see Figures 10.Q.2 and 10.Q.3. Ignore the attenuation of the gas.

    a Derive the average measured intensity in both cases.

    b Both flows have the same void fraction. What is the void fractions obtained from the measured average intensities, if the Lambert–Beer law is used on the average intensity, denoting the void fraction as ε_{gexp}?

    c Plot the void fraction ε_{gexp} as the function or the true void fraction ε_g for the case $\mu_l = 2$.

    This problem is known as 'non-linear averaging' and shows that one must be careful with interpreting this type of data.

References

[1] (a) Cartellier, A. (1992) Simultaneous void fraction measurement, bubble velocity, and size estimate using a single optical probe in gas-liquid two-phase flows. *Rev. Sci. Instrum.*, **63**, 5442–5453; (b) Groen, J.S., Oldema, R.G.C., Mudde, R.F. and Van Den Akker, H.E.A. (1996) Coherent structures and axial dispersion in bubble column reactors. *Chem. Eng. Sci.*, **51**, 2511–2520; (c) Cartellier, A. and Barreau, E. (1998) Monofiber optical probes for gas detection and gas velocity measurements: Conical probes. *Int. J. Multiphase Flow*, **24**, 1265–1294.

[2] Mudde, R.F. and Saito, T. (2001) Hydrodynamical similarities between bubble column and bubbly pipe flow. *J. Fluid Mech.*, **437**, 203–228.

[3] Juliá, J.E., Harteveld, W.K., Mudde, R.F. and Van Den Akker, H.E.A. (2005) On the accuracy of the void fraction measurements using optical probes in bubbly flows. *Rev. Sci. Instrum.*, **76**, 1–13.

[4] Xue, J., Al-Dahhan, M., Dudukovic, M.P. and Mudde, R.F. (2003) Bubble dynamics measurements using four-point optical probe. *Can. J. Chem. Eng.*, **81**, 375–381.

[5] Prasser, H.-M., Scholz, D. and Zippe, C. (2001) Bubble size measurement using wire-mesh sensors. *Flow Meas. Instrum.*, **12**, 299–312.

[6] da Silva, M.J., Thiele, S., Abdulkareem, L. *et al.* (2010) High-resolution gas-oil two-phase flow visualization with a capacitance wire-mesh sensor. *Flow Meas. Instrum.*, **21**, 191–197.

[7] Simiano, M., Zboray, R., de Cachard, F. *et al.* (2006) Comprehensive experimental investigation of the hydrodynamics of larg-scale, 3D, oscillating bubble plumes. *Int. J. Multiphase Flow*, **32**, 1160–1181.

[8] Groen, J.S., Mudde, R.F. and Van Den Akker, H.E.A. (1999) On the application of LDA to bubbly flow in the wobbling regime. *Exp. Fluids*, **27**, 435–449.

[9] (a) Kulkarni, A.A., Joshi, J.B., Kumar, V.R. and Kulkarni, B.D. (2001) Application of multi-resolution analysis for simultaneous measurement of gas and liquid velocities and fractional gas hold-up in bubble column using LDA. *Chem. Eng. Sci.*, **56**, 5037–5048; (b) Kulkarni, A.A., Joshi, J.B., Kumar, V.R. and Kulkarni, B.D. (2001) Simultaneous measurement of hold-up profiles and interfacial area using LDA in bubble columns: Predictions by multiresolution analysis and comparison with experiments. *Chem. Eng. Sci.*, **56**, 6437–6445; (c) Kulkarni, A.A., Joshi, J.B. and Ramkrishna, D. (2004) Determination of bubble size distributions in bubble columns using LDA. *AIChE J.*, **50**, 3068–3084; (d) Mudde, R.F., Groen, J.S. and Van Den Akker, H.E.A. (1998) Application of LDA to bubbly flows. *Nucl. Eng. Des.*, **184**, 329–338; (e) Mudde, R.F., Harteveld, W.K. and Van Den Akker, H.E.A. (2009) Uniform flow in bubble columns. *Ind. Eng. Chem. Res.*, **48**, 148–150.

[10] Harteveld, W.K., Mudde, R.F. and Van Den Akker, H.E.A. (2005) Estimation of turbulence power spectra for bubbly flows from Laser Doppler Anemometry signals. *Chem. Eng. Sci.*, **60**, 6160–6168.

[11] (a) Westerweel, J. (1997) Fundamentals of digital particle image velocimetry. *Meas. Sci. Technol.*, **8**, 1379–1392; (b) Raffel, M., Willert, C. Wereley, S. and Kompenhans, J. (2007) *Particle Image Velocimetry: A Practical Guide*, Springer-Verlag, Heidelberg.

[12] Hassan, Y.A., Schmidl, W. and Ortiz-Villafuerte, J. (1998) Investigation of three-dimensional two-phase flow structures in a bubbly pipe flow. *Meas. Sci. Technol.*, **9**, 309–326.

[13] Deen, N.G., Westerweel, J. and Delnoij, E. (2002) Two-phase PIV in bubbly flows: status and trends. *Chem. Eng. Technol.*, **25**, 97–101.

[14] Lindken, R. and Merzkirch, W. (2002) A novel PIV technique for measurements in multiphase flows and its application to two-phase bubbly flows. *Exp. Fluids*, **33**, 814–825.

[15] (a) Bröder, D. and Sommerfeld, M. (2003) Combined PIV/PTV-measurements for the analysis of bubble interactions and coalescence in a turbulent flow. *Can J. Chem. Eng.*, **81**, 756–763; (b) Bröder, D. and Sommerfeld, M. (2007) Planar shadow image velocimetry for the analysis of the hydrodynamics in bubbly flows. *Meas. Sci. Technol.*, **18**, 2513–2528; (c) Sommerfeld, M. and Bröder, D. (2009) Analysis of hydrodynamics and microstructure in a bubble column by planar shadow image velocimetry. *Ind. Eng. Chem. Res.*, **48**, 330–340.

[16] Bennet, M.A., West, R.M. Luke, S.P. *et al.* (1999) Measurement and analysis of flows in a gas-liquid column reactor. *Chem. Eng. Sci.*, **54**, 5003–5012.

[17] Warsito, W. and Fan, L.-S. (2005) Dynamics of spiral bubble plume motion in the entrance region of bubble columns and three-phase fluidized beds using 3D ECT. *Chem. Eng. Sci.*, **60**, 6073–6084.

[18] Warsito, W. and Fan, L.-S. (2003) 3D-ECT velocimetry for flow structure quantification of gas-liquid-solid fluidized beds. *Can. J. Chem. Eng.*, **81**, 875–884.

[19] Jin, H., Yang, S., Wang, Mi and Williams, R.A. (2007) Measurement of gas holdup profiles in a gas liquid cocurrent bubble column using electrical resistance tomography. *Flow Meas. Instrum.*, **18**, 191–196.

[20] Kumar, S.B., Moslemian, D. and Dudukovic, M.P. (1997) Gas-holdup measurements in bubble columns using computed tomography. *AIChE J.*, **43**, 1414–1425.

[21] Mudde, R.F., Bruneau, P.R.P. and van der Hagen, T.H.J.J. (2005) Time-resolved gamma-densitometry imaging within fluidized beds. *Ind. Eng. Chem. Res.*, **44**, 6181–6187.

[22] Mudde, R.F., Alles, J. and van der Hagen, T.H.J.J. (2008) Feasibility study of a time-resolving X-ray tomographic system. *Meas. Sci. Technol.*, **19**, 085501.

[23] Bieberle, M. and Hampel, U. (2006) Evaluation of a limited angle scanned electron beam X-ray CT approach for two-phase pipe flows. *Meas. Sci. Technol.*, **17**, 2057–2065.

[24] (a) Degaleesan, S., Dudukovic, M.P. and Pan, Y. (2001) Experimental study of gas-induced liquid-flow structures in bubble columns. *AIChE J.*, **47**, 1913–1931; (b) Devanathan, N., Moslemian, D. and Dudukovic, M.P. (1990) Flow mapping in bubble columns using CARPT. *Chem. Eng. Sci.*, **45**, 2285–2291; (c) Moslemian, D., Devanathan, N. and Dudukovic, M.P. (1992) Radioactive particle tracking technique for investigation of phase recirculation and turbulence in multiphase systems. *Rev. Sci. Instrum.*, **63**, 4361–4372.

[25] Seville, J.P.K., Ingram, A. and Parker, D.J. (2005) Probing processes using positrons. *Chem. Eng. Res. Des.*, **83**, 788–793.

[26] Minnaert, M. (1933) On musical air bubbles and sounds of running water. *Philos. Mag.*, **16**, 235–248.

[27] Pandit, A.B., Varley, J. Thorpe, R.B. and Davidson, J.F. (1992) Measurement of bubble size distribution: an acoustic technique. *Chem. Eng. Sci.*, **47**, 1079–1089.

[28] Boyd, J.W.R. and Varley, J. (2004) Acoustic emission measurement of low velocity plunging jets to monitor bubble size. *Chem. Eng. J.*, **97**, 11–25.

[29] (a) Boyd, J.W.R. and Varley, J. (2002) Measurement of gas hold-up in bubble columns from low frequency acoustic emissions. *Chem. Eng. J.*, **88**, 111–118; (b) Al-Masry, W.A., Ali, E.M. and Aqeel, Y.M. (2005) Determination of bubble characteristics in bubble columns using statistical analysis of acoustic sound measurements. *Chem. Eng. Res. Des.*, **83**, 1196–1207.

[30] Boyd, J.W.R. and Varley, J. (1998) Sound measurement as a means of gas bubble sizing in aerated agitated tanks. *AIChE J.*, **44**, 1731–1739.

[31] Plesset, M.S. and Prosperetti, A. (1977) Bubble dynamics and cavitation. *Ann. Rev. Fluid Mech.*, **9**, 145–185.

[32] Boyd, J.W.R. and Varley, J. (2001) The uses of passive measurement of acoustic emissions from chemical engineering processes. *Chem. Eng. Sci.*, **56**, 1749–1767.

[33] Manasseh, R., LaFontain, R.F., Davy, J., Shepherd, I.C. and. Zhu, Y. (2001) Passive acoustic bubble sizing in spaged systems. *Exp. Fluids*, **30**, 672–682.

[34] Khopkar, A.R., Rammohan, A.R., Ranade, V.V. and Dudukovic, M.P. (2005) Gas-liquid flow generated by a Rushton turbine in stirred vessel: CARPT/CT measurements and CFD simulations. *Chem. Eng. Sci.*, **60**(8–9), 2215–2229.

[35] Fishwick, R.P., Winterbottom, J.M. Parker, D.J. *et al.* (2005) Hydrodynamic measurements of up- and down-pumping pitched-blade turbines in gassed, agitated vessels, using positron emission particle tracking. *Ind. Eng. Chem. Res.*, **44**, 6371–6380.

[36] Ford, J.J., Heindel, Th.J. Jensen, T.C. and Drake, J.B. (2008) X-ray computed tomography of a gas-sparged stirred-tank reactor. *Chem. Eng. Sci.*, **63**, 2075–2085.

[37] Kraume, M. and Zehner, P. (2001) Experience with experimental standards for measurements of various parameters in stirred tanks: a comparative test. *Chem. Eng. Res. Des.*, **79**, 811–818.

[38] Bakker, A. (1992) Hydrodynamics of stirred gas-liquid dispersions, Ph.D. thesis, Delft University of Technology, The Netherlands.

[39] Clark, W.W. (2002) Liquid film thickness measurement. *Multiphase Sci. Technol.*, **14**, 1–74.

[40] Coney, M.W.E. (1973) The theory and application of conductance probes for the measurement of liquid film thickness in two-phase flow. *J. Phys. E: Sci. Instrum.*, **6**, 903–910.

[41] Geraci, G., Azzopardi, B.J. and van Maanen, H.R.E. (2007) Effect of inclination on circumferential film thickness variation in annular gas/liquid flow. *Chem. Eng. Sci.*, **62**, 3032–3042.

[42] Brown, D.J. (1978) Non-equilibrium annular flow, D.Phil. thesis, University of Oxford.

[43] Abramovitz, M. and Stegun, I. (1964) Handbook of Mathematical Functions, Dover Publications Inc., New York.

[44] Andreussi, P., Di Donfrancesco, A. and Messia, M. (1988) An impedance method for the measurement of liquid hold-up in two phase flow. *Int. J. Multiphase Flow*, **14**, 777–785.

[45] Fossa, M. (1998) Design and performance of a conductance probe for measuring the liquid fraction in two-phase gas–liquid flows. *Flow Meas. Instrum.*, **9**, 103–109.

[46] Tsochatzidis, N.A., Karapantios, T.D., Kostoglou, M.V. and Karabelas, A.J. (1992) A conductance method for measuring liquid fraction in pipes and packed beds. *Int. J. Multiphase Flow*, **5**, 653–667.

[47] Kamei, T., Serizawa, A. and Noriyasu, K. (1999) Liquid film behavior in rod bundles measured by ultrasonic transmission technique, in *Proceedings of the Conference on Two-Phase Flow Modelling and Experimentation* (eds G.P. Celata, P. Di Marcoand R.K. Shah), Editzione ETS, Pisa, pp. 1493–1502.

[48] Kapitsa, P.L. and Kapitsa, S.P. (1949) Wave flow of thin layers of viscous fluid. *Zhurn. Eksper. Teor. Fiz*, **19**, 105–120.

[49] (a) Hewitt, G.F., Lovegrove, P.C. and Nicholls, B. (1964) Film thickness measurement using a fluorescence technique. Part I: Description of the method, report No. AERE-R 4478, UKAE; (b) Hewitt, G.F. and Nicholls, B. (1969) Film thickness measurement in annular two-phase flow using a fluorescence spectrometer technique. Part II: Studies of the shape of disturbance waves, report No. AERE-R 4506, UKAE.

[50] (a) Anderson, G.H. and Hills, P.D. (1974) Two-phase annular flow in tube bends, Symposium on multi-phase flow systems, Strathclyde, Scotland. *I. Chem. E. Symp. Ser.*, No. 38, Paper J1; (b) Kockx, J.P., Nieustadt, F.T.M. Oliemans, R.V.A. and Delfos, R. (2005) Gas entrainment by a falling film around a stationary Taylor bubble in a vertical tube. *Int. J. Multiphase Flow*, **31**, 1–24.

[51] Rollefson, G.K. and Dodgen, H.W. (1944) The dependence of the intensity of fluorescence on the composition of a fluorescing solution. *J. Chem. Phys.*, **12**, 107–111.

[52] Rohatgi, K.K. and Singhal, G.S. (1962) Determination of average molar absorptivity for self-absorption of fluorescent radiation in fluorescein solution. *Anal. Chem.*, **34**, 1702–1706.

[53] Sardesai, R.G., Owen, R.G. and Pulling, D.J. (1981) Flow regimes for condensation of a vapour inside a horizontal tube. *Chem. Eng. Sci.*, **36**, 1173–1180.

[54] Aoune, A. and Ramshaw, C. (1999) Process intensification: heat and mass transfer characteristics of liquid films on rotating discs. *Int. J. Heat Mass Transf.*, **31**, 1432–1445.

[55] Schagen, A., Modigell, M. Dietze, G. and And Kneer, R. (2006) Simultaneous measurement of local film thickness and temperature distribution in wavy liquid films using a luminescence technique. *Int. J. Heat Mass Transf.*, **49**, 5049–5061.

[56] Schagen, A. and Modigell, M. (2005) Luminescence technique for the measurement of local concentration distribution in thin liquid films. *Exp Fluids*, **38**, 174–184.

[57] Schwanbom, E.A., Braun, D. Hamann, E. and And Hiby, J.W. (1971) A double-ray technique for the investigation of liquid boundary layers. *Int. J. Heat Mass Transf.*, **14**, 996–998.

Index

Hydrodynamics of Gas-Liquid Reactors: Normal Operation and Upset Conditions, First Edition.
B. J. Azzopardi, R. F. Mudde, S. Lo, H. Morvan, Y. Yan and D. Zhao.
© 2011 John Wiley & Sons, Ltd. Published 2011 by John Wiley & Sons, Ltd.